SALVAGING EMPIRE

SALVAGING EMPIRE

Sovereignty, Natural Resources,
and Environmental Science
in the South Atlantic

James J. A. Blair

CORNELL UNIVERSITY PRESS   ITHACA AND LONDON

First published 2023 by Cornell University Press

Library of Congress Cataloging-in-Publication Data

Names: Blair, James J. A., 1984– author.
Title: Salvaging empire : sovereignty, natural resources, and environmental science in the South Atlantic / James J. A. Blair.
Description: Ithaca, NY : Cornell University Press, 2023. | Includes bibliographical references and index.
Identifiers: LCCN 2022051745 (print) | LCCN 2022051746 (ebook) | ISBN 9781501771170 (hardcover) | ISBN 9781501771545 (paperback) | ISBN 9781501771187 (pdf) | ISBN 9781501771194 (epub)
Subjects: LCSH: Self-determination, National—Falkland Islands. | Natural resources—Falkland Islands. | Great Britain—Colonies— Environmental conditions.
Classification: LCC F3031 .B58 2023 (print) | LCC F3031 (ebook) | DDC 997/.11—dc23/eng/20221115
LC record available at https://lccn.loc.gov/2022051745
LC ebook record available at https://lccn.loc.gov/2022051746

To my mom, Barbara Spies Blair

Contents

Acknowledgments

This book is the outcome of a lengthy multicountry research program involving several phases of extended fieldwork at what is often called the "end of the world." It was very costly for a graduate student living in New York City with limited resources, and I have accumulated more debts than I am able to acknowledge. The research would have been impossible without generous funding. I was extremely fortunate to receive external grants and fellowships, for which I am deeply grateful, including the following: a Fulbright-IIE All-Disciplines Postgraduate Award to the United Kingdom (UK); an NSF Social, Behavioral and Economic Sciences Doctoral Dissertation Research Improvement Grant (BCS-1355717, cofunded by the Cultural Anthropology and STS programs); a Wenner-Gren Foundation Dissertation Fieldwork Grant; and a Social Science Research Council Dissertation Proposal Development Fellowship. The Graduate Center, City University of New York (CUNY), also provided extensive funding and support.

I owe a special debt of gratitude to Marc Edelman, my adviser at CUNY, who shepherded me through each stage of the research process with reliable support and surgical attention to detail. His enthusiastic embrace of the various directions I strayed, which were outside of his immediate research interests, is a testament to his expansive intellect and versatility. A special thank you is also due to Jacqueline N. Brown, Mandana Limbert, and Gary Wilder, who complemented Marc Edelman and each other with invaluable expertise and erudition. An additional thank you goes to Julie Skurski, as well as participants in the Four Field Anthropology seminar, who gave me useful feedback on early drafts.

My education at the CUNY Graduate Center occurred during a sea change. I caught the final wave of teaching and mentorship by some renowned scholars and academic rock stars whose extraordinary work and influence transcend the discipline of anthropology, including Fernando Coronil, Leith Mullings, and Neil Smith, who passed away, as well as Talal Asad and David Harvey, who entered retirement. Their prolific scholarly contributions and political commitment influenced my thinking in ways that I cannot begin to measure. My work also benefited greatly from stimulating coursework and inspiring discussion with Arthur Bankoff, Leigh Binford, Michael Blim, Jillian Cavanaugh, John Collins, Ruth Wilson Gilmore, Murphy Halliburton, Jeff Maskovsky, Don Robotham, Ida Susser, and Katherine Verdery. Like so many CUNY colleagues, I could not have navigated through this journey without the sage advice and lighthearted dedication of Gerald Creed and Ellen DeRiso. Thank you to both, and thanks also to Louise Lennihan and Rachel Sponzo for their commitment to public higher education in an era of budget cuts and austerity at CUNY. I remain very proud of my CUNY pedigree, and I also took full advantage of New York City's Inter-University Doctoral Consortium, having taken my project in new directions through the brilliant instruction of Audra Simpson at Columbia University and Ann Stoler at The New School.

This work is the product of sustained collaboration and challenging dialogue with thoughtful friends and stellar students at CUNY and across New York City: Neil Agarwal, Jure Anzulovic, Julio Arias Vanegas, Monica Barra, Christopher Baum, Carwil Bjork-James, Lydia Brassard, Matt Canfield, Emily Channel-Justice, Matthew Chrisler, Rachel Daniell, Charles Dolph, Mark Drury, Christine Folch, Julian Gantt, Rocío Gil, Zoltán Glück, Megan Hicks, Ahmed Ibrahim, Mohamad Junaid, Malav Kanuga, Madhuri Karak, Selim Karlitekin, Nazia Kazi, Esteban Kelly, Andrés León, Jenny LeRoy, Ryan Mann-Hamilton, Fabio Mattioli, Manissa McCleave Maharawal, Zeynep Oguz, Camila Osorio, Helen Panagiotopoulos, Claire Panetta, Gustavo Quintero, Kareem Rabie, Jeremy Rayner, Yonatan Reinberg, Alexandra Schindler, Daniel Schneider, Douaa Sheet, Anne Spice, Amy Starecheski, Lauren Suchman, Andreina Torres, José Vasquez, Anders Wallace, and Jonah Westerman.

I am keen and eager to extend my sincerest gratitude and appreciation to a long list of Falkland Islanders who welcomed me into the fabric of their tight-knit community. However, in order to protect the rights of my interlocutors, I have withheld names of individuals from this book. In keeping with conventions of my field, I have used pseudonyms for living participants, apart from select public figures and published authors. Islanders who read this book will likely recognize themselves or their friends and neighbors, and the absence of names is not meant to deceive them. A layer of anonymity will hopefully protect individuals from identification by outside readers who may not already know their identities. In addition to a vast number of individuals who engaged with my research in the islands, I am especially thankful for the institutional support of the South Atlantic Environmental Research Institute (SAERI), which allowed a qualitative social scientist to rub shoulders, and even penguin feathers, with a brilliant cohort of marine ecologists. On my last day in the Falklands, which also happened to be my birthday, a fisheries scientist hosted a barbecue with friends and acquaintances. There, SAERI researchers and friends set up balloons, baked a cake, and surprised me with gifts showing that they had listened closely to me and engaged with my research. I was touched by SAERI's warm embrace, and I remain very grateful for their hospitality. I also had the privilege of inheriting well-organized stacks of colonial letter books, reports, and periodicals from the Jane Cameron National Archives, as well as engaging in friendly conversation there. The Falkland Islands Community School library was another excellent source for books and periodicals. My research also benefited from in-person and electronic correspondence with an energetic and mutually supportive group of scholars who had also carried out fieldwork in the islands, particularly geographers Matt Benwell, Wayne Bernhardson, Klaus Dodds, and Alasdair Pinkerton.

In Argentina, I am especially grateful for the generous support of Rosana Guber at the Instituto de Desarrollo Económico y Social in Buenos Aires. She welcomed me into her home and permitted me to raid her "bunker," a treasure of rare and highly informative sources on Malvinas. It was a blessing to have such a gracious interlocutor with overlapping training in anthropology. Federico Lorenz and Laura Panizo also received me warmly, and I am excited to contribute to the growing scholarly interest in new approaches to the islands through history and memory. It was a thrill to be a part of such a rich and dynamic collective of friends, journalists, and radical intellectuals in Buenos Aires. Thank you to Liz Mason-Deese, Verónica Gago, Natasha Niebieskikwiat, Beatriz Sarlo, Diego Sztulwark, and Soledad Torres. This work would have been considerably more onerous without the thorough transcription provided by Mikel Aboitiz. In Río Gallegos, I had the unusual opportunity to attend the first-ever reunion of defected islanders in Argentine Patagonia. This state-sponsored attempt to reconnect broken networks of lost kin revealed surprising expressions of loyalty among Patagonian settlers and their estranged islander relatives, highlighting the importance of kinship for understanding divergent geopolitical perspectives. My ethnographic observations from that reunion were omitted from this book manuscript, but three key figures involved have since passed away, and I am grateful for the opportunity to have met them: Alejandro (Alec) Betts, Uriel Erlich, and Marcelo Vernet. Vernet—a poet, writer, and independent historian—was especially generous with his time and encyclopedic knowledge. I can still picture him grinning widely as he introduced me to others: "Este es un antropólogo y no un antropófago" ("This is an anthropologist, not a cannibal"). As with the islanders, there may be other Argentine interlocutors who figured in this research, but for reasons of confidentiality, I have omitted their names. They, too, will recognize themselves in the text, and my sincerest thanks go to them. Finally, I want to extend thanks to the Archivo General de la Nación and the Universidad de la Patagonia Austral for research support.

In the United Kingdom, the geography department in the School of Environment, Education and Development (SEED) and the Society and Environment Research Group at the University of Manchester supported the early stages of writing. I am particularly grateful to Erik Swyngedouw for his generous sponsorship, expert encouragement, and dynamic critical thinking as outside reader and mentor. I also received useful feedback at Manchester from Noel Castree, William Kutz, and Michael Martin, as well as the numerous friends I made in the geography department and copanelists at the SEED postgraduate conference. It was such a pleasure to experience Manchester alongside fellow Fulbrighter and rising star Nikki Luke, as well as several other interdisciplinary scholars and students in the US-UK Fulbright program. Manchester's anthropology department also welcomed me into their stimulating and supportive community. I had the unique privilege of drawing inspiration from the same radical environment where Max Gluckman and his students developed the extended case study method, which I had been steeped in at CUNY. Thank you to Penelope Harvey, Karen Sykes, Olga Ulturgasheva, Soumhya Venkatesan, Peter Wade, and Chika Watanabe, as well as the visiting scholar Veena Das and the many graduate students who became friends and colleagues, particularly those who organized and participated in the Royal Anthropological Institute postgraduate conference. For comments and suggestions on earlier versions of chapters, I also thank my copresenter Tania Li and workshop participants at the Department of Anthropology at London School of Economics and Political Science, particularly Gisa Weszkalnys, as well as Laura Bear, Katy Gardner, and Deborah James. Again, I have concealed the identity of a number of living interlocutors in the UK for this study. Thank you to those who participated. Lastly, the National Archives at Kew provided a state-of-the-art research setting to complement my archival work in Stanley and Buenos Aires.

Much of the writing and revision of this book took place in Washington, DC, and Los Angeles, CA. First, in Washington, it was invigorating to take a step outside of academia as a Mellon/ACLS Public Fellow, appointed as International Advocate for the Natural Resources Defense Council (NRDC). With support from NRDC's Latin America and Canada programs, I was privileged to collaborate with Indigenous peoples, biodiversity conservation scientists, and environmental activists to protect rivers, watersheds, and wetlands. Together with social movements organized by Chile's Free-Flowing Rivers Network, as well as the Cree First Nation of Waswanipi, we sought to raise awareness for more sustainable alternatives to proposed large hydroelectric dams, industrial logging, and mining projects. This gave me a unique opportunity not just to analyze problems related to settler colonialism and environmental governance from within the comfort of the ivory tower but also to advocate with partners for ways to address Indigenous dispossession and environmental injustice on the ground. I am very grateful for the committed colleagues and friends I made along the way, who kept me engaged each day after long evenings at the office working on this book: Andrés Anchondo, Josh Axelrod, Camila Badilla, Ramón Balcázar, Liz

Barratt-Brown, Andrea Becerra, Jessica Carey-Webb, Fernanda Castro Purrán, Miriam Chible, John Paul Christy, Valerie Courtois, Carolina Herrera, Courtenay Lewis, Claudia Lisboa, Manuel Maribur, Amanda Maxwell, Erika Moyer, Juan Pablo Orrego, Bernardo Reyes, Jake Schmidt, Patricio Segura, Jennifer Skene, Anthony Swift, César Uribe, and Shelley Vinyard. Many thanks, as well, to June Brown and Stephen Brown for their warmth, hospitality, and laughs.

In spite of all the challenges involved with writing a book during a global pandemic, I have been blessed to do so while teaching stellar students from diverse backgrounds in an outstanding interdisciplinary environment at the Department of Geography and Anthropology at Cal Poly Pomona (CPP). Mark Allen, Remi Burton, Kristen Conway-Gómez, Annie Danis, Amy Dao, Claudia García-Des Lauriers, Gabriel Granco, Kelly Huh, Katherine Kinkopf, Mike Reibel, D. D. Wills, Lin Wu, and Terry Young, as well as Neil Chaturvedi, Brianne Dávila, Sara Garver, David Horner, and Iris Levine at the College of Letters, Arts and Social Sciences, were all incredibly supportive. At the Lyle Center for Regenerative Studies, I have been grateful for mentorship from Mary Anne A. Akers, Lauren Bricker, Pablo La Roche, Juintow Lin, and Andrew Wilcox, as well as collaboration with Jill Gómez, Karen Mitchell, Debbie Scheider, and all my brilliant and creative students. Finally, this book may not have been possible without generous, unanticipated support from Bryant Fairley, Lydia Chen Shah, and Joselyn Yap at the CPP Center for Community Engagement.

I owe a sincere thanks to Cornell University Press for believing in this book project. In particular, I am grateful for the warmth and intelligence of Jim Lance and Clare Jones, as well as the very generous readers, Marcos Mendoza and Gisa Weszkalnys, whose identities were disclosed after the manuscript was approved. Thanks also to Karen Hwa, Jennifer L. Morgan, and Michelle Witkowski for production and copy-editing as well as Enid Zafran for indexing. I have tried to incorporate all suggestions, but any errors or shortcomings are, of course, my own.

Beyond these institutional research connections at home and in the field, I am pleased to have joined a broader constellation of emerging scholars working on sovereignty, natural resources, and environmental science in anthropology and geography. I am grateful to the participants of stimulating conference sessions, including the 2015 American Anthropological Association (AAA) panel "Extractive Infrastructures and Energy Assemblages"; the 2016 AAA panel "Development's 'Half-Life': Diverse Moments in the Circulation of Extractive Capital"; the 2017 American Association of Geographers panel "Data Infrastructures, Nature, and Politics"; the 2017 AAA panel "Unsettling the Nation-State: Emerging Ethnographies of Argentina"; the 2018 AAA panel "The Political Ecology of Enclosure in Global Perspective"; and the 2019 roundtable "Seeking Truth amongst the Powerful: Ethical and Legal Issues of 'Studying Up'"—all offered rich and thought-provoking papers and discussion. I am especially thankful for thoughtful feedback on early drafts from Jenny Goldstein, Maron Greenleaf, Stephanie McCallum, and Eric Nost.

Last but certainly not least, I thank my family: Maru, Abigail, Tom, Keramet, Edward, Barbara, Jun, Joe, Mary Ann, and Elinor—all shining examples of compassion, hard work, and dedication. My aunt Terry Spies, a professional cartographer, even contributed this book's maps on short notice. As my grandfather Raymond Spies used to say, "I'm humble and I'm proud of it." Without your love and support, it never would have been possible to complete this book. Thank you so much.

Portions of text in this book have been transformed but have appeared previously in:

Blair, James J. A. 2013, March 16. "Loud and Clear." *The Economist*. http://www.economist .com/news/americas/21573581-islanders-seek-sway-world-opinion-voting-stay -british-loud-and-clear.
——. 2013, June 21. "Referendum Rewound." *The Economist*. http://www.economist .com/blogs/americasview/2013/06/falkland-islands.
——. 2013, August 8. "Self-Determined." *The Economist*. http://www.economist.com /blogs/americasview/2013/08/argentina-falklands-and-un.
——. 2013, March 11. "Sending Their Message." *The Economist*. http://www.economist .com/blogs/americasview/2013/03/falkland-islands-referendum.
——. 2014, February 28. "Treasure Islands?" *The Economist*. http://www.economist.com /blogs/americasview/2014/02/oil-and-gas-falklands.
——. 2016, August 4. "Brexit's South American Ripple Effect." *NACLA Report on the Americas*. https://nacla.org/news/2016/08/04/brexit%E2%80%99s-south-american -ripple-effect.
——. 2016, December 20. "The Limits of Environmentalism at Earth's End: Reindeer Eradication and the Heritage of Hunting in the Sub-Antarctic." *Engagement* (blog). https://aesengagement.wordpress.com/2016/12/20/the-limits-of-environmentalism -at-earths-end-reindeer-eradication-and-the-heritage-of-hunting-in-the-sub -antarctic/.
——. 2017. "Settler Indigeneity and the Eradication of the Non-Native: Self-Determination and Biosecurity in the Falkland Islands (Malvinas)." *Journal of the Royal Anthropological Institute* 23 (3): 580–602. https://doi.org/10.1111/1467-9655.12653.
——. 2019. "South Atlantic Universals: Science, Sovereignty and Self-Determination in the Falkland Islands (Malvinas)." *Tapuya: Latin American Science, Technology and Society* 2 (1): 220–36. https://doi.org/10.1080/25729861.2019.1633225.
——. 2022. "Data Gaps: Penguin Science and Petrostate Formation in the Falkland Islands (Malvinas)." In *The Nature of Data: Infrastructures, Environments, Politics*, edited by Jenny Goldstein and Eric Nost. By permission of the University of Nebraska Press. Copyright 2022 by the Board of Regents of the University of Nebraska.
——. 2022. "Tracking Penguins, Sensing Petroleum: 'Data Gaps' and the Politics of Marine Ecology in the South Atlantic." *Environment and Planning E: Nature and Space* 5 (1): 60–80. https://doi.org/10.1177/2514848619882938.

Abbreviations

AGN	Archivo General de la Nación (Buenos Aires)
BAS	British Antarctic Survey
BGS	British Geological Survey
BHP	Broken Hill Proprietary Company Limited (BHP Billiton)
BOT	British Overseas Territory
BP	British Petroleum
EEZ	exclusive economic zone
EIA	environmental impact assessment
EIS	environmental impact statement
Ex-ESMA	Espacio Memoria y Derechos Humanos (formerly Escuela de Mecánica de la Armada)
FCO	Foreign and Commonwealth Office
FIC	Falkland Islands Company
FICZ	Falklands Interim Conservation and Management Zone
FIDC	Falkland Islands Development Corporation
FIG	Falkland Islands Government
FIOHEF	Falkland Islands Offshore Hydrocarbons Environmental Forum
FIPASS	Falklands Interim Port and Storage System
FOCZ	Falkland Islands Outer Conservation Zone
FOGL	Falkland Oil and Gas
FPSO	floating production and storage offloading
GDP	gross domestic product
GIS	geographic information system
GLS	global location sensing
GPS	Global Positioning System
ITQ	individual transferable quota
JCNA	Jane Cameron National Archives (Stanley)
LNG	liquefied natural gas
MIOR	Misión Internacional de Observación del Referendo
MLA	member of the Legislative Assembly
MPA/MPC	Mount Pleasant Royal Air Force Base/Mount Pleasant Complex
NATO	North Atlantic Treaty Organization
NGO	nongovernmental organization
NSF	National Science Foundation

NTR	nontechnical risk
PRP	Permanent Residency Permit
PTT	platform terminal transmitter
QUANGO	quasi-autonomous nongovernmental organization
RAF	Royal Air Force
RIOM	Referendum International Observation Mission
ROV	remotely operated underwater vehicle
SAERI	South Atlantic Environmental Research Institute
SATCO	South Atlantic Construction Company
STS	science and technology studies/science, technology, and society
TED	time elapsed for disaster
TLP	tension leg platform
UN C24	United Nations Special Committee on Decolonization (Committee of 24)

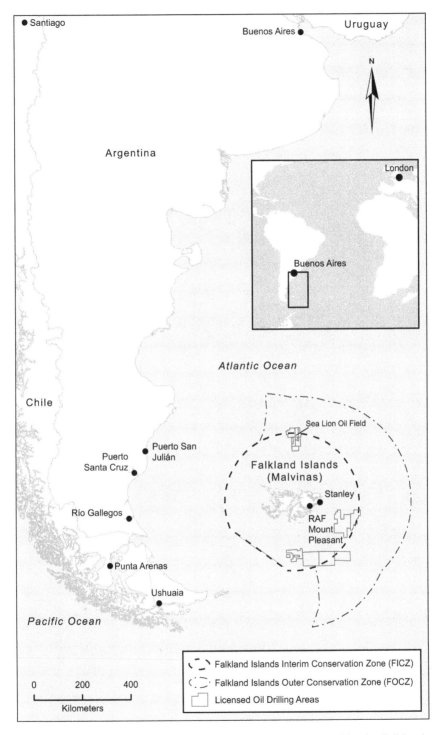

FIGURE 1. Reference map featuring oil drilling areas licensed by the Falkland Islands Government (FIG) in the South Atlantic.

Map by Terry Spies (data available online: http://dataportal.saeri.org/).

SALVAGING EMPIRE

BRITISH TO THE CORE

In March 2013, residents of the Falkland Islands (in Spanish, Malvinas) held a referendum on whether or not they wished to remain British.[1] Around that time, upon arrival at Mount Pleasant Royal Air Force Base, which functions as an international airport, I noticed that the baggage claim area looked starkly different than it had looked during an earlier research visit. Previously, the only décor consisted of a crumbling papier-mâché penguin and signs warning against invasive species. In preparation for the referendum, the Falkland Islands Government (FIG) had covered the walls with blown-up photograph placards.[2] They featured local charismatic megafauna, such as penguins, albatrosses, and sea lions, but the new display centered on a photo of six Falkland Islander children, leaping from a grassy field with arms stretched high (see figure 2). They looked almost as though they were emerging, autochthonous, from the South Atlantic territory their British ancestors settled.[3] The words "Our Islands, Our Home" were painted across the blue-sky backdrop.[4]

On the bumpy ride from the Mount Pleasant base to Stanley, the islands' only town, the shuttle driver discussed preparations for the referendum. Stores, houses, and cars had been plastered with Union Jack flags and "British to the Core" stickers. The other passenger in our van was a helicopter pilot from New Zealand whom the UK government hired to assist in the eradication of rats and reindeer on South Georgia Island, situated between the Falklands and Antarctica.[5] While we talked, the driver's young daughter, who was wedged between my front passenger seat and her mother's driver seat, tapped two metal objects together, producing a staccato rhythm. She motioned for me to look at the source

FIGURE 2.   "Our Islands, Our Home" placard and pamphlet cover.

Source: Falkland Islands Government.

of the sound: her unlikely instrument was a pair of bullet shells. She continued playing them as we drove past a stretch where Zimbabwean "de-miners" were clearing active land mines still embedded within the barren, hilly landscape decades after Argentina's 1982 military occupation. The minefields had been fenced off from the archipelago's abundant sheep, but travel in "Camp" (from the Spanish *campo*, meaning the countryside outside of Stanley) still risked getting bogged in peat. This carbon-rich soil, from which residents had long cut turf to burn in kitchen stoves, has been substituted by gas for heating. Turbines now harness the near-constant westerly winds for electricity.

Our driver remarked that offshore oil exploration was generating hope.[6] In 2010, FIG-licensed oil firms made their first commercial discovery (estimated to be the equivalent of 1.7 billion barrels of oil in place with more than 500 million barrels recoverable) in the Sea Lion well of the North Falkland Basin.[7] We began to approach the new oil yards at the edge of Stanley, overseen by the Falkland Islands "totem pole" (see figure 3), a column of signs listing the hometowns of British military officers, tourists, and scientists, as well as their distance in miles (Stanley is situated at latitude 51°S, and London is located nearly 8,000 miles away, at latitude 51°N as shown in the reference map in figure 1). A common feature of military bases around the world, Falkland Islanders have appropriated the "totem pole" to mark their British imperial heritage and stake their claim to what I call *settler indigeneity* in one stroke.[8] With the Queen's head on the reverse side, the "totem pole" appeared on the official referendum coin.[9]

Descending into Stanley, we entered a steep grid of brightly roofed houses, huddled together on the hillside, which offer refuge from the incessant, howling gusts of wind. Cats and ornamental gnomes greeted us from bountiful vegetable gardens. Finally, beyond the proliferation of tourism shops, hotels, pubs, war monuments, and government buildings, lay Stanley Harbour. There, the rusted shipwreck of *Lady Elizabeth* has remained since the iron barque drifted into the Falklands under a Norwegian flag on March 13, 1913 (see figure 5).[10] It was damaged on a rock at the entrance to Berkeley Sound to the northeast of Stanley while sailing from Cape Horn on the way to deliver lumber from Vancouver, Canada, to Delagoa Bay (now Maputo Bay) in Mozambique. Declared unseaworthy, the corroded wreck was salvaged and repurposed as a floating warehouse by the Falkland Islands Company (FIC) until 1936, when it was put ashore in Whalebone Cove. Today, the *Lady Elizabeth* is regularly eclipsed by bustling traffic from cruises, container ships, squid-fishing jiggers, British Antarctic Survey (BAS) ice-strengthened ships, offshore patrol vessels, and oil barges.

This description of the stimulating airport shuttle ride provides a snapshot of the islands, which have been in the process of transformation in the decades since the violent 1982 conflict between the UK and Argentina.[11] While the war certainly has not faded from memory—land mines and military exercises had become aspects of everyday life—the placards, flags, stickers, and "totem pole" indicate a resolute sense of local British identity. In a less overt manner, the home gardens, commercial fishing, oil drilling, scientific research, and biosecurity campaigns suggest more subtle ways in which islanders cultivate a living and fortify their dominion. As they engage in new forms of environmental management, the islands' settlers have asserted a bolder sense of indigeneity. This book describes how Falkland Islanders and their trustees have sought to salvage the British Empire: not only with defense forces but also by extracting natural resources.

The Falkland Islands (Malvinas) are a former British Crown Colony, yet they are handled as an ongoing "special and particular colonial case" in international law.[12] The islands are located about 300 miles (480 km) away from Argentine

FIGURE 3. Falkland Islands "Totem Pole."

Photo by James J. A. Blair.

Patagonia's coast, and Argentina's government has long claimed sovereignty over the territory and its surrounding resources, so it remains on the UN's list of "non-self-governing territories."[13] The Falklands had been just one piece of the empire on which the sun never sets during the height of the UK's global dominance in the nineteenth and early twentieth centuries. However, in the decline of the British Empire following World War II, as most former colonies in Asia, Africa, and the Americas gained independence through decolonization, the Falklands remained under British sovereignty, designated a British Dependent Territory in 1981.[14] It was renamed a British Overseas Territory (BOT) in 2002. The South Atlantic archipelago, comparable in area to the US state of Connecticut, consists of East and West Falkland and 778 smaller islands, only a fraction of which are inhabited. According to a 2021 census, a permanent military garrison at Mount Pleasant Royal Air Force Base comprises 347 people. The total resident population is 3,541, 80 percent of whom live in Stanley. Around 70 percent of total residents consider their national identity to be Falkland Islander, British, or a combination of both. English is the standard language, but local vernacular includes Anglicized Spanish syncretisms, such as Camp, as I describe in chapter 2. The census counted 86 different nationalities with a significant influx of migrants from St. Helena, Chile, and the Philippines. Individuals with temporary immigration status make up 31 percent of the usual resident population.

Taking an anthropological approach to geopolitics, this book is a historical ethnography that explores how the settlers of the Falklands have constructed themselves as "natives" through everyday forms of environmental governance. Unlike comparable settler colonies predicated on "the elimination of the Native," no living Indigenous population was present on the islands at the time of European colonization.[15] Most of the Falkland Islanders are white settlers, making their invocation of self-determination different from that of former colonial subjects with Indigenous rights. I argue that by claiming self-determination and consenting to British sovereignty, the Falkland Islanders have formed a settler colonial protectorate. This protectorate has empowered the islanders to own land, control territory, extract resources, conserve the environment, and ultimately to salvage the last bits of the British Empire from the tide of decolonization.[16]

An Emphatic "Yes"

The 2013 referendum offered a window into some of the ways Falkland Islanders salvage empire. Stanley residents greeted international media crews by parking Land Rovers across the harbor in rows to spell out the word YES (see figure 4), and hundreds paraded through town in a rally that included SUVs, motorbikes,

FIGURE 4. Land Rovers spelling out the word "YES" across from Stanley during the 2013 referendum.

Photo by James J. A. Blair.

and horses. In addition to the ubiquitous Union Jack, posters featured FIG Member of the Legislative Assembly (MLA) Dick Sawle—one of the originators of the referendum—as John Bull or Uncle Sam, instructing his constituents to vote "Yes." Local sentiments during this moment ranged from pride and hubris to uncertainty and anxiety.[17]

When I visited the islands during the week of the vote, I pitched a story to newspaper editors to help fund my travel. I secured an assignment from *The Economist* and wore two hats as an anthropologist and journalist. Being both a researcher and a reporter put me in close quarters with British and Argentine scholars and journalists, as well as observers from Canada, the US, Brazil, Chile, Mexico, New Zealand, and Uruguay.[18] To demonstrate democracy and ensure high voter turnout in Camp, the FIG arranged flights to eight outer islands and assembled five teams of Land Rovers. I spent much of the voting day on one of the unpaved mobile polling routes. Farmers invited me into their settlement houses and offered me tea and biscuits as we discussed what the referendum meant to them. Unsurprisingly, none of my interviewees veered from the "Yes" decision, but some residents of younger generations did raise the possibility of future independence, perhaps by redistributing oil revenue to reimburse the UK for defense costs. In this sense, islanders viewed oil as a resource to fund a more self-sufficient protectorate with taxes and royalties that might keep the military around. Most islanders expressed support for the continued presence of British forces while also seeking to expand trade and travel through South America.

Upon returning to Stanley, I took refuge from the squalls of freezing rain and snow. I awaited the ballot count in the Town Hall alongside Argentine public intellectual Beatriz Sarlo, who had come out in support of the islanders in a letter signed by her and sixteen other prominent Argentine figures.[19] The turnout was high at 92 percent, and the ballot count went relatively smoothly, apart from audio problems and a miscalculation of the percentage of "Yes" votes (the FIG chief executive announced 98.8 percent but meant 99.8 percent). There were just three naysayers among 1,517 valid votes. When the near-unanimous results were revealed, crowds gathered under Stanley Green's Whalebone Arch to toast "Her Majesty and the Falklands" and sang "Rule Britannia." Decked out in custom-made Union flag suits and dresses, islanders embraced one another in inebriated elation. Even if the vote was a foregone conclusion, some voters admitted feeling afraid and nervous when the polls closed. Others stood speechless in tears of ecstasy. That week, islanders told me that they "felt like [they] had died and gone to heaven." Ironically, a few days after the vote, this divine sensation subsided, as their adversarial Argentine neighbors were blessed with media attention of their own. Plumes of white smoke appeared at the Vatican as an Argentine cardinal, Jorge Mario Bergoglio, was chosen to be the next pope.[20]

The 2013 referendum was not designed to tease out facets of national identity or even realize hopes for a clear political future. The FIG planned the vote as a public mandate that could be used as a diplomatic tool to affirm British sovereignty in response to constant pressure from Argentina. Rather than a decolonial defense of Indigenous land, the islanders' self-determination claim may be understood as a manifestation of the persistent ideology that the critical theorist Stuart Hall called "authoritarian populism."[21] Hall described authoritarian populism as a conjuncture of capitalist state domination with popular consent.[22] In this sense, the 2013 Falklands referendum was part of a global trend of authoritarian populism that has transformed the contemporary world order by bringing far-right demagogues into seats of power, from Viktor Orbán in Hungary and Vladimir Putin in Russia to Donald Trump in the US, Jair Bolsonaro in Brazil, and Boris Johnson in the UK.[23] The 2016 Brexit referendum for the UK to leave the European Union, which Johnson championed, may ironically have negative long-term repercussions for the Falklands' export economy with the introduction of tariffs, but both plebiscites were rooted in the authoritarian populist rhetoric of former British Prime Minister Margaret Thatcher.[24] As Hall keenly observed, Thatcher constructed her nationalist project of neoliberalism by appealing to "the people" in the 1982 war to defend the Falkland Islanders' wishes for self-determination against Argentina's military dictatorship.[25] For Thatcher, the Falklands had become a powerful metonym for imperialist nostalgia, transmitted through a smokescreen of patriotic glory for democracy's triumph over totalitarianism.[26] This book tells the story of how the islanders' claim of self-determination came to perpetuate sovereign control over land, territory, and resources in the afterlife of the British Empire.

Sovereignty Saga

Despite the islanders' emphatic referendum vote, contestation over the islands' sovereignty endures. Argentina's Malvinas sovereignty claim has long remained an enormously popular national cause.[27] The promise of an ostensibly just "clean war" to recover the islands for Argentina in 1982 momentarily suppressed outrage at the state terror of the Dirty War (1976–1983), in which tens of thousands of *desaparecidos* (disappeared people) were kidnapped, tortured, and murdered.[28] Library bookshelves swell with dramatic accounts from battles of the 1982 conflict, and military history buffs may not be satisfied with this book's main focus on more mundane matters of local belonging, natural resource management, and environmental science.[29] Nonetheless, for unfamiliar readers, I offer a brief outline of the sequence of events in the war followed by remarks on the broader geopolitical significance of the ongoing sovereignty dispute in the South Atlantic.

On April 2, 1982, an Argentine task force invaded the Falklands, overtaking the few British Royal Marines stationed there.[30] The next day, Argentine military forces seized South Georgia. Before British prime minister Margaret Thatcher responded with force, Alexander Haig, US secretary of state under Ronald Reagan, shuttled between London and Buenos Aires to negotiate a potential settlement. On April 25, the British recuperated South Georgia. Haig's mission ended, and the US tilted diplomatically toward the UK, placing economic sanctions on Argentina. On May 1, British forces attacked Stanley's airstrip. Peru's president Fernando Belaúnde Terry attempted to negotiate peace with Argentine general Leopoldo Galtieri, but this bid failed when a British nuclear-powered submarine controversially sank the Argentine cruiser *Belgrano* outside the established Total Exclusion Zone on May 2, killing 323 people. Argentine forces returned fire on May 4, hitting the British frigate *Sheffield* with an air-to-surface missile and destroying a British Harrier plane. As UN peace negotiations commenced, British special forces raided the outer Pebble Island, destroying several aircraft.

On May 21, a British task force landed at Port San Carlos, located on the opposite side of East Falkland Island from Stanley. After much bloodshed, the British Parachute Regiment took control of the settlements at Darwin and Goose Green, where 115 Falkland Islanders (including 43 children) had been confined to a hall for nearly a month.[31] Argentine and British forces sustained respective air and sea losses as the British advanced on Stanley in an iconic 56-mile "yomp" with full loads over three days. On June 12, fierce battles occurred on the mountains surrounding Stanley: Mount Longdon, Two Sisters, and Tumbledown. The British emerged victorious in Stanley on June 14 when Argentine general Mario Menéndez surrendered. In total, it is estimated that 255 British, 649 Argentine, and three Falkland Islander lives were lost in the relatively swift conflict.[32]

The motivations for the war and its impact on British and Argentine culture and politics are well studied.[33] The Argentine writer Jorge Luis Borges famously remarked that "the Falklands thing was a fight between two bald men over a comb."[34] Other dramatic accounts of the war have been depicted in popular television shows like *The Crown*. The British victory in the South Atlantic established Thatcher as the formative figure responsible for implementing widespread neoliberal economic policies with the apparent consent of "the people." To mobilize military forces, Thatcher referred to the Falkland Islanders as "kith and kin" of a similar British "island race," which I describe in chapter 3 as an authoritarian populist appeal to whiteness as well as a way to salvage what remained of an imperial diaspora in the decline of the British Empire.[35] Having proven that she was made of metal, the Iron Lady leveraged the UK's patriotic political mood to secure the Conservative party's power and fight so-called enemies within.[36] This included both organized miners' unions and criminalized Black Britons.

The Argentine state has continued to assert the popular national cause for the Malvinas, even after General Galtieri's military junta surrendered and democracy was restored.[37] During the postwar *Desmalvinización* period, child conscripts became subaltern veterans in a liminal struggle for state benefits and social programs for themselves and their grieving families.[38] Struggling to overcome economic crisis from debt and inflation, the Argentine government under Presidents Néstor Kirchner and Cristina Fernández de Kirchner revived the Malvinas sovereignty cause as a core value of Peronism: arguably an earlier authoritarian-democratic form of populism.[39] Seeking to renovate Argentina's damaged national narrative, Fernández de Kirchner's government attempted to pass bills that threatened oil companies exploring near the islands with legal action and banned ships that fly a Falklands flag from docking in ports of the province of Buenos Aires. The Argentine Foreign Ministry has continued to argue that the islands and marine resources are within its waters. As a result, Argentina has gained considerable support from the international community. Year after year, the UN Special Committee on Decolonization (also known as the Committee of 24 or C24) has backed Argentina's position.

To materialize these separate, parallel narratives of sovereignty, two different museums opened during my fieldwork in 2014. Each sought to commemorate and honor the divergent legacies of the Falklands/Malvinas.[40] In Buenos Aires, Fernández de Kirchner's government launched the Museo de Malvinas, which is located controversially in the campus of Ex-ESMA (Escuela de Mecánica de la Armada, or Army Mechanical School), a former detention and torture center made infamous during the military dictatorship. The Falkland Islanders opened their own new Historic Dockyard Museum in Stanley, which includes a moving recording of residents who witnessed the violent battles of 1982 as children. The two museums express their respective national geopolitical claims not only through memories of war but also via the scientific authority of marine ecology and natural resource governance, which this book explores in ethnographic detail. For instance, the Museo de Malvinas displays an elephant seal skeleton prominently with maps of its migratory patterns and animated underwater films, designed to naturalize Argentine national sovereignty over the South Atlantic. Meanwhile, the Historic Dockyard Museum also includes bones and skulls of marine mammals as well as taxidermized seabirds perched beside sundry artifacts of maritime heritage and resources: models of early cutters or schooners, sealing narratives, an explanation of the life cycle of "Falkland Calamari," and maps of proposed offshore oil drilling.[41] In general, the Museo de Malvinas seeks to resurrect early-nineteenth-century stories of imperial dispossession and rebellions of South American gauchos (cattle herders) against early British settlers, which I describe in chapter 1. This contrasts with Stanley's Historic Dockyard Museum, which focuses primarily on quotidian social practices and colonial customs of Falkland Islanders as a people.

Political theorists have long debated sovereignty in the South Atlantic, yet the Falkland Islanders are a sociopolitical entity for which we do not have clear language.[42] The unsettled dispute rests on a fundamental tension between international legal principles of territorial integrity and self-determination in decolonization.[43] The rights claims not only of the islanders but also of Argentina work against our preconceptions of how these underlying principles generally apply. The key diplomatic distinction continues to lie between recognition of the interests of the Falkland Islanders on the one hand and protection of their wishes or desires (deseos) on the other. On December 16, 1965, the UN General Assembly passed Resolution 2065, which promoted peaceful bilateral negotiations between the governments of Argentina and the UK, "bearing in mind . . . the interests of the population of the Falkland Islands (Malvinas)."[44] Excluding "wishes" from the equation and labeling the islanders a "population" rather than a "people" limited their right to self-determination, whereas including these terms may have obstructed Argentina's claim to sovereignty through territorial integrity.

After much pressure from the Falklands lobby, which I outline in chapter 3, the UK's position on preserving the "wishes of the islanders" in any settlement became critical to the breakdown in negotiations leading to the violent 1982 conflict.[45] In her memoir, Thatcher wrote, "We looked carefully at the resolutions and at the charter. The battle on self-determination is between those who believe that 'interests' can be paramount (in which case governments can decide what is in the best interests of the people) and those who say that the peoples [*sic*] expressed wishes are the proper guide to these interests and the proper yardstick for self-determination."[46] Despite wavering in the years preceding the war, the British position has since centered on the "paramountcy" of the islanders' wishes.[47]

Refusing to accept the islanders' wishes as "a people," Argentina's government has repeatedly referred to them as an "implanted population" (*población implantada*).[48] Under Fernández de Kirchner's presidency, the Argentine Foreign Ministry created a Malvinas Secretariat, dedicated specifically to critiquing what they viewed as the UK's unjust "usurpation" and systematic occupation of the islands.[49] From this perspective, the islanders were glossed as mere pawns for the UK's exercise of imperial control over the territory for strategic military advantages and natural resource wealth.[50]

To defend their wishes to remain British, the islanders have hinged their own argument on the apparent historical absence of a colonial clash with a living Indigenous population in the Falklands. In contrast, they sharply critique the history of genocide in Argentina. The FIG points out that Argentina, which they call an "aspiring colonial power" with "an expansionist agenda," is a settler nation itself, and that most of its citizens have fewer generations of continuous residency than some Falkland Islanders. The FIG's MLAs insist that the Islanders are "a people" in their own right, and not a downtrodden colony paid to stay in the Falklands by the British government.[51] Despite being well defended by the Mount Pleasant military base, Falkland Islanders describe Argentines as bullies. They accuse their neighbors of intimidation and even economic terrorism for creating obstacles to shipping and travel.[52] In this sense, most islanders view the sovereignty dispute as an impediment to progress. One Stanley resident described the Falklands metaphorically as "an airplane on due course, but every now and then you hear noises from the engine: those noises are Argentina."[53] No matter how much pressure the Argentine government applies, Falkland Islanders are unlikely to budge from their firm stance on British sovereignty.

In short, this sovereignty saga has deferred indefinitely the fulfillment of either Argentine national interests over territory or the Falkland Islanders' wishes to determine their own future in the South Atlantic. Readers may draw their own conclusions, but this book is not a defense or apology for any particular position in the sovereignty dispute. It is not an exhaustive account of armed conflict or an authorized war history. Nor is it a plea or proposal to advocate for a new diplomatic solution to resolve the ongoing imbroglio. Instead of arguing in favor of a national or international cause, this book engages critically with contradictory aspects of these contending claims. Taking an anthropological approach against methodological nationalism, it seeks to make strange these familiar geopolitical assertions in the South Atlantic.[54] I invite readers to step back from the blood spilled on the battlefields and the normative order of international law. Let us redirect our focus instead on the collectivity of people at the center of the dispute by probing the colonial roots and current predicaments of a twenty-first-century settler society. We will do so through historical investigation of the nativist narrative of settler indigeneity among Falkland Islanders, as well as ethnographic description of how their assertion of British sovereignty has taken material form through natural resource governance in the South Atlantic.[55]

Environmental Wreckage and Imperial Salvage in a Perpetual Frontier

Beyond offering the first full ethnographic monograph of society and environment in the Falkland Islands, this book makes contributions to four main scholarly fields: (1) Atlantic history; (2) empire and imperialism; (3) settler colonial studies; and (4) social scientific research rooted in environmental anthropology and critical geography, particularly work at the conjuncture of political ecology and science and technology studies (STS). The following section contains more academic jargon than the rest of the book, so general readers are welcome to skim or move to the next section.

First, this book is an Atlantic history of the imperial present.[56] Atlantic historians have largely restricted their scope of analysis to the early modern era (1500–1800), a stretch during which European imperial powers asserted global aspirations through intense maritime activity within the Atlantic Ocean.[57] The virtual borders of such a "world" may have dissolved through industrialization, nation building, and globalization.[58] However, the Atlantic remained the main circuit of British capital and the slave trade in the nineteenth century.[59] In 1833, the same year that the UK abolished the legal institution of slavery throughout most of its colonies, giving way to freedom struggles in the Caribbean, the British permanently established their colony of the Falklands in the South Atlantic.[60] This case of delayed—yet still ongoing—British possession of an overseas territory allows us to not only excavate the Atlantic World's colonial heritage but also interrogate what imperialism looks like today.

Second, our point of departure for analyzing how empire has endured is to examine various forms of resource extraction as the accumulation of what historical anthropologist Ann Stoler calls "imperial debris," and more specifically in our case, wreckage.[61] The Falklands are often portrayed as remote and isolated, but even as a colonial outpost of the British Empire, the archipelago has long been a commercial hub. Before the opening of the Panama Canal in 1914, much of the traffic and goods transportation between the Atlantic and Pacific Oceans went through the islands. Distressed ships, including the *Lady Elizabeth* in Stanley Harbour described in my opening vignette, took refuge there for supplies or repair (figure 5). Before agriculture became central to the islanders' economy, ship chandlery and wreck salvage helped to define the South Atlantic as an extractive frontier. Shipwrecks were treated as marine resources with an extended use life in a colonial circular economy. Settlers extracted ships' cargo, repaired the wrecks at extortionate rates, or reused them as floating warehouses for storage of other goods. If provisions were not readily available, islanders wielded their local authority to detain vessels. In other instances, they ventured outside the rule of law and resorted to outright piracy, theft, and plunder.[62]

South Atlantic settlers thus perceived the stranded wrecks surrounding them as products of "second nature" from which to generate profit.[63] Here, "wreckage" refers not only to the spectacle of disaster for seafarers but also the opportunity of salvage for settlers.[64] Most of this book is not about literal shipwrecks; treasure hunting enthusiasts should look elsewhere for triumphant tales of rescue missions, maritime archaeology, and wreck discoveries.[65] Nonetheless, rooted in this local history, the conceptual metaphor of wreckage is helpful for framing

FIGURE 5. Wreck of *Lady Elizabeth* in Stanley Harbour.

Source: Brian Gratwicke (available online: https://commons.wikimedia.org/wiki/File:Lady_Elizabeth_shipwreck _IMG_6593.jpg).

how extractive capitalism has allowed imperialist projects to prosper in the South Atlantic. This *environmental wreckage* exemplifies the creative modalities of frontier capitalism that anthropologist Anna Tsing has called "salvage accumulation."[66] Tsing uses the term primarily to study the global commodity chain of matsutake mushrooms, comestible fungi that grow in the "ruins of capitalism," particularly deforested industrial timber plantations.[67] Even though the Falklands are virtually treeless, the local economy has been built on the wooden ruins of colonial wreckage. In addition to the numerous uses for shipwrecks, themselves, some houses in Stanley were constructed out of wood originating in Tierra del Fuego, where Yagán (or Yámana) Indigenous people cut and hauled timber for starvation wages of a biscuit per day.[68] As historians Peter Linebaugh and Marcus Rediker have analyzed, such "hewers of wood and drawers of water" have long been overlooked as "the laboring subjects of the Atlantic economy."[69]

By focusing on how the Falkland Islanders not only salvage capital but also assert British sovereignty through these historical transformations of the South Atlantic environment, this book thus extends scholarship on imperial ruins and "ruination" that stresses the continuity of empire.[70] According to anthropologist Gastón Gordillo (2014, 253), ruins may become "time capsules" that condition the possibility for contested politics of Indigeneity, territory, and resources.[71] Linking colonial ruins to current forms of industrial destruction, Gordillo draws connections between and among the rubble of Jesuit missions, stranded steamships, abandoned trains, and forests razed for agribusiness in Argentina. Similarly, this book traces lines of continuity from salvaging shipwrecks, herding feral cattle, sealing, and whaling to sheep raising, commercial fishing, and oil drilling. It examines how the South Atlantic has undergone successive regimes of intensive land use and resource production, as well as how these extractive enterprises have become irrevocably linked to contested sovereignty claims. Following Gordillo and critical thinker Walter Benjamin's notion of modern progress as a sedimentation of catastrophic events, I thus interpret "wreckage" broadly—not as a relic of the past but rather as an ongoing process that informs the imperial present in the South Atlantic.[72] In this sense, environmental wreckage is not just a mode of producing surplus value during an earlier colonial period; it has also been a way to continue salvaging empire while upholding modern liberal values of economic growth and self-determination.

By using the term "empire" here, I am not name calling or hyperbolizing, but instead reflecting the historical record and empirical fact of imperial domination over land, territory, and resources. As a British Overseas Territory, the Falklands are governed through what Stoler has called "degrees of imperial sovereignty."[73] According to Stoler, "'Degrees of sovereignty' describes a principle of governance that convenes a contested political relation. It resituates focus on an embattled space and a longer temporal stretch, in which gradations of rights, deferred entitlements, and incremental withholding or granting of access to political and economic resources shape the very conditions that imperial formations produce and productively sustain."[74] This political concept is influenced by the critical insights of legal scholar Antony Anghie, who has analyzed the role of positivist jurisprudence and international law in facilitating imperial domination by promoting Western norms of sovereignty and self-determination.[75] Anghie argues that the Mandate System and the League of Nations resulted in protectorate forms of dependency, administered through colonial and postcolonial sovereignty in peripheries like the South Atlantic. Nonetheless, political scientist Adom Getachew has drawn on the historical writings of anticolonial critics and nationalists across the Black Atlantic to show how these powerful figures reinvented self-determination in "postcolonial cosmopolitan" projects of national and regional world making.[76] Building on these critical theories of imperialism, this book shows how Falkland Islanders have attempted to salvage the remains of empire through a white settler reappropriation of self-determination.[77]

Third, this book engages settler colonial studies, influenced by Indigenous political critique, to consider how *imperial salvage* functions not just through dispossession of land, infrastructure development, and resource extraction, but also claims of peoplehood that follow a particular logic of settler indigeneity.[78] Critical scholars of settler colonialism have reconceived self-determination as a phase of state policy that perpetuates colonial regimes of land tenure through gestures of multicultural recognition.[79] They draw parallels between self-determination and other legal processes of "eliminating and excluding Natives" in the seizure of land.[80] For instance, Mohawk anthropologist Audra Simpson has described how consent often performs the trickery of a "ruse" that makes theft of Native land seem like freedom, even after dispossession.[81] This book extends this nuanced critique to analyze the reappropriation of self-determination by settlers. In the case of the Falklands, self-determination may be understood not strictly as a politicization of Indigeneity but as a "native" category for asserting settler exclusivity based on local residence and environmental governance.[82] In this instance, by claiming self-determination, settlers have sought to salvage empire from Argentina's threat of decolonization through continuous occupation and resource extraction, as well as disaffiliation from the imperial legacy of direct alien rule over Indigenous peoples.

Across the Atlantic World, the extraction of wealth and power from shipwrecks accelerated settler colonialism through dispossession of coastal Indigenous peoples.[83] My own field of American anthropology was founded on Franz Boas's romantic antiracist mission to treat Indigenous cultures, languages, artifacts, and bodies as forms of wreckage in their own right that needed to be preserved through "salvage ethnography."[84] Discussing racial dynamics of intelligence and culture in his influential book *Mind of Primitive Man*, Boas drew on a South Atlantic story of what he called "Darwin's Fuegian who was educated in England and returned to his home [Tierra del Fuego] where he fell back into the ways of his primitive countrymen."[85] I revisit this story in chapter 1 and show how O'rundel'lico (also known as James "Jemmy" Button)—the Indigenous individual whom Robert FitzRoy captured on May 11, 1830, as part of his imperial scientific project—did not simply regress to a "primitive" state after returning to South America.[86] As I will describe, O'rundel'lico eventually participated in an overlooked anticolonial uprising with other Yagán people who refused to accept wealth inequality and abusive colonial work conditions in a missionary settlement located on the Falklands' Keppel Island. Even though there is no historical record of a living Indigenous population at the initial moment of European colonization in the Falklands, the missionary settlement in the islands was used as a staging ground for the conquest of South America, comparable to the key role residential schools played in the cultural genocide of Indigenous children in North America.[87] This book situates the islanders' assertions of self-determination and settler indigeneity in this broader regional context of colonial violence and dispossession in the South Atlantic.

Fourth, rather than preserving Falkland Islanders as an isolated culture in a naïve ethnographic community study, this book uses critical tools from political ecology and STS to interrogate how settlers have salvaged empire by forming global connections with multinational corporations and environmental scientists to capture and conserve resources.[88] The 2013 referendum, held amid a boom in commercial fishing and offshore oil drilling in the South Atlantic, made the islands an apt site for using methods from environmental anthropology and critical geography to examine the relationships between and among authoritarian populism, resource "extractivism" and environmental governance.[89] Without either narrowing the focus to the sovereignty dispute or omitting it from the scope of analysis, this book examines ethnographically how the geopolitical controversy figures into everyday engagements with the landscape, natural resources, and the environment. It does so by conceptualizing the South Atlantic as a *perpetual frontier*: an emergent arena for natural resource governance and environmental science where the islanders' contested peoplehood and commercial interests remain in a constant state of deferral.

Despite the promise of wealth, natural resources—particularly oil and gas—are also widely regarded as a peril. Research on modernization in oil booms indicates that the sudden surplus in revenue during oil development poses considerable challenges to democratic governance. Prominent economists and political scientists have argued that resource-reliant postcolonial nations are "cursed" with economic stagnation, environmental degradation, political conflict, and authoritarian regimes.[90] Political-ecological and ethnographic research, however, has called into question the limits of the "resource curse" hypothesis, describing how predicaments of the poor mesh uncomfortably and in locally specific ways with elite enclaves.[91] This book explores such unequal social formations in the Falklands by focusing on the relationship between the islanders' self-determination claim and their governance of land, territory, and resources.

In his influential book *Carbon Democracy*, Timothy Mitchell suggests that carbon energy has conditioned the possibility for democratic politics.[92] Not all petro-states are liberal democracies with market economies, but according to Mitchell, former imperial powers have promoted self-determination in order to obtain consent to being governed as protectorates for energy production.[93] Local governments at extraction sites have then become "Native agents" of empire, embedded in what geographer Michael Watts calls the "oil complex": a technopolitical network of engineers, scientists, shareholders, state functionaries, and military personnel.[94] These insights resonate in the context of the Falklands, the difference in this case being that the islanders' self-determination claim, as well as their mineral resource licenses, are under constant scrutiny from Argentina. Amid the sovereignty dispute, this book examines how islanders have sought to develop new ways of securing access to and control over natural resources and the environment as a means to ultimately salvage empire.[95]

Resource extraction may be motivated by material interests, but scientists, planners, and engineers also coproduce standardized packages of political worldviews in making knowledge claims about nature.[96] By examining how the sovereignty dispute forms assumptions for experts assessing environmental impact, this book analyzes the situated knowledge with which resource managers and scientists preserve and protect the South Atlantic.[97] It enters the "technological zones" through which expert procedures produce a discourse of confidence and conservation that enhances extractive industry interests.[98] Rather than considering alternatives to making the South Atlantic a sink for oil pollution, I analyze how risk assessment reproduces what Red River Métis/Michif scientist Max Liboiron describes as "colonial entitlement" to nature, resources, land, and life.[99]

Both the UK and Argentina have declared a climate emergency based on abundant scientific evidence of global warming, and just south of the Falklands, Antarctica's vast ice sheets are breaking off. This book shows how these competing powers nonetheless seek to reinforce their respective claims to sovereignty over resources—including fossil fuels—with complicit support from environmental scientists. I consider how sustainability science has made such extractivism seem "green," as entrepreneurs and environmentalists try to refashion the

South Atlantic as an ecotourism attraction and a globally important place for biodiversity conservation.[100] Ultimately, the book examines how settlers have shifted their relations with the environment from the "ecological imperialism" of exterminating native pests to assertions of indigeneity in non-native species eradication and habitat restoration.[101]

The 2013 referendum showed near-total consensus on British nationality and citizenship, but questions remain. How do the islanders negotiate asymmetries in land access, language use, race, and ethnicity as they recast themselves as "a people" with resource rights? And, as they try to account for the risks involved with extracting resources, how are the islanders naturalizing their colonial heritage of environmental wreckage as stewards of the South Atlantic with scientific claims to authority over biodiversity conservation? With independent nationhood deferred indefinitely, the islanders are trying to control their own future through resource management.

I argue that by consenting to British sovereignty through self-determination, the Falkland Islanders have crafted a settler colonial protectorate. This political formation has allowed settlers to secure land, extract resources, manage the environment, and assert indigeneity. Ultimately, it has authorized Falkland Islanders to salvage empire in the South Atlantic.

Argonauts of the South Atlantic

Remote archipelagos have, of course, been fruitful sites for ethnographic research since the field of anthropology's very beginnings.[102] Rather than analyzing the internal logic of an isolated "people without history" in an out-of-the-way place, this book examines how, despite its location on the global periphery, the South Atlantic has established powerful connections on multiple scales.[103] Observing how Falkland Islanders have presented their self-determination claim as evidence of stability for regional trade partners, multinational oil firms, and investors in the City of London, I have sought not just to represent subaltern subjects but also to "study up" among experts and elites.[104]

The research informing this book incorporated twenty months of participant observation, analysis of colonial letters and reports, and more than 100 interviews conducted with (1) townspeople, farmers, migrants, resource managers, scientists, planners, engineers, and business elites in the Falklands; (2) government functionaries, scientists, and defected islanders and their displaced descendants in Argentina; and (3) oil executives, repatriated islanders, and government representatives in the UK. During the austral winter of 2012, I carried out field research in Buenos Aires, Tierra del Fuego, and the Falklands. For local perspectives on the Malvinas cause in Argentina, I attended relevant events and established affiliations with local anthropologists and historians. I then traveled to the islands for two weeks of exploratory fieldwork. I visited the Falklands for a second time during the March 2013 referendum on self-determination. I observed balloting both in town and in the countryside along one of the five unpaved mobile polling routes. I administered surveys before, during, and after the vote and then used my survey as a basis to structure interviews and observations. My observations during the referendum were published in *The Economist.*[105] Upon my return to the US, that affiliation gave me media accreditation to attend and report on relevant UN meetings, where I interviewed petitioners from the islands and Argentina as well as diplomats from other member states.[106]

I returned to the islands for extensive field research from January to May 2014, and then again from September to December 2014. Most of my research took place in Stanley (see figure 6), but I seized every opportunity to collect data in Camp. I visited multiple Camp settlements by automobile, ferry, or interisland flight. In addition to interviewing residents during the referendum on the mobile polling route, I went lamb marking, attended a "bachelor and spinster's ball" held in a shed, and took part in Camp Sports on West Falkland Island, which

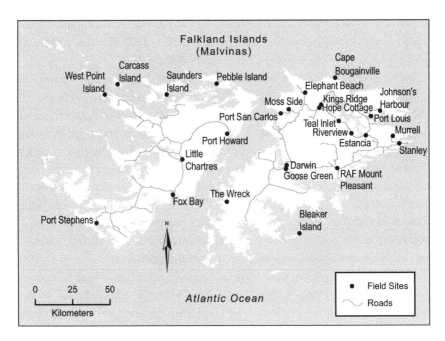

FIGURE 6.   Field sites in the Falkland Islands (Malvinas).

Map by Terry Spies (data available online: http://dataportal.saeri.org/).

included sheep shearing, peat cutting, and other rural athletic contests. Much of my fieldwork also involved participating in the field research of marine ecologists, particularly tracking penguins with locational sensors, which I discuss in chapter 6.

To capture how debates around infrastructure and resource management are forming new relations of power on the islands, I observed and analyzed plans for development and financing. I interviewed a wide variety of actors in the government and private sector, examining especially closely the plans for a new commercial port and temporary dock (see chapter 5). With office space in the South Atlantic Environmental Research Institute (SAERI), I had an inside perspective on environmental impact assessments of the proposed oil development, and I participated in biannual meetings of the Falkland Islands Offshore Hydrocarbons Environmental Forum (FIOHEF).

In rural settlements, I observed changing relations with the landscape. Stone ruins offered traces of gaucho corrals, and fences indicated layered property relations. These kinds of observations were complemented by work in the local archives, where I examined colonial reports, gazettes, diaries, letters, periodicals, and early maps for historical perspective on resource governance. Finally, to grasp how the islanders are using their self-determination claim as leverage for enhancing the prospects of oil, I conducted interviews with resource managers— one of whom was my generous housemate—as well as visiting consultants. Once relations were established with participants, I gathered life histories with Falkland Islanders as well as migrants and contract workers.

Between visits to the islands, I carried out a shorter phase of fieldwork in Argentina from May to August 2014. In Buenos Aires, I primarily conducted archival research at the National Archives. To gain a better understanding of Argentina's contemporary claim, I interviewed functionaries of the Ministry of Foreign Affairs. I also interviewed scientists involved in the government's "Pampa Azul" (Blue Pampa) campaign, which is designed to showcase environmental research in the South Atlantic (see chapter 6). In the Patagonian province of Santa Cruz, I met defected islanders and their descendants. Interviewing these actors and recording their life histories has provided for a more robust analysis of histories of migration and settlement, experiences with the dispute, and opinions on resource management in the South Atlantic.

I began analyzing data and writing in early 2015 in the UK, where I also conducted targeted interviews and further archival work. Through networks established during my initial research, I met and interviewed repatriated islanders connected to the Falklands lobby in London. At the National Archives in Kew, I examined colonial reports and gazettes, some of which are omitted from the islands' archives. I visited the London offices of firms involved in oil exploration, where I interviewed top executives, and I met oil affiliates and supply chain consultants in Aberdeen. I was also invited to participate in a workshop on data analysis and marine spatial planning in the South Atlantic at the University of Cambridge. This became part of a wider critical dialogue that my research generated with scientist interlocutors.

Unlike much of the British or Argentine scholarly research on the Falklands/ Malvinas, I have conducted this project from the relatively unsentimental position of being an American researcher. In other research contexts, I have engaged reflectively with local social movements in the role of an advocate for marginalized communities and their environments.[107] During the fieldwork for this book, I made friends with many individuals and sympathized with their devotion to the islands. I also grew a deep fondness for this unique place and its stunning landscape. However, I sought to offer ethnographic insights that might strengthen social equity and environmental governance without directly consulting or accepting benefits from industry or state actors.[108] As I discuss in chapter 6, there is not only a prior history of hostile relations between Argentines and islanders but also at least one instance of a whistleblower who raised concerns about potential impacts of extractive industries on penguins and other wildlife, received personal threats from political and business elites, and ultimately became exiled from the Falklands.

Despite careful consideration of these ethical problems, the sovereignty dispute and industry gatekeeping presented several dilemmas. I never felt unsafe, and my hosts were always hospitable, but I have had to avoid being entangled in the imbroglio. Functionaries of the Argentine Foreign Ministry tried to persuade me to send them private documents; I declined to do so. Royal Air Force military personnel hinted that I had been monitored. Ministers of the UK Foreign and Commonwealth Office asked for advice on lobbying US Congress. Oil managers threatened to destroy my phone until I showed them that it was not recording. The Falkland Islands Government invited me to apply for a tender to conduct a social impact monitoring program on the proposed oil development—a tempting opportunity to apply my research concretely—but it would have risked potential fines and imprisonment in Argentina, so I did not submit. Newspaper or magazine editors drew on my research and reporting to publish stories about the Falklands that followed a line of neoliberal multiculturalism and imperialist nostalgia that I did not endorse. I even received a veiled threat of potential defamation action from a high-ranking official if I published anything about their involvement in certain scandalous topics.

The sovereignty dispute also impacted the logistics of my multicountry research itinerary. For example, when my debit card went missing in Patagonia, I thought I had succeeded in explaining to the Bank of America agent where the

Falklands were located until I discovered later that they had sent a replacement card to an address in Djibouti! Their explanation was that "Malvinas," not "Falkland Islands," appeared on Google Maps. To reach the islands from South America, there was one flight per week from Punta Arenas, Chile. After conducting research on the Argentine side of Patagonia, I had to travel to Chile to then fly to the islands. However, when I arrived at the bus station, I was told that the customs agency workers were on strike for a week, which forced me to miss my scheduled flight. There are surely inferior places to be stranded than the magnificent glaciers and mountain ranges of Patagonia, but the uncertainty surrounding border crossing made it difficult to enjoy the layover. Argentine customs agents questioned me when they discovered passport stamps labeled "Falkland Islands"—or worse, date stamps that did not match because previous Argentine customs officers had purposefully neglected to mark my tourist visa upon reentrance from the islands, presuming it was domestic and not international travel. It was a relief to get past the border checkpoint where, as if to taunt me for missing my flight, a sign read: "Las Malvinas son Argentinas" (The Malvinas are Argentine; figure 7).

FIGURE 7.   Border crossing near Río Turbio, Argentina.

Photo by James J. A. Blair.

Chapter Summary

This book is structured in three parts that guide readers through the process of imperial salvage in the South Atlantic. Despite the apparent absence of a classic colonial encounter between European colonists and prior Indigenous inhabitants, the story begins by addressing the violence of dispossession that nevertheless occurred before arriving at the wreckage of extractive industries and the possibility of survival through biodiversity conservation. This is not a standard ethnography in which all of the historical content is frontloaded in the first chapter, though chapter 1 does provide a foundation of "history from below."[109] As an Atlantic history of the imperial present, the sedimentation of the past is layered throughout each chapter of the book.

Drawing on colonial letters, diaries, and ordinances as well as interviews with descendants of early settlers, the first chapter outlines the history of settlement and the movement of people between the islands and Patagonia. A cattle frontier built on conditions of debt peonage resulted in the alienation and exile of South American gauchos after permanent British occupation. Addressing the legacy of a fraught missionary settlement in the Falklands, the chapter revisits the story of the Indigenous Yagán figure O'rundel'lico. The second chapter analyzes how islanders have adhered to and customized British imperial property ideologies of enclosure and improvement of the commons. It describes how the Falklands' property regime has transformed land tenure from a usufruct colony founded on debt bondage to an absentee-owned monopoly of community sheep ranches and the current system of privatized small family farms. Chapter 3 then analyzes social contradictions inherent to the colonial subjection of settlers and the islanders' assertion of peoplehood. It traces the local construction of a white ethnic community: locally born "kelpers." It then describes how, since the 1982 war, Falklands society has become an imperial diasporic mixture of islanders, British contractors, and military personnel as well as migrants. Despite apparent multiculturalism, citizenship remains contingent on degrees of imperial sovereignty, and strict immigration laws and customs restrict access to the "Falklands Way of Life."

Shifting to environmental wreckage, chapter 4 describes how the islanders have built upon past extractive economies to craft new resource regimes. It relates speculation surrounding plans to develop the Sea Lion oil field with past strategies to capture wealth by salvaging shipwrecks, sealing, whaling, penguining, and commercial fishing. As hopes rise for an oil bonanza, local visions of becoming a commercial energy exporter threaten to eclipse previously established modes of resource production. The fifth chapter examines how islanders have used infrastructure as a means of resolving the temporal paradox between retrospective peoplehood built on a shared past and prospective personhood constituted in an uncertain future of oil production. It pays particular attention to the construction of the aptly named "Noble Frontier" temporary dock facility, built to accommodate oil exploration, as it relates to mired plans for a permanent deepwater port. It then examines shifting debates over the engineering design of the Sea Lion drill site. By attending to the incremental process of oil development, the chapter throws new light onto not only petro-capitalism but also settler colonialism, which tends to emphasize durable structures over infrastructural events.

Chapter 6 analyzes the enclosure of the South Atlantic as a knowledge frontier. It examines how the UK and Argentina secure their respective claims to resource-rich maritime territory through scientific research. Centering on the Data Gap project, financed jointly by the FIG and its offshore oil licensees, the chapter describes how marine ecologists tag penguins with tracking sensors and curate their data as part of a new geographic inventory of the South Atlantic. Data representing the penguins' foraging patterns feed into environmental impact assessments and are intended to influence where oil firms plan to locate drilling and fuel transfer sites, but transparency measures serve geopolitical and corporate interests at the possible risk of local ecological damage. Finally, toward a theory of settler indigeneity, the final chapter examines quotidian practices to ensure biosecurity and preserve biodiversity. It discusses the implications of Darwin's cross-species comparisons during his voyage of the *Beagle* and then tracks how the historical disgust that Darwin and early settlers felt toward Yagán Natives of Tierra del Fuego influenced the desire to expel nonhuman bodies and objects in various eradication campaigns. I contend that by "colonizing" *with* native species in ecological restoration projects, settlers aim paradoxically to assert indigeneity while salvaging British heritage.

Fundamentally, this book is a historical ethnography that examines how settlers have asserted indigeneity in dynamic relation with the South Atlantic environment and explores how Falkland Islanders and their trustees try to salvage empire through resource management. By examining geopolitical aspects of the sovereignty dispute in ethnographic detail, the pages that follow relate the politics of energy and environmental science to the history of empire and international law.

Part 1

DISPOSSESSION

SETTLER SAFE ZONE OR COLONIAL STAGING GROUND?

Archaeological research indicates that humans were present in the Falklands centuries before European colonization.[1] However, there is no historical evidence of an initial colonial encounter between European settlers and Indigenous people in the islands upon the Europeans' arrival. This irregularity made the South Atlantic seem like a safe zone for imperial expansion, free from the risk of rebellions by Indigenous and enslaved people that were common in other colonial frontiers. Indeed, a visitor to the islands reported the following in a British military publication:

> The settler on the Falkland Islands need not fear the many disappointments and almost insurmountable difficulties experienced by the hundreds who embarked their all in the Swan River scheme—He need not dread, on his return home from a journey, to find his wife and children murdered by the ferocious and blood thirsty savage, as has lately been the case in Van Diemen's Land—neither need he fear to hear the warwhoop of the Indian burst upon his ears, as he is assembled round his domestic hearth with his family, as was not long ago the case in the back settlements in America—he need not fear as in the African Settlements, the murderous attack of the Caffre—neither has he to reside among a number of slaves, against whose rising he has not a moment real protection. He has only steadily to pursue his aim, certain of never being in want, and with every prospect of acquiring wealth.[2]

However, European colonization of the South Atlantic archipelago did not actually occur in such a free and fearless fashion. Far from creating a hermetically sealed settlement, the founding of the colony was marked indelibly by uprisings of toiling laborers and Indigenous peoples of South America, who became linked to the islands through the interventions of commercial enterprises and evangelical missions.

Despite their unwavering British imperial heritage, Falkland Islanders now liken themselves to an Indigenous people facing a constant colonial threat—from Argentina. "They want to take over our islands and take all our wealth. The same that they did with the Mapuche Indians," one elected counselor told me. Islanders back their self-determination claim with British residents of up to nine generations. They point out that Argentina's claim to sovereignty through territorial integrity belies its violent history of national expansion through dispossession. Yet, despite its own history of forced displacement of Indigenous peoples, Argentina's government has continued to dismiss the islanders as an "implanted population."

This chapter describes how the settlement of the Falklands and the colonization of Patagonia became mutually constitutive. As a strategically well-positioned colonial frontier, the archipelago served as a launching pad for the conquest of Patagonia and the dispossession of prior Indigenous inhabitants of South America more generally.[3] In turn, the islands' settlers felt ripples of resistance in rebellions of overexploited workers and Indigenous peoples crossing the Southern Cone and South Atlantic. Far from a utopian settlement, perceived as safe from the unsavory aspects of colonialism, the Falklands have been embroiled in violent upheaval since their founding.

Who Colonized First?

Given the implications for contemporary sovereignty claims, the European "discovery" of the islands is the subject of considerable scrutiny and hot dispute among historians and scholars of international law and politics.[4] Falkland Islanders and their British trustees have long held that the English captain John Davis first sighted the archipelago on his ship *Desire* in 1592 (hence the islanders' official slogan "Desire the Right").[5] However, Iberian maps dating from up to fifty years earlier feature a comparable archipelago called Sansón, suggesting that Magellan (or Esteban Gómez, one of his deserters) encountered the islands on his famous 1520 expedition.[6] Dutch sailor Sebald de Weert likely observed the islands in 1600 soon after encountering another Dutch crew that had shot and killed twenty-five to forty Selk'nam (or Ona) people and captured six of their children in 1599.[7] This may have been the first instance of violent bloodshed inflicted upon Indigenous peoples leading to genocide in Tierra del Fuego. Finally, British commander John Strong may have been the first European to actually set foot on the islands in 1690 on the *Welfare*. He named the sound separating the two main islands Falkland, which later became the English name for the full chain.[8]

Even though they have no part in the current sovereignty dispute, the French were, in fact, the first Europeans to form a settlement in the islands. To build a settlement, French officer Louis Antoine de Bougainville assembled an expedition of sailors and merchants from Saint Malo who gave the archipelago the French name Iles Malouines, from which the Spanish name Malvinas derives. Significantly, in addition to these residents, de Bougainville's colony consisted primarily of displaced Acadians already at home in a North American settler colonial milieu. From a hemispheric perspective, in its earliest recorded settlement, the European colonization of the islands was thus an extension of the conquest of the Americas.

In early 1764, de Bougainville and his crew sailed into what is now called Berkeley Sound on East Falkland Island (in Spanish, Isla Soledad), claimed possession of the islands for Louis XV, and constructed a fort, St. Louis (hereafter, Port Louis).[9] The French colonists built a sustained colony with wood from Patagonia and abundant oil and meat from seals and sea lions. They also collected penguin eggs, hunted geese, and harvested root vegetables and peat for cooking and heating.[10] Despite bright prospects for developing a commercial sealing and whaling outpost, the French colony was short-lived. In 1767, de Bougainville delivered the settlement to the Spanish, who renamed it Puerto Soledad.

However, before the Spanish took possession of de Bougainville's colony, the British commander John Byron (grandfather of poet Lord Byron) had already independently surveyed another section of the archipelago at Port Egmont on the northwestern Saunders Island in 1764. Without apparent knowledge of the French settlement, Byron's crew occupied Port Egmont and planted a vegetable garden that a garrison led by Captain John McBride inherited two years later.[11] The British settlement grew under McBride's successor, Captain Anthony Hunt, until the Spanish removed the garrison forcefully but without any casualties in 1770.[12] After much combative diplomacy, the British reoccupied Port Egmont with Spanish permission in 1771, only to abandon it again in 1774.[13]

The Spanish managed to take over the French fort in 1767 and secure exclusive control over the islands from 1774 to 1811, while the British were preoccupied with the American Revolution and the Napoleonic Wars. Nonetheless, the islands were never hospitable to the Spanish settlers. The first of twenty governors, Felipe Ruiz Puente commanded an original garrison of over 500 personnel, of which roughly 130 would remain in Puerto Soledad, with these consisting of a fluctuating population of soldiers, seamen, seventeen remaining Acadians, as well as servants and slaves.[14] The settlement grew with more than 2,000 head of cattle and up to thirty-eight buildings made out of stone, imported wood, or turf.[15] But, lacking sustained vegetable or peat cultivation, the Spanish relied on annual provisions from Río de la Plata. Struggling to supply adequate rations, they suffered from shortages and used the settlement primarily as a military outpost and penal colony.[16] If they had not inhabited the islands at the moment of European colonization, the first Indigenous South Americans did inhabit the Malvinas during this period—as prisoners.[17]

Irrespective of the sovereign military power, transient sailors, particularly North American sealers, occupied the islands' coasts throughout the early colonial period. From 1811 to 1820, the islands served as an ungoverned haven for sealers, whalers, and *gauchos*—debt-bonded cattle herders—of various nationalities. On November 6, 1820, David Jewett, an American naval officer turned privateer commander assisting in the South American independence campaigns, sailed the frigate *Heroína* into the former Puerto Soledad (Port Louis) and raised the Argentine flag among the sealers.[18] The short-lived colony of Buenos Aires that existed in the Malvinas between Jewett's act and permanent British occupation and settlement is the territorial legacy on which present proponents of the populist Argentine national sovereignty cause pin their hopes.

The Lost Colony of Buenos Aires

In addition to sealers and whalers, feral cattle flourished throughout Isla Soledad (East Falkland Island) during the power vacuum of the early nineteenth century. While Jewett did not remain in the islands after planting the Argentine flag, Luis (or Louis) Vernet, a merchant originally from Hamburg, seized the opportunity to make a longer-term investment. Vernet and his business partner, Jorge Pacheco, signed a contract on August 25, 1823, for usufruct rights to cattle from the government of Buenos Aires.[19] On December 18 of that same year, the government gave Pacheco a concession for a tract of land for Vernet to build a colony on Isla Soledad.[20] After an initial expedition failed in 1824, Vernet enlisted three foremen and twenty-three gaucho *peones* (peons) to recover his losses in 1826.[21]

In 1829, Buenos Aires named Vernet its political and military commander of the Malvinas,[22] but his colony was more of a business enterprise for raising cattle and selling hides than a nation-building project for the nascent Argentine republic. In this respect, the islands resembled the mainland closely during this period. Instituting his own paper currency, a common practice among some *latifundistas* (large estate owners), Vernet paid the gauchos one to two "dollars" per head of cattle that they caught and slaughtered.[23] The work of making horse-riding gear such as lassos and saddle pads, constructing corrals, minding cattle, building huts, and cooking beef was included in this contract. In addition to the gauchos, some of whom came from South America, Vernet's colony included five Charrúa Indigenous individuals who had been imprisoned in Montevideo: Manuel González, Luciano Flores, Manuel Godoy, Felipe Salazar, and "M." Lattore. Moreover, although it often goes unmentioned in the historiography, this colony was built on contraband slave labor. Vernet bought at least thirty-one *"negros y negras,"* enslaved people captured previously during the war with Brazil, from an associate in Patagones on August 7, 1828, along with horses and other supplies.[24]

"Gaucho," as current Falkland Islanders are keen to point out, refers to *occupation* rather than national origin. While instrumental to state formation in Argentina, the gaucho did not become symbolic of authentic Argentine national identity until the turn of the twentieth century.[25] Gauchos were usually classified racially as *mestizos* (mixed race), although various colonial letters group the Charrúa captives together with them as either gauchos or as *indios* (Indians, in English).[26] Similar to their status in other cattle frontiers of the Americas, Vernet's hired gaucho cattle herders were subjugated in a system of debt peonage and marginalized as "barbarous" vagrants or bandits.[27] Vernet's colony of around 100 individuals also included Dutch, German, and British colonists. He tried, but failed, to persuade then-UK minister to Buenos Aires Woodbine Parish to recruit more British settlers and "capitalists abroad," to replace the merchants of Buenos Aires, whom Vernet berated for having "no spirit of enterprise."[28]

Vernet envisioned his colony not just as a gaucho satellite for cattle herding but also as a potential commercial center for fishing and a launching point for conquering Patagonia (as indicated in the anonymous quote above, it was perceived as a safe haven from potential confrontations with Indigenous peoples). He wished to capture wealth from the sealing and whaling activities already under way by establishing settlements of at least three families and twelve men on each island, possibly with a naval warship to enforce regulations.[29] Vernet also gained control of Staten Land, a wood source closer to the South American coast. He and his wife Maria enjoyed the cosmopolitan comforts of a library of books in Spanish, German, and English and quite a variety of other amenities.[30] They held footraces and dances, entertaining guests with music played on their imported piano.[31]

The colonial enterprise seemed to be relatively harmonious for Vernet until he tried to enforce Buenos Aires's regulations on fishing and sealing. In 1831,

Vernet detained three sealing schooners, and after taking them to trial in Buenos Aires, he was never able to return to the islands. Ironically, according to one of the last accounts of Vernet's settlement, Maria sang "Di Tanti Palpiti" in the company of the crew of one of the detained schooners. The aria begins with the words "Oh patria! dolce, e ingrata patria! alfine a te ritorno!" (Oh fatherland! Sweet and ungrateful fatherland! At last I return to you!).[32]

1833 and the Legend of "El Gaucho" Rivero

It is often overlooked that the US Navy, not the British, destroyed Vernet's settlement in 1832. With Vernet stuck in Buenos Aires awaiting trial, there was little he could do to protect his settlement when Captain Silas Duncan—commander of the *Lexington* reporting to the bellicose US president Andrew Jackson—sacked Port Louis in response to Vernet's detention of the three American sealing vessels.[33] General Juan Manuel de Rosas sent a garrison to maintain stability in the schooner *Sarandí*. However, its crew (most of whom were themselves Americans or British) staged a mutiny shortly after arriving in the islands. They assassinated Rosas's commander, Esteban Mestivier, thus leaving the colony defenseless. Just over one month later, on January 2, 1833, the British ship *Clio* arrived to decisively reclaim the islands for Britain.

As with the struggles leading to the formation of the Argentine Republic, gauchos played a central role in the violent aftermath of the event that Argentina's government views as the British "usurpation" of the islands.[34] On August 26, 1833, Antonio Rivero (also referred to as Antook), one of the gauchos formerly employed by Vernet, led seven others, including gauchos José María Luna, Juan Brasido (or Brasilio, nicknamed "Rubio"), and the five aforementioned Charrúa Indigenous prisoners, in a violent uprising, in which Vernet's former managers, French foreman Jean (or Juan) Simon, Irish steward William Dickson, and Scottish superintendent Matthew Brisbane, as well as two other inhabitants named Ventura Pasos of Buenos Aires and Anton Vaihinger (called "Wagner," a German), died.

More than a century later, Argentine intellectuals mythologized this event to usher in a populist brand of nationalism that left-Peronists have repeatedly seized upon since.[35] Here, the national identities of the two parties became dichotomized as "Argentine gauchos" on the one hand and "British colonists" on the other. Depending on one's political sympathies, historians have interpreted "El Gaucho" Rivero's uprising alternately as an apolitical crime of murder, a rebellion in defense of Argentine sovereignty against British rule, and an act of potentially revolutionary resistance against unfair labor conditions, carried over from Vernet's debt peonage regime.[36] Testimonies that the gauchos wanted to be paid in silver or gold rather than Vernet's devalued paper currency support the latter.[37] In any case, this revolt demonstrates that permanent British settlement of the islands did not occur free from violent struggles between and among colonizers, bonded laborers, and Indigenous peoples.

During my fieldwork, researchers and journalists, particularly those collecting evidence to support Argentina's sovereignty cause, lamented the restricted access to Port Louis, the primary settlement during the pre-1833 settlements of the French, Spanish, and Luis Vernet under Buenos Aires (see figure 8). However, when I ventured to the now-family-owned working sheep farm with a group of Stanley-based "ramblers" who had scheduled a visit ahead of time, the current landowner welcomed us cordially. Rolling up to the gate in his four-wheeler, he handed out maps and pamphlets identifying many of the same turf houses and enclosures that I had first encountered in the dusty reports of Vernet's estate.[38] There, I had the opportunity to take photographs and walk along gaucho-era ruins, including a stone corral, the French commandant's house and "secret tunnel," gauchos' cottages, original church and Spanish buildings, a store and gun battery, a house, a magnetic observation station, and the site of a Spanish fort. The settlement's graveyard features the tombstone of Vernet's former deputy, Brisbane, who was one of the victims of the Gaucho Rivero uprising.

FIGURE 8.   Port Louis settlement.

Photo by James J. A. Blair.

Hailing from Perth, Scotland, Brisbane had been part of James Weddell's crew, where he encountered Yagán people during one of his voyages in Tierra del Fuego near Cape Horn in 1823. Two years later, Brisbane returned to Tierra del Fuego, where he led a sealing vessel. Brisbane apparently developed a decent rapport with the Yagán people, though he complained about their pilfering and the prices they demanded for otter skins—indicating they had learned their lesson from previous instances of colonial swindling and dispossession.[39] Nonetheless, Brisbane bartered with and recruited Yagán people to work on his ship. He was resourceful and survived being shipwrecked at least three times. In 1830, Brisbane and his crew spent three months surviving out of a small boat they constructed out of the wreckage of one of those disasters, until a group of Manek'enk (or Haush) Indigenous people approached them with bows and arrows, sending Brisbane back to Port Louis. In 1831, he even survived being wrecked on the remote South Georgia Island, once again building a boat out of the wreckage of his salvaged sealing vessel.[40]

Upon returning to the Malvinas, Brisbane might have thought his job working at the colony store for Vernet would be an uneventful reprieve from the constant danger of shipwrecks or confrontations with Indigenous peoples in other colonial frontiers, but it was no safe zone. First, he was detained by Duncan and brought to Buenos Aires on the *Lexington*, and then, after being released back to the Malvinas, he was slain by Gaucho Rivero and his comrades. At Port Louis, a rather biased epitaph demonstrates how Brisbane has become symbolized in contradistinction to his assailant Gaucho Rivero as a martyr for the British. It reads: "In memory of Mr. Matthew Brisbane who was barbarously murdered on the 26th, August 1833."[41] If the British claim of peaceful settlement free of colonial violence was indefensible, they could still salvage symbolic virtue from Brisbane's death by memorializing it this way.

Charles Darwin visited the Falklands twice with FitzRoy aboard the *Beagle*. First, arriving just after British settlement in March 1833, they immediately encountered a coastal environment littered with wreckage. Darwin remarked, "It is quite lamentable to see so many casks & pieces of wreck in every cove & corner: we know of four large ships in this one harbour."[42] On a second visit, one year later in March 1834, upon disembarking his vessel, the founder of evolutionary biology found himself among a small population of Englishmen, gauchos, and other "runaway rebels and murderers."[43] Darwin described the Falklands as "ruined" and "worth nothing."[44] The Gaucho Rivero rebellion had already occurred, and FitzRoy encountered the dead body of Brisbane, whom he had admired for his fearlessness in the face of wreckage and danger.[45] I will return to Darwin's account in chapter 7, but these direct impressions indicate how the naturalist and his captain were impressed by the ruins of multiple rounds of violent dispossession in the islands.

The British eventually captured Rivero and his comrades, bringing them to England for trial in early 1834 (only to be released in Montevideo a year later).[46] The rest of Vernet's hired gauchos actually stayed in the islands under British rule. Vernet viewed the incidents involving the *Lexington* and El Gaucho Rivero as unjust trespasses on his usufruct rights to cattle and land.[47] Rejecting debt bondage, Rivero and his fellow rebels appeared to value a more egalitarian system of pasturage rights, common to this period, without any monopoly on violence (see chapter 2).[48] Argentina's left-Peronist political class now frames both of these contradictory figures as fallen heroes of their sovereignty cause, and their stories are central to the dramatic reenactments and exhibits of national pride in the Museo de Malvinas in Buenos Aires. No matter how these historical actors are remembered, the violent acts of the *Lexington* gunboat, the mutineers of the *Sarandí*, and El Gaucho Rivero offer undeniable evidence that the South Atlantic was no colonial safe zone. Rather, permanent settlement occurred amid considerable violent dispossession and armed conflict.

Integration and Alienation of South American Gauchos

After permanent British settlement, island residents of South American origin were integrated into the main colony as a labor force. In 1842, the main settlement and seat of British colonial government moved from Port Louis to Port Stanley and subsequently expanded into the rural Camp. Twenty-two of the thirty-three residents of Vernet's settlement had remained at Port Louis after the British reclaimed the islands in 1833.[49] Even after the uprising of Antonio Rivero, gauchos continued to play a central role in maintaining the livelihood of the British settlement, which initially consisted of a small naval garrison. The gauchos labored just as they had done under Vernet's regime, capturing cattle and producing meat and hides.

While most of the holdovers from Vernet's colony were men, as were the naval officers, three women remained: Antonina Roxa, Carmelita, and Gregoria Madrid. The latter two were African descendants; Carmelita had three children with three different men in the 1830s before tragically drowning herself in 1845, and Gregoria Madrid married a settler and lived in Stanley until her death in 1871. We know more about Antonina Roxa, who was described as an "Indian Princess" from Salta via Río de la Plata.[50] A local icon, not only was Roxa's role important in the domestic sphere of the settlement, but she was also considered a skilled gaucho and a "person in her own right."[51] In Stanley, she owned land and multiple houses, and accumulated a substantial herd of cattle and other livestock. After marrying and divorcing an American sailor, Roxa became the first resident to swear an oath to the Crown in 1841. For this reason, Falkland Islanders have framed Roxa as a subaltern symbol for the islanders' allegiance to British sovereignty as an inclusive, even feminist, cause. In the 1850s, she remarried a gaucho from Montevideo and acquired, leased, and worked on Camp settlements at Hope Place and Port San Carlos in East Falkland Island before her death in 1869.

As the colony grew in the mid- to late nineteenth century, settlers debated how labor recruitment would impact its embryonic sense of Britishness—with xenophobic undertones. The islands' absence of so-called aborigines and the relatively few settlers made some British colonial commissioners interested in converting the settlement into a penal colony and naval station.[52] Speculative investors moved instead toward advancing the commercialization of the cattle trade, which did occur with the establishment of the Falkland Islands Company (FIC) in the 1850s (see chapter 2). However, ideas to attract tens of thousands of small farmers never materialized, and racialized fear—similar to that expressed in the anonymously written passage at the beginning of this chapter—influenced how white settlers viewed the impact of importing "Indian gauchos" on their safety.[53] These are the early conditions upon which the contemporary settler colonial protectorate, explored further in chapter 3, was built.

Thus began a lasting racially charged debate about population control: who would be deemed "desirables" and who "undesirables"? The FIC's founder, Samuel F. Lafone, and his agent proposed recruiting five-ninths of their workforce from the Shetland Islands, where residents were considered to be "hardy and industrious" and "religious men" with "good morals." One-ninth would have to be from Río de la Plata because of their superior knowledge of cattle management. The FIC proposed that two-ninths come from southern Chile, where "the people are more docile and humble than the inhabitants of the River Plate." And the final ninth would be "intelligent Basques and Catalans" working in the saladeros of Montevideo.[54]

Collapsing this calculated division of labor by race and ethnicity, the early British colonial governors Moody and Rennie enforced their own hierarchy, which disadvantaged South Americans more generally. Passing an Alien Ordinance in 1849, the government deported South American workers—classified as Aliens, Banda Oriental (referring to what is now Uruguay), Spaniards, Indians, Spanish Indians, Indian gauchos, or Spanish gauchos—whose behavior they deemed "barbaric" or whom they suspected of spreading subversive, anti-British views.[55] "Aliens" and "Resident Strangers" were listed separately from "whites" and the remaining few "coloured" people in colonial reports.[56] To prevent Stanley from becoming "overrun by Spaniards and Indians to the great annoyance and inconvenience of the inhabitants," the colonial government and FIC required that cattle be concentrated out of reach from town in the southern plains of East Falkland Island, called Lafonia after the eponymous FIC founder.[57] By 1858, the FIC had "no Spaniard in its employ" at Stanley.[58] Finally, to displace the gauchos with a preferred population, the government encouraged immigration by residents of Buenos Aires of northern European descent.[59] When the economy shifted from cattle to sheep, colonial managers ultimately found little use for gauchos and recruited Scottish or English shepherds instead.[60]

The gauchos did not passively accept their replacement with British laborers. As their population decreased, a second, less mythologized Gaucho Rivero–like incident occurred when the gaucho Manuel Gil murdered FIC Camp manager John Rudd. There are multiple theories about Gil's motives, but based on the reactions of Governor James George Mackenzie, it is clear that there was widespread fear among the British managerial class that "all the Spaniards resident in the islands" were complicit, if not directly involved, in conspiring for the manager's death.[61] The lack of institutional control over violence in Camp is one of the reasons for the vanishing of the cattle frontier in the Falklands, and the gauchos with it. However, while the islands may no longer be a cattle frontier, the settler colony that took shape in its wake introduced new agroindustrial practices that have created a lasting frontier condition for different regimes of extractive capital.

O'rundel'lico and the Constant Gardiner

In his eloquent and incisive study on power and history, *Silencing the Past*, Michel-Rolph Trouillot destabilized conventional narratives about the Haitian Revolution.[62] In one essay, Trouillot exposed how different interpretations of "The Three Faces of Sans Souci" have silenced an untold war within the war. To Europeans and most Western guild historians, Sans Souci is a palace in Potsdam, Germany, with no apparent connection to the long-overlooked, first, and only successful slave revolution in human history. To Haitians and their visitors, Sans Souci is the palace in Milot where Henri Christophe, the formerly enslaved person who became emperor of Haiti, shot himself. However, few realize that Christophe built this resilient palace in nearly the exact same location where he had ended the life of a rival revolutionary named Sans Souci. The latter was an African-born enslaved person who had risen through the ranks as a guerrilla fighter under Toussant L'Ouverture and rebelled against Christophe's primarily Black Creole hierarchy after defeating the French, with an army of his own. By excavating this untold story from the ruins of the palace, Trouillot showed how every mention of Sans Souci, the palace, effectively silences the history of Sans Souci, the man. He wrote, "Silences are inherent in history because any single event enters history with some of its constituting parts missing. Something is always left out while something else is recorded."[63]

In the history of the Falklands and Patagonia, the multifarious figure of Allen Gardiner has had a comparable silencing effect in the South Atlantic. Maritime historians associate *Allen Gardiner* with the name of the schooner that assisted in the establishment of a series of South American Missionary Society stations. Scholars of Patagonia draw inspiration from proto-ethnographic studies, such as *Uttermost Part of the Earth* by E. Lucas Bridges, son of the missionary Thomas Bridges, who disembarked the *Allen Gardiner* in Tierra del Fuego and founded settlements at Ushuaia and Harberton.[64] There, Bridges chronicled the "dying" language and culture of the Yagán (or Yámana) and Selk'nam (or Ona) peoples in their "original condition," and converted the Indigenous people into loyal Christians. Other landowners, such as José Menéndez and mining investor Julius Popper, hired clans of "Indian hunters."[65] Considering that there is no historical evidence of a "precontact" population of Indigenous people in the Falklands, it is common to treat the archipelago's settlement as a distinct narrative from Patagonia's violent colonization process. However, Allen Gardiner stands out in time to bridge these contradictory imperial histories, demonstrating how they were in many ways mutually constitutive.

The South American Missionary Society (formerly the Patagonian Missionary Society) actually had three successive ships named *Allen Gardiner*—a schooner (1854–1873), a yawl (1874–1884), and a steam yacht (1884–1896)—all in memory of the society's British founder.[66] Former Royal Navy commander Allen Francis Gardiner had taken a keen interest in delivering "civilization" to the Natives of Tierra del Fuego and Patagonia since before Robert FitzRoy encountered and transported four Yagán Indigenous people to and from England with Darwin on the journey of the *Beagle*.

After establishing the society in 1844, Commander Gardiner assembled a crew to search for one of the famed Yagán individuals whose story Franz Boas drew upon to develop salvage anthropology: O'rundel'lico, who has been known more commonly as James "Jemmy" Button—alluding to what environmental anthropologist Laura Ogden calls the "colonial fantasy" that the teenage boy was sold by his own people for a button.[67] Ogden draws on the archives of the US explorer Charles Wellington Furlong, as well as anthropologist Anne Chapman, who points out that, contrary to most accounts, "a large shining mother-of-pearl button" was not payment for O'rundel'lico's custody.[68] FitzRoy gave the button to a relative who was with O'rundel'lico, but without any agreement that he would be "purchased." Chapman found it more accurate to state that O'rundel'lico was captured and held hostage without consent.

In addition to O'rundel'lico, FitzRoy had already taken eleven Kawésqar (or Alakaluf) people hostage, most of whom escaped, besides three children, two of whom were "returned."[69] The remaining hostage, Yokcushlu (named Fuegia Basket), was treated as a dehumanized pet, and she eventually joined O'rundel'lico and two additional hostages—El'leparu (also known as York Minster) and Boat Memory (an embodiment of lost wreckage whose Indigenous name is unknown)—as captives on a journey of the *Beagle* to England in 1830. There, Boat Memory died of smallpox, but the other three hostages attended boarding school and even met King William IV. Treated as samples for scientific racism, their heads were measured by a phrenologist before departing on the *Beagle* on December 27, 1831, with FitzRoy and Darwin back toward Tierra del Fuego, where they were released at Wulaia on the west coast of Navarino Island on January 19, 1833.[70] Darwin and FitzRoy encountered O'rundel'lico once more in 1834, naked and thin, yet far from "regressed," for he had not lost any memory of the English language and manners he had been forced to learn (see figure 9). As Chapman describes, O'rundel'lico was content with the abundance of animals and fish available, and he had no desire to return to England.[71]

Commander Gardiner made two journeys to Tierra del Fuego without encountering O'rundel'lico, whom he hoped might serve as a potential interpreter.

JEMMY IN 1834.   JEMMY BUTTON IN 1833.

FIGURE 9. O'rundel'lico (also known as James "Jemmy" Button) in 1834 (*left*) and 1833 (*right*).

Source: FitzRoy 1839. Reproduced with permission from Van Wyhe 2002.

Gardiner had already endured tense colonial conflicts, "seeking heathens" to convert in South Africa, the Andes of Argentina, Chile, and Bolivia, as well as Indonesia.[72] In 1841, Gardiner left his wife and two children on East Falkland Island for ten months while he sailed to Chilean Patagonia to spread the gospel to Tehuelche people. He went back and forth between England and South America, and he managed to organize the Patagonian Missionary Society in 1850 to reconnect with O'rundel'lico and catechize the Yagán. Yet, when Gardiner first encountered Yagán people, he found that they were not satisfied with the items he exchanged for fish and shell necklaces; they were more interested in clothes.[73]

Gardiner's second visit to Tierra del Fuego in 1850–1851 ended in his death by starvation. Prayers, complemented with gifts of buttons and knives, seemed to have made a better impression on the Yagán people. However, fearing an attack that never transpired, Gardiner killed "one or several Indians" and had to flee, according to Thomas Bridges.[74] Upon reflection, Bridges removed blame from the Yagán people, who had become understandably irritated by the missionaries' lack of generosity with the wealth they had possessed yet failed to distribute in a fair manner. Having fled to an inhospitable place called Spaniard Harbour at Aguirre Bay, one of the missionaries' ships, the *Pioneer*, was severely damaged on a rock. The missionaries salvaged the wrecked bow and stretched canvas over it to use as a sleeping apartment.[75] They tried to find food they had stashed at Banner Cove, the previous location, but they fled again, fearing an attack from relatives of the Yagán people whom Gardiner had murdered. After surviving through much of the winter on seabirds, shellfish, and seaweed, Gardiner and his mates eventually died of scurvy and starvation. The Patagonia Missionary Society had ordered a six months' supply of provisions to reach the missionaries, which Lafone, the FIC founder, had sent. However, that ship was also wrecked on Staten Island and never reached them.[76] On one of the last entries of his diary, Gardiner outlined a plan for future missionaries: "To convey a few of the natives to the Falklands, to teach them English, and learn their language, and to provide a brigantine or schooner of a hundred tons' burden as a mission vessel."[77]

Indigenous Refusal and Decolonial Salvage

Determined to carry out Gardiner's final instructions, the society built a schooner and named it in honor of the murderer they considered a martyr. In October 1854, Captain William Parker Snow sailed the *Allen Gardiner* from Bristol, England, taking possession of the Falklands' northwestern Keppel Island to construct a station on February 5, 1855. Stanley residents and their British colonial administrators expressed mixed feelings about the society's plans to extract Indigenous Yagán from Tierra del Fuego and "civilize" them remotely in a new settlement on Keppel Island. Shipping agent Admiral B. J. Sulivan, supported the society's plan, reasoning that it could reduce the fatal confrontations between wrecked sailors and Indigenous people on the coast.[78] In this sense, the Patagonia Missionary Society sought to "save Indians" by converting them and "save castaways" in need of refuge from potential wreckage and confrontation.[79] Others, including Colonial Chaplain Charles Bull and Governor Moore, contended that the scheme amounted to "kidnapping" or slavery, if the true motive was to force the Natives to work with cattle at Keppel.[80]

Even the *Allen Gardiner*'s own Captain Snow, who had previously struggled to suppress the slave trade, had reservations about "any abduction of these poor ignorant natives."[81] Unlike Commander Gardiner, Snow succeeded in locating O'rundel'lico on November 2, 1855. Yet, even though this Yagán person did understand English and grasped the missionaries' intentions, Snow reported that O'rundel'lico refused to let them bring his son with them to Keppel, as they had requested. Instead, he and a hundred or so others demanded gifts, and, again, they were not satisfied with the presents, given how much wealth they knew the English to possess.[82] Snow suggested that the missionaries should go to them and "not inveigle them to us," especially given the strict Alien Ordinance in the Falklands.[83]

In spite of Snow's dissent, to accelerate the settlement of Keppel, the society sent General Secretary Rev. George Pakenham Despard to revive the operation in 1856. Despard clashed with Snow over the scope of the enterprise at Keppel, leading to Snow's dismissal. As Snow suspected, Despard devised plans to recruit Natives not only for religious instruction and "civilization" but also as laborers for wood cutting in Tierra del Fuego, as well as gardening and cattle raising at Keppel for profit so as to be self-sustaining.[84] Yagán lumber workers in Tierra del Fuego were compensated with fishing lines, knives, clothing, and scarlet comforters, which they may have found more satisfying this time around.[85] However, used clothes that missionaries distributed also carried diseases that would cause fatal epidemics across the Yagán communities.[86]

Among Despard's crew was another Allen Gardiner, son of Commander Allen Weare Gardiner. At first, this Gardiner encountered the same wall of Indigenous refusal that Snow hit when asking Yagán parents for consent to take their children back to Keppel Island, even after promising clothes, livestock, and vegetables.[87] Gardiner even dodged threatening pursuits and an apparently imminent attack from sixty canoes of neighboring Indigenous communities.[88] The Patagonia Missionary Society reached a "breakthrough" in 1858 when Allen W.

Gardiner followed through on his father's fatal wish and persuaded O'rundel'lico to come to Keppel in the *Allen Gardiner*. With consent from his community, O'rundel'lico had in turn convinced the society to amend its proposal to take families, including his first wife Lassaweea (also known as Jamesina) and three of his children, rather than only boys on their own, as missionaries had sought.[89] Having already endured a "civilizing" process nearly three decades earlier, O'rundel'lico was seen as critical to the mission at Keppel. However, like Allen Gardiner's many forms, the subjectivity of O'rundel'lico was changing and unpredictable. Despite his literacy with British social rules, manners, and customs, the missionaries grew disappointed in O'rundel'lico's "idleness" and "lack of humanity or curiosity."[90] O'rundel'lico refused to do the onerous work Despard assigned to him, such as "ditching, gardening, fencing, painting, carpentry, shoe-mending, and tailoring."[91] Despard was also surprised when O'rundel'lico demanded that they take his whole family back home when Lassaweea was falsely accused of stealing. The missionaries may also have underestimated O'rundel'lico and his relatives' skills in the art of deception, a central dynamic of slavery and subversion in the Atlantic World.[92]

Seeking to "Civilize and Christianize" more Yagán individuals and return O'rundel'lico and his family, catechist J. Garland Phillips and Captain R. S. Fell sailed the *Allen Gardiner* to and from Wulaia (or Woolya) in Tierra del Fuego in 1858 and 1859. O'rundel'lico considered this site home, and the missionaries were familiar with it from previous shuttles. This time, the society wanted to establish a more permanent presence. They left the *Allen Gardiner* to garden, collect wood, and establish a house of worship, but disputes ensued over unfair compensation for Indigenous labor and further accusations of theft. O'rundel'lico and his relatives were offended at the meager provisions of "musty" biscuits and old rags that were only provided as wages after hauling trees.[93] Requests for raises of more biscuits and better clothes were denied, and O'rundel'lico refused to allow the missionaries to take his daughters to Keppel, though they did bring nine other relatives.[94] They spent nine months gardening, raising livestock, doing housework, learning English, teaching Despard Yagán, and going to church. However, as they prepared to return to Wulaia, the missionaries conducted a search for missing equipment, which deeply offended two of the residents: Macall-wenche (known as Billy Button) and Schwaiamugunjiz (or Squire Muggins).[95] The *Allen Gardiner* stopped at Port Stanley, where the Keppel residents were greeted approvingly by "flocks of cottagers" who sent them off with presents.[96]

Upon arrival at Wulaia, another search took place, upsetting Macall-wenche and Schwaiamugunjiz, who resisted. Missing knives, handkerchiefs, and a harpoon were discovered, but the two who had been accused got into their canoes and went ashore with their spouses.[97] As the crew left the *Allen Gardiner* to cut wood and garden, about seventy canoes carrying up to three hundred Natives gathered. O'rundel'lico again expressed disappointment in the lack of gifts that the missionaries had to offer. Then, on Sunday, November 6, 1859, as the crew gathered in the mission house to begin church service, Captain Fell, Phillips, and the rest of the crew were suddenly overpowered with clubs and stones and brought to their death. Only the cook Alfred Cole survived, by remaining on board the *Allen Gardiner*.

In an act of *decolonial salvage*, the Natives plundered the ship. They recovered pieces of iron from deck lights, poles, the wheel, and sails. According to Cole, the *Allen Gardiner* was reduced to a "mere wreck," with nothing left except the hull and masts.[98] Cole made it to the shore, where he survived on berries, mussels, and limpets for four days until a group of Yagán people dressed in clothes taken from the *Allen Gardiner*, including Captain Fell's blue coat, took him in and fed him.[99] At Wulaia, Cole was greeted by O'rundel'lico, who clothed him and treated him well. Distressed at the lack of correspondence after a long delay, Despard sent American former sealer Captain William Horton Smyley to search for the missionaries. Smyley was originally contracted by Lafone to find the first Commander Allen Gardiner, and on that previous trip he had encountered the corpses of some of that first crew, but not Gardiner.[100] This time, Smyley found the *Allen Gardiner* afloat but stripped of its parts.[101]

Smyley escorted Cole and O'rundel'lico back to Stanley for trial. There, Cole blamed the event on O'rundel'lico for not getting "as much as he thought he had a right to, and that he was at the head of the whole proceedings."[102] When cross-examined, O'rundel'lico said that he had never wanted to stay at Keppel, and acknowledged his comrades were angry at having their bags searched, but he accused the Selk'nam (Ona), who inhabit the main island of Tierra del Fuego, of perpetrating the violence.[103] Rev. Bull and Governor Moore exculpated O'rundel'lico of responsibility for the "deplorable tragedy" and viewed the "plunder" in which he took part as righteous indignation for their "enforced and irksome residence" at Keppel.[104] The trial thus supported claims that O'rundel'lico and his peers had been involuntarily abducted, reinforcing Stanley residents' uneasiness about the missionary settlement at Keppel. O'rundel'lico's role in the uprising and his pardon secured his place in history, not only as the captive survivor of multiple fierce procedures based on a highly flawed social-evolutionist framing but also as a principled Indigenous leader who refused to accept unfair labor conditions and wealth inequality during the establishment of the Keppel residential school and the colonial intrusion of Tierra del Fuego.

Despite the disaster at Wulaia and Governor Moore's strict prohibition of the involuntary importation of more Natives to the Falklands, the society continued using Keppel Island as a base for its missionary work. Despard resigned, and

O'rundel'lico never returned. However, the younger Ookokko, one of the only residents of Wulaia who reportedly expressed concern at the time of the uprising, became the society's new "industrious" darling. He and his 16-year-old wife, Cammilenna, apparently participated in the Keppel Island mission project at their own will. Ookokko was more willing than O'rundel'lico had been to act as an interpreter, and he helped garner support for the Society not only from fellow Yagán but also from Stanley residents like Charles Bull.[105] The missionaries even brought other Indigenous groups to Keppel, including Tehuelche people from the Argentine province of Santa Cruz in Patagonia.[106] In 1865, Despard's replacement, Rev. Waite Hockin Stirling, brought another cohort of Yagán boys to England for intensive training in Bristol, including Wamme-striggins (nicknamed "Threeboys"), the son of O'rundel'lico, who had made the same transatlantic journey with FitzRoy decades earlier.[107] Stirling's intention was to dispel myths of the Yagán as subhuman beings, yet the voyage ironically reproduced this stigma through some of the same scientific racist procedures that Boas used to make antiracist claims in salvage anthropology, such as anthropometric head measurements.[108] The heads of Wamme-striggins and another Yagán recruit, Uroopa, were measured to be "above the averages yielded by the population of Bristol and its neighborhood" in terms of cerebral capacity.[109] On their return to the Falklands in 1867, one of the Yagán recruits, Uroopa, fell ill with consumption (tuberculosis) and died on the repaired *Allen Gardiner*. Before his passing, Stirling christened Uroopa as "John Allen Gardiner," the first baptized Native of Tierra del Fuego.[110] Thus, in a final act of historical silencing, Allen Gardiner's name was made eternal, even though O'rundel'lico, who died from a devastating epidemic, would become known only as "Jemmy Button."[111]

Wamme-striggins also died at sea, three weeks after Uroopa's passing, and both of their bodies are buried in Stanley: physical remains of an Indigenous presence, embedded in the soil and immersed in the wreckage of the Falklands.[112] By the turn of the twentieth century, hundreds of Yagán people—including children taken without the consent of their parents—had gone to Keppel, Thomas Bridges had begun compiling his dictionary of Yagán, and the society had established mission stations at Ushuaia and elsewhere in Patagonia. Despite significant setbacks, the settlement at Keppel Island effectively served its purpose as a remote base for the society, and with the opening of the western islands of the Falklands to settlement, ordinary Falkland Islanders became more integrated with their neighbors at Keppel. The society continued using Keppel as an agricultural hub for "the industrial education of the natives"—involving coerced child labor in an isolated, disease-ridden, unforgiving environment—leaving most of the Christian indoctrination and "civilizing" activities to the Ushuaia branch.[113] The *Allen Gardiner* and its nautical reincarnations continued shuttling the missionaries and Yagán people between Tierra del Fuego and Keppel until the society sold the farm in 1911.

The captivating saga of Allen Gardiner, O'rundel'lico, and the South American Missionary Society has become the stuff of horror or legend, depending on one's perspective. Yet, as with Sans Souci, each recording leaves out key parts of history. Christian evangelists made heroes out of Commander Gardiner and his successors, and historical fiction writers have tried to tell the tale of the redoubtable "Jemmy Button," with mixed results.[114] Given the disquiet among Stanley residents and their British colonial governors, it is critical to remember the history of the Keppel Island settlement and the insurrection at Wulaia not just as a romanticized "morality tale" about "Jemmy Button the tragic antihero"—as Ogden astutely observed from the archive of Furlong—but also as a possible act of political resistance against a cruel use of colonial power.[115] The missionaries' removal of the Yagán people and the commercialization of their unpaid labor defied post-abolition antislavery laws.[116] Treated as "hewers of wood and drawers of water," these people were proletarianized or enslaved in their own land to cut down trees that were exported to the Falklands to supply lumber for sale in Stanley.[117] The settlement on Keppel Island ultimately played a similar role in cultural genocide as the horrific residential boarding schools of North America, and the results were not just "civilization" and Christianity, but exploitation, disease, and death.[118] In this context, it behooves us to reconsider the possibility that the so-called massacre at Wulaia was a legitimate anticolonial resurgence against dispossession.[119] Beyond the spectacle of bloodshed, O'rundel'lico and his community's reverse salvaging of the *Allen Gardiner* for the benefit of Natives rather than settlers offers a vision of Indigenous refusal.

This chapter provides a broad historical overview, ranging from claims of discovering the Falklands/Malvinas to colonial experiences of dispossession in the South Atlantic. Here, dispossession took an unusual colonial form in comparison with other histories of conquest in the Americas. While there may be no historical evidence of an immediate encounter between European colonizers and an Indigenous population in the islands, imperial claims to the territory as *res nullius* (land belonging to nobody) were always in dispute.[120] Competing forms of rule over the frontier led to multiple violent conflicts. Temporary settlements of the French, British, and Spanish failed to establish permanent colonies in the eighteenth century. The merchant Luis Vernet had a grand vision for a colonial project, which began to take form as a colony of Buenos Aires. However, when

Vernet tried to enforce regulations on sealing, the US destroyed his settlement, clearing way for the British to reclaim the islands in 1833.

Control over the far-flung cattle frontier remained tenuous under British rule, when Antonio "El Gaucho" Rivero led an uprising against Vernet's former colonial managers. Ironically, the links between the Falklands/Malvinas and Patagonia began to take hold just as the gauchos became alienated and displaced from the South Atlantic. Beginning in the 1880s, a wave of shepherds seeking to invest their savings and own their own land migrated to "the Coast." This coincided with and in some ways complemented the second "Desert Campaign," in which Argentine General Julio Roca led bands of mercenaries to annihilate the Indigenous peoples of Patagonia and Tierra del Fuego. The British colonial governor even "kindly welcomed" Roca during a visit in Stanley in 1901.[121] While the FIC and other absentee landowners had already enclosed the vast majority of the Falklands (see chapter 2), the Argentine Republic, particularly Carlos Moyano, a founding governor of the province of Santa Cruz, encouraged nonlandowning islanders, such as the son of the slain FIC manager Rudd, to stake new claims to land in Patagonia. Some of these migrants amassed considerable wealth by starting their own businesses on the FIC model. Others never owned land and continued to be peons or managers on the farms of Patagonia. From 1882 to 1895, more than twenty-six families and sixty-one additional settlers migrated to South America. The migrants established farms throughout what is now the province of Santa Cruz, helping to found the towns of Río Gallegos, Puerto Santa Cruz, and Puerto San Julián, as well as strengthening connections with Punta Arenas, Chile, and the missionary settlements in Tierra del Fuego.[122]

Even though South American gauchos were racialized, alienated, and displaced by British shepherds within the islands, Rivero's rebellion has become symbolic of Argentina's claim to sovereignty. In 2012, the Argentine government passed Decree 256 in Rivero's memory. The so-called Gaucho Rivero Law (*Ley Gaucho Rivero*) banned ships scheduled to stop in the Falklands and flying a British flag from using the ports of the province of Buenos Aires. The impact this law has had on containerization is discussed in chapter 4, but it has been most effective in hampering the Falklands' tourist industry. After the law was passed, most cruise lines changed their itinerary to avoid the Falklands or the South Atlantic more generally, having received threats and protests led by the left-nationalist group Movimiento Patriótico Revolucionario Quebracho in support of the legislation. These threats caused twenty-one cancellations that would have brought 23,000 potential visitors to the Falklands in 2012–2013 (ironically, Ushuaia's tourist industry in Tierra del Fuego was also impacted, triggering counterprotests). The cruise tourism industry bounced back after the protests died down starting in 2014, but when cancellations continue to occur "due to weather," and the sun is shining in Stanley, local residents usually assume that the haunting politics of Gaucho Rivero have something to do with it.[123]

Finally, the historical fact of the Keppel Island settlement still works to silence the possibility that there was a precolonial Indigenous population in the archipelago. Whenever arrowheads, canoes, or middens are discovered in the Falklands, the immediate assumption is that the Yagán who were taken to Keppel left the items, and the idea that they might be artifacts of an earlier era is dismissed.[124] The Indigenous presence at Keppel has thus ironically strengthened settlers' claim that Europeans were first to occupy the islands, even though archaeological findings indicate otherwise.[125] The Falkland Islanders in Stanley on East Falkland Island were somewhat insulated from much of the ugly history between the missionaries and the Yagán people on Keppel Island and Tierra del Fuego. Nonetheless, the story of the South American Missionary Society is inseparable from the settlers' nativist narrative.[126] As they expanded their domain westward, settlers modeled sheep farming stations after the Keppel settlement, and while it has been uninhabited, some islanders have been eager to convert the island into a tourist destination to showcase both cultural heritage and biodiversity.[127] Even though the colonial government expressed discomfort with forced removal of Indigenous peoples, by aiding the missionaries, the islanders and their trustees were ultimately complicit with the systematic abduction and colonization practiced at Keppel.

In sum, the South Atlantic was never a settler safe zone, free from the threat of violent insurrection against dispossession. Subaltern historical figures like Gaucho Rivero, Antonina Roxa, and O'rundel'lico have been portrayed as either barbarous or civilized to support or decry a particular national claim in the ongoing geopolitical dispute. The legacy of failed colonialists such as Vernet, Brisbane, and Gardiner also became wreckage of empire, which current claimants have tried to salvage and celebrate as fallen heroes who sacrificed themselves for a larger sovereignty cause. While there may not have been an initial encounter between colonizer and colonized in the Falklands, the area became an enduring site of contestation and a staging ground for imperial expansion.

COMPANY ISLANDS

During my fieldwork in Stanley, I spent most weekday mornings at the Jane Cameron National Archives (JCNA), where I had a three-hour window to squeeze as much meaningful historical data as possible out of stacks of colonial reports and letter books. There, I also met Joan Spruce, a senior local historian responsible for various historic preservation projects, including the previous iteration of Stanley's tourist museum and a booklet on the presence of corrals and gauchos in the islands. Noticing my avid interest in the minutiae of the islands' colonial past, she informed me of her newest hobby: cataloging unrecognized "built heritage" sites in the Camp.[1]

To get a better sense of what this research entailed, Joan invited me to her house in Stanley. I rode my bicycle over one afternoon and found in her possession a compendium of human-made turf walls, stone corrals, and other sites. Starting with a topographic map, Joan divided the islands neatly into twenty-nine sheets, each forming a section of a book project. Together with her husband and two other islander couples, Joan had combed all corners of the islands by Land Rover in order to identify seventeen stone corrals, more than sixty turf corrals or walls, shanties, "try pots" (large casks for boiling penguins down for oil; see chapter 4), unmarked graves, ruins of farm buildings, stone enclosures, whale skeletons, and other sundry artifacts. In addition to taking photos and marking their maps, Joan and her team recorded the GPS coordinates of structures. When the sites were too hard to reach, they enlisted the nongovernmental organization Falklands Conservation, the FIG Public Works Department, or friends to take geodata bearings. Upon her request, I provided Joan with photographs of a human-created "chasm" that served as a moat to prevent sheep from going onto the beach on Carcass Island.[2] These structures had become naturalized as part of the built environment in the rural Camp.

Raised in a fully analog environment on one of the remotest islands in the archipelago, Joan marveled in the technological capabilities of her computer's digital software. She devoted a digital folder to each of the twenty-nine sections of the islands containing geographical references, descriptions, and footnotes. But most fundamental to her project was Google Earth software, which Joan used not only to mark points collected from on-site observations but also to search for visual evidence of new sites from satellite images. Before the digital era, Joan's team would just walk into the FIG Mineral Resources Department and ask for aerial photographs to find places of interest. At the time of my visit, a vast array of GIS coordinate pushpins covered the satellite image of the islands' landscape on Joan's customized Google Earth map, each with a specific title such as "Swan Inlet small corral." I watched Joan, seemingly in a trance with her eyes glued to the screen, push and scroll her mouse to zoom in on the satellite image, wait for the resolution to improve, and then squint to see if she could make out any possible sign of a turf corral. Joan told me that she got such a thrill out of finding new sites thanks to this method that she could do this forever, despite the considerable costs.[3]

The rapid expansion in development due to revenue from commercial fishing, tourism, and oil exploration, including new laydown yards and roads, has brought urgency to Joan's project. She eventually published a book called *Falkland Rural Heritage* for builders or casual explorers; it and the accompanying twenty-nine-sheet map set are available to the public at the local museum.[4] Joan hopes that publishing this record will influence the FIG Historic Buildings Committee to protect more structures—or at the very least restrict islanders from pulling rocks out of corral walls to fashion makeshift barbecue pits. Joan's effort may have a lasting influence. In response to her reputation as the local expert on corral ruins, construction workers had already called her to seek permission to bulldoze a mound of turf to build a road. When she looked up the GPS coordinates and found that the site was a turf corral dating from before the permanent British occupation of 1833, Joan convinced the foreman to redesign the road so that it would squiggle around the perimeter instead.

Joan's devotion to her hobby raises questions about the role of cultural heritage and patrimony in claims over sovereignty and property. Why would a Falkland Islander who vehemently supports British sovereignty, like nearly all of her local peers, be so interested in preserving artifacts such as corrals that the Argentine government cites as evidence of their prior occupancy and British usurpation?[5] In fact, unlike most military history buffs and veterans who visit the islands' battlefields to pay homage to fallen soldiers, Joan is more interested in pre-1833 artifacts than relics of the 1982 British war victory.[6] She periodized the project so that it would only cover sites erected before the start of the military conflict, and she was equally interested in recording vestiges of the French and Spanish settlements, as well as that of Luis Vernet, the merchant named political and military commander of the Malvinas on behalf of Buenos Aires. *Falkland Rural Heritage* does tend to downplay the presence of gauchos as "small in number."[7] The book disassociates them from their origins in the continent, stating that "being a gaucho was not limited to South American nationals and the term just meant men who worked with cattle."[8] Nonetheless, as evidence of her impartiality, Joan pointed to her record of a turf rectangle at Sussex Farm, which she suspects served as an early prison for the Spanish colony or Vernet's settlement. Rather than silencing the past of gauchos and corrals, Falkland Islanders like Joan have attempted to salvage their history and incorporate these indelible marks on the landscape into their own narrative of enclosing and improving upon it.

Toward a more comprehensive understanding of property relations within the islands, this chapter analyzes historical transformations of enclosure and improvement that have changed the face of the archipelago. Even without encountering a precolonial Indigenous population when they arrived in the Falklands/Malvinas, British colonialists privileged themselves for naturalizing land rights without consent based on a classical liberal understanding of property rooted in the theory of political economist John Locke.[9] To eke out a living and ultimately gain considerable profit on the remote colonial frontier, British settlers embraced Lockean tropes of resourcefulness through enclosure and improvement of the commons. Falkland Islanders rest their contemporary claim to British sovereignty on the modern liberal notion of self-determination, but they defend their desires through adherence to a deeper notion of self-sufficiency. In turn, these liberal political economic concepts inflect how islanders like Joan assert a shared cultural heritage and sense of place—even through attachments to the legacy of gauchos and corrals—embedded in the landscape. Before moving on to addressing the overlapping maritime claims, which led to new resource wealth from commercial fishing and oil exploration in the late twentieth and early twenty-first centuries (outlined in part 2), this chapter analyzes the localized terrestrial conditions that made the colony a significant producer of cattle hides and then sheep wool in the nineteenth and twentieth centuries.

To grasp more fully the significance of the islanders' property relations for their emergent bundles of rights claims, it is important to consider the historical stages of dominion and ownership in the Falklands/Malvinas. This chapter examines how property, particularly in the rural Camp of the islands, has undergone various transitions of monopolization and subdivision of land. The settler colonial property regime has ranged from (1) a usufruct cattle frontier founded on debt bondage to (2) a feudalistic absentee-owned system of community sheep ranches to (3) the current system of privatized small family farms. Throughout this history, *enclosure* has taken various forms, from cattle corrals and sheep fencing to land subdivision. While colonial administrators have continually sought to make settlers self-sufficient, the ways islanders appropriated the *improvement* concept have fluctuated from paternalistic dependence on the Colonial Office and particularly the Falkland Islands Company (FIC) on the one hand to surprisingly subversive projects of mutual aid on the other.

An anthropological approach that treats property not just as ownership and rights over pieces of land but rather as relations between and among persons with respect to soil is helpful for understanding social rank and class.[10] Tania Li's *The Will to Improve* is instructive for using such anthropological methodology to make familiar Lockean property ideologies strange.[11] Based on fieldwork in Central Sulawesi, Indonesia, Li found that, by rendering political contestation "technical," colonial and neoliberal development programs were actually constituted by that which they exclude. Improvement—not just the enlightenment idea of human progress, but the commitment to make labor productive and land profitable—is a social force. It encloses and collapses contradictory claims to belonging and indigeneity.

Enclosure is the historical condition of possibility for adding value to land through agricultural development. Anthropologist Liza Grandia has shown how Guatemalan elites dispossessed Q'eqchi' Maya people from their land and labor through corporate trade agreements. This resulted in various forms of enclosure, ranging from conflicts over cattle ranching and logging to the establishment of protected areas for biodiversity conservation and carbon credit trading.[12] Building on these anthropological approaches to property, this chapter is therefore guided by two questions: how were property relations constituted historically in the Falklands/Malvinas, and how are they currently being reconstituted politically?[13]

From this anthropological standpoint on property, this chapter examines how shifting ethics of enclosure and improvement have justified the extraction of wealth and resources in the Falklands/Malvinas.[14] I show how, in some respects, settlers fulfilled the hopes of British colonial administrators to craft a colony of self-sufficient individuals, yet in other instances, the islanders customized the improvement doctrine as a solidary struggle against the FIC's monopolistic control over enclosed land.

Specters of the Gauchos

Joan was not the first Falkland Islander I encountered who decried Argentine aggression and simultaneously obsessed over South American gaucho culture and history. On my first visit to Camp, a retired sheep farmer, Dallas, noticed me inspecting the two coats of sheep's wool hanging in his yard and invited me into his shed. There, he confessed his labor of love: crafting horse-riding gear out of leather in the style of Rioplatense gauchos. With palpable pride, Dallas demonstrated his skills of threading together intricate whips and saddle attachments. From the outside, no one would have suspected that the interior walls of this quaint British-style sheep farmer settlement house would be lined with gaucho adornments. This was not his only secret: Dallas said that his most important item is the stool he sits on to do his leatherwork. Lifting his seat cover, he revealed that it was full of Budweiser beer cans. He insisted that I have one, and we discussed his nostalgia for living off the land and getting around by horseback.

Having grown up in England, Dallas confessed that, when he took the job as a shepherd in the 1960s, he thought the Falklands were in Scotland! Eight thousand miles away and over fifty years later, he not only feels that he belongs in the Falklands but, given his passion for gaucho heritage, he has grown a strong affinity to South America as well. While he expresses contempt for Argentina and its government, he vacations regularly in Chile and Uruguay, where he attends rodeos and meets other horse-riding gear aficionados.

Corrals and gaucho heritage provide a reminder of what the islands might have been like if they had not become a British colony. The French and Spanish introduced livestock to the islands in their temporary settlements of the late eighteenth century. When the Spanish left the islands unattended in 1811, feral cattle flourished throughout East Falkland Island. As I discussed in chapter 1, this provided a new source of "wild" cattle, which merchant Luis Vernet viewed as an exciting business opportunity. Again, after Vernet had acquired usufruct rights to cattle on East Falkland Island (Isla Soledad), making horse-riding gear, such as lassos and saddle pads, and constructing corrals were included in his contract with the gauchos.

While some islanders soften or distort the Rioplatense origin of social habits and customs within the islands, there is no denying the important place of gaucho cattle herding in the countryside and the corrals mark that legacy. Stanley's archives contain a quite detailed inventory of the corrals and structures of Port Louis, a project that a group of curious year nine islander pupils carried out in 1991. The class project offers a clear sense of Vernet's vision for land tenure: he sought to cut up the settlement into relatively equitable lots in harmony with natural boundaries like Pig's Brook, while still consolidating larger tracts for wealthier colonists willing to help populate his colony.[15] Branching out beyond the settlement at Port Louis, Vernet divided the territory into eleven sections and constructed five corrals, which dot the landscape of Isla Soledad. The chain of corrals facilitated the killing of 5,553 cattle between 1826 and 1831.[16] The corrals were designed to capture cattle by *volteada*, a method of coaxing wild cattle by driving them into an already tame herd. However, the gauchos' captain, Simon, instructed his peons to hunt cattle using lassos instead; this less efficient method subjugated the gauchos in a system of debt peonage.[17] It is also what Gaucho Antonio Rivero cited as the principal motive behind his own insurrection, in addition to the general state of upheaval after the USS *Lexington*'s destruction and unfair payment in worthless paper currency.[18]

The introduction of wire fences and Scottish shepherds would eventually displace the gauchos, even after more of them were recruited when the British took control of the settlement in 1833. But corrals remain present throughout the islands as vestiges of the peons and their relations with the land. They also conjure Vernet's vision for a prosperous colony built on the export of cattle products (and potential fishing and sealing). Under formal British colonial rule, a firm Lockean private property ideology would all but evacuate the islands of Vernet's conception of usufruct rights and ultimately dismantle the gauchos' frontier pasturage regime—a familiar story that resonates throughout the Americas.

Settler Colonial Diglossia

In addition to the corrals, a particular language ideology keeps the gaucho presence active in the landscape through islanders' speech habits.[19] Despite nearly two centuries of continuous British occupation, islanders continue to use and embrace loanwords deriving from Spanish, Indigenous languages, and the historic gaucho culture.[20]

As unusual as it may seem, Falkland Islanders who consider themselves "more British than the British," and who view Margaret Thatcher as their liberator, refer to each other as "chay," an Anglicized syncretism for *che*, the Patagonian term of affection immortalized as the nickname of Argentine revolutionary Ernesto "Che" Guevara. Indeed, the wardrobe of one of my interlocutors in the islands included a T-shirt featuring El Che's iconic portrait. Another wore a printed T-shirt with the word "Poocha!" in bold (an Anglicized syncretism of the Spanish *pucha*, a polite version of *puta*, meaning in this context "shoot" or "gosh"). Gaucho language is very much in fashion among Falkland Islanders, and it is not exclusive to islanders of multiple generations; I have known British, Chilean, and St. Helenian migrants to be quite fond of employing "chay," "poocha, man!"

and a variety of other syncretisms. These speech habits simultaneously index their affection for the colonized place and assert their own belonging.

In a parallel ethnographic situation in the Pacific, Miki Makihara has described how Chile forcibly proletarianized Indigenous islanders in Rapa Nui through the auspices of the "Easter Island Exploitation Company"; this is especially apparent in the company's transformative introduction of the Spanish language.[21] Makihara calls the language ideology that formed in Rapa Nui "colonial diglossia." Linguistic anthropologists use the term diglossia to describe a sociolinguistic arrangement that either differentiates varieties of the same language in a single speech community or compartmentalizes endangered minority languages in instances of bilingualism.[22] In the case of Rapa Nui, colonial diglossia features a "high" standard use of Spanish and a "low" vernacular Indigenous language. As the next section explores, a similarly monopolistic colonial company did play a central role in the population's shift of language and culture, but the language use of Falklands/Malvinas residents reveals different relations of power from those of Rapa Nui. Falkland Islanders' speech habits demonstrate a "high" standard use of English and "low" vernacular use of Spanish (as well as some words, such as "chay/che," likely deriving from Indigenous Tehuelche).[23] Unlike Rapa Nui, where the "high" colonial state language transformed the use of a "low" vernacular Indigenous speech, in the Falklands, the formerly dispossessed "low" South American Spanish and Indigenous language forms continue to penetrate and punctuate the settlers' own vernacular use of English. To be consistent, I refer to this particular language ideology as *settler colonial diglossia*.[24]

While the use of motorbikes and Land Rovers has distanced new generations of islanders from the cattle frontier gaucho culture, colloquial expressions among islanders of Anglicized Spanish syncretisms for horse-riding gear and features of the land derived primarily from South American Spanish persist. As linguist Yliana Rodriguez has analyzed, syncretic place names used within the islands, which stem from gaucho heritage, differ from the Spanish toponyms for the islands that have been developed as part of the geographic imaginary of the Malvinas in Argentina.[25] Because Spanish toponyms deriving from gaucho heritage are terms that Falkland Islanders, themselves, employ, it may be tempting to interpret them as speaking *against* a strictly British national heritage. In fact, functionaries of the Argentine Foreign Ministry consider Anglicization evidence of deliberate implantation of the population.[26] But to islanders, the survival of these syncretisms is part of what they call the "Falkland Islands way of life," an indigenized settler notion of peoplehood, which I analyze in chapter 3. Attachments to place not only enclose areas; they also enable global connection.[27] These speech habits show how boundaries are porous, and it is worth considering how such place names and features of the land compose part of the environmental wreckage that settlers have salvaged and repurposed in a new sociolinguistic formation of the South Atlantic.

Settler colonial diglossic expressions offer a window onto the instability of imperialism in its failure to fully enclose colonial territories. The domains in which code-switching thrives are mainly the masculine, heteronormative, working-class conservative barbecues of the rural Camp. Buffered from the British contractors and more cosmopolitan bourgeois-bohemian residents of Stanley, some Campers drop "chay" into each and every sentence, as though the contexts for code-switching were irrelevant. Between and among Falkland Islanders in the Camp, few remarkable things are just "awesome." The proper way to phrase it would be "Poocha! That's awesome." One interlocutor explained that the use of syncretisms is a way of acknowledging and reinforcing that he is talking to another islander (usually, but not exclusively, one born in the islands or having descended from multiple generations).

After a long day of lamb-marking in one Camp settlement, I discussed the significance of diglossia with my hosts, a mixed Falkland Islander/British expatriate family and Santiago, a Chilean Spanish teacher who had accompanied me. Sipping on pisco and then Chilean Carmenere, our Falkland Islander host told me that she was humiliated during her education in the UK when an English teacher told her it was nonsense to say that she went down to the "arrosure" (derived from the Spanish *arroyo*, meaning stream) to catch some zebra trout (*Aplochiton zebra*, a native fish species). But this was the language she acquired growing up in Camp on East Falkland Island. Interestingly, almost all of the Spanish-originating place names are located in East Falkland because the main period of gaucho activity occurred before the West was settled. Yet, now that the gauchos have been displaced, settler colonial diglossia is common throughout all areas of the Camp.

An especially intriguing variant of settler colonial diglossia may be found in a remote settlement of West Falkland Island. The farm's elderly owners, Humphry and Lou, are a fascinating Falkland Islander-Argentine couple. Several members of their family attended boarding school, farmed, and were even born in Argentina; their daughters were stranded there during the 1982 war.[28] Humphry and Lou had a unique perspective on both the islands and Argentina, which they referred to as "the Argentine" (an antiquated English usage translating "La Argentina"). They do not take offense by, and even sympathize with, Argentines who insist on calling the islands "Malvinas." Lou has fond memories of growing up in the Belgrano neighborhood of Buenos Aires, and Humphry went to school there during the rise of Perón. They are shocked at the so-called *villas miserias* (shantytowns) that dot Buenos Aires today.

Within their settlement, Humphry and Lou call horse pens "corrals" and streams "arrosures," but the accent veers closer to Argentine *castellano*: for example, pronouncing the "y" in "arroyo" as a hard "j." Yet Humphry and Lou adamantly support the Falkland Islanders' claim to self-determination. And, critical of the legacy of colonialism within Argentina, Humphry points out that, while "che" may have entered into Falkland Islands' vernacular speech via the primarily Rioplatense gauchos, most Argentines forget or deny that the term likely derives from the Tehuelche word for "you."

Humphry spent eight years managing the expansive Condor sheep estancia, spreading from near Río Gallegos, Patagonia, to the Straits of Magellan (Humphry uses Spanish syntax for this and other place names). Condor was under the same British ownership as Port Howard, one of the largest farms developed by the Falkland Islands Company. To understand how these transregional expressions of place formed historically, it is crucial to understand how this company gained a monopoly on land through enclosure of the islands: at first for cattle herding, but ultimately for sheep ranching.

Monopolizing the Colony

Subsequent to British settlement, the aforementioned English merchant Samuel Fisher Lafone caught wind of the islands' potential riches, apparently through direct correspondence with Vernet.[29] Based in Montevideo, Lafone owned and managed a number of *estancias* and *saladeros* (cattle ranches and salting facilities). Seeking to expand his business, he bought from the first British colonial governor Richard Moody (1842–1848) exclusive rights to the wild cattle in the southern peninsula of East Falkland and adjacent islands in 1846 (the same date as the Enclosure Acts in England), marking the first moment of formal land expropriation and enclosure under British rule.[30] This area is still known as "Lafonia," even though Lafone himself never actually set foot on that land; he managed it through a local agent as an absentee landowner.

In 1850–1852, Lafone established by Royal Charter the Falkland Islands Company, in which he became a major shareholder.[31] The company was ostensibly set up to persuade British settlers to populate the colony as partners. However, like Vernet, Lafone was principally a businessman. His prioritization of ranching over colonization underscored what would become a durable tension between and among three social forces: (1) the FIC and other absentee landowners' drive for revenue, (2) the British colonial government's administration, and (3) occupying settlers' immediate concerns with political stability and economic subsistence.[32]

Merchant George Thomas and his brother John Bull Whitington, as well as colonists John Markham Dean and Captain Robert Christopher Packe, formed rival clans in opposition to the FIC for rights to land, cattle, and salvaged shipwrecks.[33] To edge out the opposition, Lafone's brother-in-law John Dale, the FIC's first colonial manager, recruited and imported new groups of gauchos from South America to tame and kill wild cattle. Colonists refashioned the corrals that dotted the landscape into physical stakes for making claims to frontier territory in the Camp.[34] To establish its local headquarters, the FIC erected a now iconic stone corral in Darwin, the islands' first main settlement outside of Port Stanley, in 1859.[35]

FIC managers built up the company's workforce but delayed supplementing labor with enclosure and cultivation of land.[36] Due to the lack of shelter in the great plains of Lafonia, they found that cattle and their pastures had "deteriorated."[37] Corrals, sod walls, gorse hedges, and weekly rodeos helped to confine wild cattle to some extent, but there were tremendous gaps in boundaries, and frequent disputes over killing cattle ensued.[38]

Ultimately, the FIC's cattle venture struggled to meet its goals, instead ushering in a monopolistic land tenure regime focused on sheep farming. As discussed in chapter 1, parallel to this shift in livestock, the FIC recruited shepherds from the Scottish Highlands and elsewhere in Britain to displace the South American gauchos. In 1867, the FIC's outlook brightened when, under the direction of Colonial Manager Frederick E. Cobb, the company hired Wickham Bertrand of New Zealand as Camp manager. Bertrand helped Cobb to solve the islands' scab disease, which had prevented sheep from growing wool (see chapter 7). Wool then became the staple commodity on the islands while wild cattle went virtually extinct. The sheep population increased from 65,000 in 1870 to a peak of 807,000 in 1898.[39]

Along with the improvement in agriculture came more expansive practices of enclosing land with hundreds of miles of wire fences.[40] Fencing kept sheep from dispersing, which caused scab infection and pasture degradation. The material for fencing was costly, so while enclosure propelled the FIC, it exacerbated uneven development for smallholders.[41] Installing fences was also hard work, as one laborer, writing under the pen name "El Vagabundo," described in this poem submitted to the January 1913 issue of *Falkland Islands Magazine*:

At the Making of the Fence
Most men sticks to dead things, said things, read things,
    Bits o' people's lives an' never thinks o' life themselves;
But I am all for seen things, green things, clean things,
    Things you see in ships an' not in books on musty shelves.

I wish I lived in old times, bold times, gole times,
  Times o' 'Enry Morgan w'en 'e sailed the Spanish Main;
I'd go a-buccaneerin', veerin', steerin',
  An all me crew a-cheerin' as they 'eaved the moorin' chain.
An' the diamonds would be chinkin', blinkin', winkin',
  In the blazin' sunlight streamin' through the 'atches cracked an'
    worn;
While the crew would all be boozin', snoozin', carousin'
  An' the Jolly Roger flyin' at the mizzen 'acked an' torn.
But there aint no use desirin', firin', admirin',
  Longin' for the drinkin' an' the fightin' an' the fun;
So I swings me bloomin' 'ammer, slam' 'er dam'er,
  'Ere comes the boss (Gawd cuss 'im!) 'an me job aint 'arf begun.[42]

Longing for the earlier days of colonialism before the entrenchment of the FIC, "El Vagabundo" curses the onerous work of enclosure. Seeking a fresh start, many smallholders and workers like this one began migrating to Patagonia during this period, as discussed in chapter 1.[43] Among those who stayed in the islands, new forms of enclosure and private property refined the islanders' speech habits: sheep fencing introduced a new series of colloquial terms that blended with the gaucho-derived terms.[44]

Making Settlers "Self-Sufficient"

While it certainly played a key role, enclosure alone did not suffice for the British to establish a profitable colony. Empire-building, virtually from scratch—in a landscape without trees—required that the ground beneath one's feet serve as the primary resource. Converting the land into a resource meant adaptation and improvement of the soil. Thus, unlike their Spanish counterparts, British settlers extracted abundant peat for heating fuel, and, as discussed in chapter 1, vegetable garden plots were some of the only traces of the initial temporary British settlement of the 1760s at Port Egmont on Saunders Island.

Governor Moody, who administered the main settlement's relocation on East Falkland Island from Port Louis to Stanley, commented extensively on native soil and grass types. Noticing that the maximum capacity for improvement would derive from the rich soil and thick plant life of estuaries and valley floors, he guided settlers on how to reuse animal feces, peat ash, and kelp seaweed for manure, as well as the proper ways to till the soil in these areas.[45] Similar to British imperialists elsewhere who propagated notions of the colony as "the useful

garden," Moody encouraged settlers to experiment with horticulture and cultivation of the native tussac grass (see chapter 7).[46] Introducing a temporary lease
system of land tenure, Moody envisioned intensive enclosure and improvement
as the proof necessary to attract new settlers and investors to join in the colonial effort. Nevertheless, with vast tracts of land at the disposal of a consistently
sparse population, reports show that uncontrolled grazing was ubiquitous
throughout the islands for many years to come.[47] Viewing the land as a literal
loan, one outside consultant described the settlers' laissez-faire stocking regime
as "drawing on their principal as represented by the soil and pasture" instead of
prospering off of their "interest."[48]

Relatively slow to adapt land for durable pasture, settlers acted promptly to
breed more profitable livestock. Farmers had imported sheep from "The Plate,"
or Río de la Plata, in South America, which of course had been introduced previously from Europe. Still, British colonial administrators marked this stock as
a racially inferior *Mestizo* (mixed) variety. To "improve" the South American
sheep and yield higher rates of profit from their fleece and flesh, settlers crossed
the animals with "hardy" English rams.[49] With the shearing shed as their laboratory, settlers used wool bales as petri dishes for experimenting with different
breed combinations. In 1860, a quarter of the colony's 8,000 sheep were of a
"pure" English breed, which British governors boasted about as having been a
"profitable flock," "admirably adapted to the climate."[50] One bullish governor asserted that crossing the Mestizo sheep with Leicester breeds doubled the fleece
weight and increased the value of wool exponentially.[51] After the FIC accumulated and bred up to 40,000–50,000 of such sheep by the early 1870s, smallholders imported fewer Ríoplatense sheep and instead began to source the majority
of their stock internally from the company.[52]

Once wool from "improved sheep" became a staple commodity, British colonial administrators felt adequately prepared to assess the colony's potential for
more independent growth. Here, capitalist agrarian improvement was a gauge
for the settlers' self-sufficiency. In 1867, one governor wrote with cautious optimism, "It is not, I hope, unreasonable to anticipate that this Colony, improving
as it is in many important aspects, will eventually provide by local taxation for
the maintenance of a modern establishment, and cease to be a burden on the
mother country."[53] On the one hand, administrators such as this one cited improvement as scalable value, worthy of further assistance from Imperial Parliament. On the other hand, loyalty to the UK and its property ideology reproduced
dependence on government aid for further colonial improvements and many private enterprises.[54]

Mutual Improvement

Through the establishment of a General Improvement Society, Governor George Rennie (1848–1855), a sculptor by training, attempted to mold the settlers into a more independent collectivity. To demonstrate confidence and skill in adapting the soil and climate, the society initiated what would become a long tradition of an annual horticulture show. I had the pleasure of attending the horticulture show of 2013, which coincided with the referendum on self-determination. The event, held in a church hall, featured tremendous vegetables divided into various sections with prizes like "most outstanding of the potato classes." One side of the hall was devoted to cut flowers, home produce (jams, marmalades, chutneys, and pickles), eggs, butter, and other dairy produce. Across the end of the hall was a bounty of home produce—baked breads, cakes, cookies, and savories. Along the other wall were the decorated cakes (a main attraction), as well as a children's section and, of course, vegetables. There were more than forty different categories of vegetables, ranging from white potatoes to red potatoes to purple potatoes and potatoes with red eyes. Finally, at the far end of the hall were potted plants, including herbs. Decent soil and shelter are scarce in the islands, but the impressive array of plants and foodstuffs at the annual horticulture show reflects a historical propensity of settlers to produce fresh vegetables.

In 1852, Governor Rennie distributed twelve allotments of one acre each for "the working classes to build on or cultivate gardens."[55] Gardening of root vegetables and select leafy greens caught on among Stanley residents, eliminating the need to distribute food rations to settlers. Remarking on the General Improvement Society a decade later, Governor William Robinson reported:

> European vegetables thrive exceedingly well. The Falkland Islands Improvement Society (a Society recently formed for the encouragement of market-gardening) held its first meeting in April last. I never saw, in any part of the world, finer potatoes, turnips, cauliflowers than were then exhibited. Every house in Stanley has its plot of garden ground attached, and, owing to the number of vessels that call here during the year, the cottages find a ready and profitable sale for their surplus produce.[56]

In this sense, by becoming more self-sufficient, individual households would appear to have fulfilled the ecological-imperial promise of what environmental historian Alfred Crosby called a Neo-Europe.[57] This was not a modest "From each according to their ability, to each according to their need" arrangement, for, beyond providing for themselves, improvement laid the foundation for capitalist development. With a surplus of fresh produce, islanders were able to sell further provisions to the booming ship supply business amid increased traffic from the California gold rush before the US transcontinental railroad was completed.

However, this entrepreneurial spirit made settlers feel stifled by the limits of growing produce or raising livestock on their own properties. In 1859, thirty residents of Stanley petitioned for a commonage ordinance that would provide a more egalitarian distribution of land and access to grazing and pasture on the Government Farm.[58] Moreover, in Camp, discord between managers and laborers sharpened. Employers viewed plentiful gardens as evidence that workers were spending excess time and capital on horticulture at their expense.[59] Sensing the mounting pressure from opposition in Camp, managers ridiculed their workers and rejected grievances. They dismissed shepherds as idle, overpaid, and tardy.[60]

By 1889, the government-led General Improvement Society had rebranded itself as the member-run Stanley Social Club and Mutual Improvement Association. The club continued to put on horticulture shows, and in 1890, the friendly competition extended into the annual Sports Week in Stanley and Camp, consisting of rural recreation including sheep shearing, horse racing, dog trials, peat cutting, footraces, and other events. The Sports Association still holds such festive events each year at the end of the shearing season (see figure 10).

Yet here the focus was no longer on the individual self but rather on the collectivity. At the turn of the century, settlers started to understand improvement not just as a process of making nonhuman animals and land interest-bearing but also as a moral and political project of mutual aid. In their weekly meetings, the Mutual Improvement Association read a series of papers, lectures, and debates on such topics as "compensation," "mutual improvement," "our civilization," "universal disarmament," "the unemployed problem," "socialism" (on multiple occasions), "reform or socialism," and "socialism—what is it?" (the latter discussed Malthusian theories on overpopulation, comical in such a thinly peopled territory).[61] The mutual improvers congratulated each other on efforts to help the poor and young with subsistence provisions, all while privileging British moral standards of etiquette over the so-called wicked Gaucho propensities of the earlier era of "barbarism."[62]

The general can-do ethic of the emergent working class influenced improvements not just to private landholdings but also to public works in Stanley. With a newfound sense of urgency, government and residents developed roads and sanitation, water supply, the construction of a jetty for the general public, compulsory education, and a savings bank.[63] Nostalgia for this era is on display in the newly opened Historic Dockyard Museum in Stanley. Images depicting sheep

FIGURE 10.   Shearing competition (author in back). Camp Sports West.

Photo courtesy of Charlie Stuart.

shearing and dipping, lamb-marking, gathering, peat, and agriculture develop-ment have descriptions like "living off the land," "resourceful," and "practical." The Horticultural Society has an active Facebook page geared toward recruiting audiences or new members. The level of competition is high, and much debate surrounds the rules of categorization: Is rhubarb a vegetable? How big a pot should be used for herbs? What kind of manure is the winner using? Although years ago, enough horses, cows, or sheep abounded to provide ample manure, fewer livestock remain near Stanley now. An organizer of the horticulture show explained, "If you want a ton of manure from Goose Green, you can get it for a case of Carlsberg [imported beer]—a ton of shit for a case of Carlsberg—and I don't drink Carlsberg, so I say it's trading shit for shit." Self-reliant horticultur-alists have also taken up composting.

Trading beer for manure may seem humorous or unremarkable, but it is a clear example of the unofficial "barter system" that has been prevalent in the Falklands since the early-twentieth-century interest in mutual aid. In addition to providing for themselves as individuals, most multigenerational islanders use a nonmonetary reciprocal gift exchange system to support one another. Debts for favors are frequently settled through a gift of animal meat. I first observed this custom when my host on the outer Saunders Island invited me to accompany him by boat to Hill Cove on West Falkland Island to pick up some dog food. A man in a Land Rover was awaiting our arrival on the coast, and after exchanging a few words—but no money—he gave us the bag of dog food. In return, my host unloaded something mysterious under the cloak of a tarp. I asked him later what it was, and he told me it was "meat that we killed"—a mutton.

Here, rather than meat meant either for personal consumption or a commodity exchanged as payment for the dog food or potential profit, the mutton was a gift, that is, an inalienable form of wealth imbued with the value of a debt owed, much like the kula shells or potlatch blankets anthropologists have observed people tokenizing and exchanging in noncapitalist economies around the world.[64] This arrangement, which Falkland Islanders refer to as the barter system, is akin to a "moral economy": collective actions common among peasants and people working in rural areas that have alleviated subsistence crises throughout history and even transnationally.[65] Here, we might recall the key role of barter and demands for reciprocity to alleviate wealth inequality in struggles over dispossession between the Yagán people of Tierra del Fuego and the missionaries from chapter 1. The islanders' ongoing barter system has its roots in the notion of mutual improvement that earlier generations of working-class shepherds took up to challenge unequal access to land and unjust exploitation of labor. Nonetheless, whether they were self-sufficient or mutually supportive, for much of the twentieth century, settlers remained dependent on the FIG and, increasingly, the FIC.

A Company Colony

By the mid-twentieth century, the FIC owned and enclosed up to nearly half (46 percent) of the land in the islands. The company established a monopoly on wool trade, shipping, retail, chandlery, banking, accommodations, and entertainment.[66] Through this hold on production, distribution, and consumption, the FIC had overpowered the local British colonial governors in local influence. Acquiring the West Store in Stanley, a one-stop shop for nearly any everyday necessity, the company reconfigured the port village as a company town, and the islands more generally as a *company colony*. Journalist Ian Jack put it succinctly in 1978:

> The men of the camp . . . live in tied houses on company land. They shop in the company store for goods delivered by company ship, and have bills deducted from company wages. Many of them use the company

as a bank, the wool they shear from the company sheep goes to Tilbury, again by company ship, where it is unloaded at the company wharf, stored in the company warehouse and sold on the company wool exchange in Bradford. By means of directorships and shareholdings, and by owning the only means of transport and marketing the Falkland Islands Company extends its influence over the islands' few other landlords. For better or worse, the Falklands are company islands.[67]

With even more widespread influence in Camp than in Stanley, the FIC built schools and stores, screened movies, established a network of horse-riding traveling schoolmasters, and redistributed rent-free housing and pensions for life with thirty years of service. Even though most FIC profits were paid to absentee shareholders in the UK (86 percent from 1919–1938, for example), the firm would remain hegemonic locally until the late 1970s.[68] FIC ownership changed hands in 1972, when Dundee, Perth, and London Shipping took the firm over, and when Charringtons fuel and oil company acquired it a year later in 1973, but few changes to operations occurred until the 1982 war.

While landed property, sheep farming, and the overall economy would become more diversified in the 1980s and 1990s—as explored in the next section—many islanders still see the FIC as the biggest offender for overexploiting labor and resources. Despite their expansive social programs, one elderly former FIC employee remarked, "I didn't get nothing out of them—only what I pinched—and I wasn't the only one. Even some of the bosses were thieves." For local residents like this one, the FIC remains a cagey outside presence, justifying petty theft. Even though the company's monopoly has largely dissolved, giving islanders more opportunities to manage their own farms, the FIC now recruits most of its labor from countries such as Chile, St. Helena, Peru, Poland, and the UK for two-year contracts. While they are content to be free from the shackles of the company, xenophobic members of Stanley's white working class view the new multicultural FIC as a different kind of threat, which is explored in more depth in chapter 3.

During my fieldwork, I discussed some of these local dynamics with Larry, the FIC's director, who was also the head of the Falklands Chamber of Commerce. Born in Hong Kong, educated in the UK, and having lived in South Africa, Larry practically embodies the British Empire. Explaining where he feels he belongs, Larry started to say, "I don't have any . . ." and then corrected himself: "Home is here for me. I've been here since 1986. It's the longest I've been anywhere in the world. . . . I found my way down to the Falklands on a three-year contract, and basically stayed here."

An imposing presence, Larry is gregarious bordering on bombastic. He is a far-right Thatcher-Reaganite who is not afraid to speak his mind. Larry boasted of being quoted by the BBC, comparing the Falkland Islanders to Palestinians under siege. Taking on an almost accusatory tone, he lambasted Americans' lack of moral principles for not supporting the islanders' self-determination claim. When I asked him about the similarities and differences between the way the British handled Hong Kong and the Falklands, he thought about it for a few seconds and then pinched his pale cheek and said, "Yellow skin."

At one point in our conversation, the phone rang. Larry told someone the name of a specific part that needed to be ordered. I was surprised that Larry was responsible for such low-priority matters, but he explained that the phone he answered gets passed around, so it just depends whose week it is. Nestled into the company's warehouses, Larry's office is isolated but far from luxurious. Papers are spread out in layers all over his desk. When I visited, his secretary was absent, as were chunks of the wall, exposing electrical wiring. A humble headquarters for the forceful director of such a historically dominant institution, the office speaks to the awkward way in which the FIC was unseated in the islands' power structure.

Although the FIC does not hold as much agricultural land as before, Larry confirmed that the company still has fairly extensive holdings around Stanley. A land bank of about 400 acres positions the FIC advantageously as a local agent for supporting oil exploration, with space for development of storage facilities and the elusive potential deepwater port in Port William. As described in chapter 5, the FIC also won the contract to supply labor for the construction of oil infrastructure through the joint venture. Among other things, this kept the FIC pitted against islander-owned shipping agents and fishing companies.

However, in contrast to the early years of the company, when the British colonial government kept the FIC in check through penalties on killing cattle, Larry considered his relationship with the locally elected FIG Legislative Assembly "very positive."[69] He considered the UK-appointed chief executive "one of his best friends." Larry rejected the idea that today the islands are a company town—now, the pie is bigger. There have been fewer squabbles with the private sector, and Larry characterized himself as an advocate against monopolies in the islands. "I believe in free trade. We should free things up rather than restrict people's ability to do things," Larry said. He did not want the FIC to be a "dead hand" on the Falklands, as islanders have viewed it in the past. And he pointed out that there are a number of locally owned companies that make more money than the FIC, particularly in fishing.

Ultimately, Larry saw the Falklands' private sector and government as being "all in the same boat," united geopolitically in opposition to Argentina. To illustrate the FIC's role in the sovereignty dispute, Larry recounted his trip to Ushuaia in Tierra del Fuego. When he announced he was from the Falklands, his receptionist gave him a hard time but accommodated him anyway. In the hotel restaurant, he ordered the best bottle of wine available. He was delighted that the waiter put on white gloves to serve him, and with the exchange rate, the bottle only cost £10. However, to avoid controversy, Larry was put on a separate bus tour from other guests. The tour guide passed by Ushuaia's central Malvinas memorial without mentioning it, but Larry would not miss the opportunity to be a gadfly. When the guide told him that Ushuaia was founded in 1870, Larry countered that the company he works for was founded long before that in 1850. Waiting for the guide to ask which company it was, he responded proudly, "The Falkland Islands Company."

Subdivision of Land

While broader impacts of the war on the islands are discussed above, the most immediate outcome of the war was a large-scale breakup of the monopoly on land by the absentee landowners of the FIC into plots owned by individual families. In 1976, the UK government sent a survey team to the Falklands made up of researchers from The Economist Intelligence Unit and other FCO affiliates and chaired by Lord Edward Arthur Alexander Shackleton. Son of the famous Antarctic explorer Sir Ernest Shackleton, the chair was a geographer and Labour Party politician. The results of the survey were three comprehensive volumes (including an updated 1982 synopsis immediately following the war) outlining the islands' economy, resource prospects, and strategies for development.[70] Adopting the tone of colonial reports in other British territories, the studies describe locally born islanders as "the indigenous 'kelpers'" (see chapter 3) or even the declining "indigenous stock" with whom new would-be settlers would have to avoid friction.[71]

Yields of profit from FIC wool exports had also declined, and Shackleton warned of the "grave danger of regression to a primitive form of land utilisation as practised during the early phase of colonisation based on wild cattle."[72] In order to eliminate stagnation and prevent a return to the ostensible anarchy of gaucho culture, the reports outlined, among other recommendations, potential improvements of grassland and breed and the establishment of new rural education programs for more efficient sheep-farming methods. All three volumes, including the report published after the war, indicated that offshore resource exploitation in particular would require cooperation with and even the participation of Argentina, so before pursuing fishing or oil licensing in earnest, the reports suggested a "radical" land tenure reform program for wider ownership and reinvestment of profits.[73] At the time of the surveys, islanders ran only ten farms, representing 4.8 percent of total area, as family businesses. Moreover, only three out of the ten were owner-occupied.

Consistent with the Lockean concept of property, Shackleton viewed enclosure as a necessary precedent for improvement of the colony and its land. Drawing on previous reports that pointed to fencing as a primary means of pasture renovation and improvement, Shackleton's team considered the impacts of plowing, rotating, sod-seeding, and oversowing on various soil and grass types.[74] Concerned that the interests of distant monopolistic shareholders may not be appropriately attuned to local knowledge and experience, the reports recommended breaking up large company-managed community ranches into small family-run farms. Shackleton described Camp managers as "paternalistic," and attributed "the relatively low level of enterprise and engagement of the indigenous population" to internal "dependence" of the working class on the government and FIC.[75]

The General Employees Union, formed in 1943, had negotiated a Camp Wage Agreement, which established a minimum wage, piece rates, adjustments, long-term awards, and an unusual bonus system (based on the price of wool rather than productivity). But the now-ubiquitous practice of contracting professional gangs for sheep shearing for four to five days, two to three times per year, had just begun to erode the rural working class. Shackleton viewed this flexible labor supply of shearing gangs as a sign of progress. To enhance contract shearing with a more intensive system of sheep farming, the reports promoted a titling process administered through a Quasi-Autonomous Non-Governmental Organization (QUANGO) that would increase private property ownership for occupants through the dissolution of community farms.[76] Shackleton's report echoed earlier policy drafts, but the UK parliament received his suggestions with heightened concern in the context of the islands' stagnant wool economy.[77] Recommendations from Shackleton's first two reports went into force before the conflict with Argentina. In 1979/1980, the FIC sold its Green Patch settlement, an experimental farm built on East Falkland Island in the 1920s, to the FIG. The FIG then divided the farm into six sections, each sold at a subsidized price to a family of settlers.[78] Displaced working-class farm employees, who did not gain ownership of subdivided land, moved to Stanley, migrated externally, or bought relatively small, unclaimed islands in the archipelago.

The Port Stephens subdivision was the only farm privatized through an employee buyout rather than government expropriation and sale. First purchased by the Dean family in 1869, the farm became an FIC settlement in 1945, and during my fieldwork it was managed by a family whose residence there dates to 1909.[79] The settlement is the farthest southwest on West Falkland Island. In the main settlement house, a plaque with a Disney cartoon Goofy character hung on the mudroom door, welcoming visitors to "God's Own Country." The owner explained to me that when he was giving orders to one of his workers on the farm, they asked him, "Who are you? God?" And he replied, "Yes, I am God, and this is God's Own Country." Unsurprisingly, this divine manager viewed the subdivision as the inevitable result of unsustainable labor costs and the strengthening of British trade unions, including that of the islands, in the 1970s. Selecting his closest in command, he divided the 250,000-acre, 50,000-sheep farm into five sections (his was somewhat larger), each stocked with 10,000 sheep. While the FIG usually provided mortgages for subdivisions, in this case, the FIC financed it.

Since the end of absentee ownership, the government has crafted policy promoting greater differentiation in the scale and size of settlement farms. Some islanders reflect that the subdivision of land sacrificed "settlement culture" for an ineffective economic model. In the privatizing shift away from FIC monopoly, the FIG encouraged instead what it called an "enterprise culture."[80] However, Bernhardson points out that by conceding ownership of pastoral land to "native islanders," tenure reform actually restored the nineteenth-century geographical formation of settlement.[81] The difference was that, as in enduring latifundio systems that continue to operate in other export-oriented economies, the debt peonage of the initial cattle herding period became inverted, now with a shortage of labor rather than bosses.[82]

According to some of my interlocutors, the subdivision transformed the Falklands more than the war itself. It brought about a significant change in how people view the land, because in the 1970s, most Camp residents did not own a patch. Absentee landowners owned roughly 90 percent of thirty-six farms. Inversely, about 90 percent of the now ninety-seven farms are locally owned (see figure 11).[83] The new landowning middle-class status group of small farmers supplemented the subdivision with new techniques of enclosure, including wind- and solar-powered electric fencing, as well as mechanized shearing, shepherding with motorbikes, and a new abattoir.

Under the new self-regulated property regime, rural to urban migration toward Stanley has decreased the Camp population substantially. For example, Port Stephens's population went from forty-two to four.[84] Moreover, privatization was only partial. Even though it had sold all of its farms by 1992, the FIC still held mortgages. Along with the emergent QUANGO Falkland Islands Development Corporation (and now Falkland Landholding), the FIC would remain a leader in the private sector for financing leases and loans and subcontracting public works. During my fieldwork in 2014, the FIG was crafting a Rural Development Strategy that would use tax breaks to incentivize settlers to start

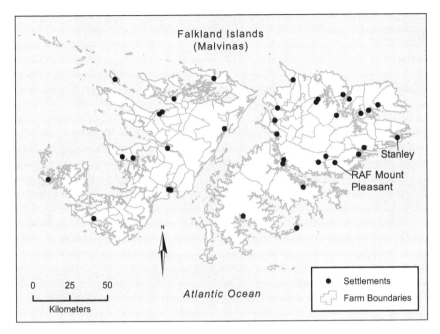

FIGURE 11.   Falkland Islands farm boundaries map (updated 2020).

Map by Terry Spies (data available online: http://dataportal.saeri.org/).

businesses in Camp. Nonetheless, the prospects of wealth from oil, fishing, and tourism continued to draw islanders into Stanley.

Lafonia's all-but-abandoned landscape now bears the ruins of the FIC settlements: telephone poles installed for bare-wire windup phones, spots where sheep grazed at fence corners, sod or steel shelters built for sheep to avoid hypothermia, and clumps of gorse that served as miniature corrals for gauchos. Remote settlements reveal areas to stack peat or pile ash, empty dog cages, dilapidated dip yards, and sheds for hanging mutton or milking cows, as well as the signature "palenkey" (from the Spanish *palenque*), a wooden frame or post used to hang mutton or beef carcasses.

From the Goose Green settlement, one of the largest of the former FIC Camp settlements near Darwin, a boggy path between minefields leads to Bodie Bridge, the world's southernmost suspension bridge. It once connected the vast farms of Lafonia to the FIC's original headquarters for wool production. Long abandoned, the bridge is falling apart. Although visiting contract workers delight in taking the risk to venture across it on weekend adventures, local residents view this as trespassing. From Joan's registry of rural heritage to small farmers' new forms of enclosure and improvement, islanders are selectively seeking to preserve ruins of their settler colonial past and suspend British sovereignty in the present.

Corrals, built primarily before the permanent British settlement, remind us of an initial cattle frontier settlement based on usufruct rights to land through debt bondage of gauchos. Although there was some continuity after British occupation, the cattle venture declined, and the dominant colonial FIC developed a different paternalistic land tenure system focused on sheep farming. Despite widespread enclosure through sheep fencing, uncontrolled grazing prevented the improvement of grasslands. Experiments in sheep breeding sustained the wool trade, but colonial administrators' efforts to make the settlers self-sufficient clashed with a working-class reappropriation of improvement as mutual aid. Nonetheless, the FIC enjoyed a near-total monopoly on production, distribution, and consumption in the islands, which endured until a dramatic subdivision of land following the 1982 war.

Noticing the settlers' propensity to garden and save against the odds, one of the key recommendations given to the UK after the 1982 war was the commercialization of vegetable production.[85] Initially, the idea was that the presence of a more permanent garrison would offer a new consumer base. Owned and managed by a particularly vocal islander who lost an eye in the conflict, the market garden Stanley Growers now provisions small cruise ships and the occasional oil exploration crew with "ultra fresh" veggies, as well as supplying fishing vessels with stock of "a reasonably high standard." UK expatriates and others in town decry high prices for local residents, but the islands' remoteness makes produce, particularly that which is imported wholesale, more costly.

Some residents lament that, with the wider distribution of cash from fishing, oil, and tourism, the barter system may be disappearing. But as David Graeber has shown convincingly, the common notion perpetuated in classical economics textbooks that barter is an origin from which market economies evolve naturally is a myth.[86] In the Falklands, I have observed that while mutton or lamb may not be swapped as frequently as it might have been earlier (at one point in time, meat gift deliveries by plane were announced on the radio for unsuspecting residents), such exchanges have actually proliferated in different forms—including but not limited to a case of beer. If a person leaves their car in a Camp settlement overnight while visiting friends, there is a fair chance that it will be filled up with gasoline in the morning. Residents created a Facebook group called the "Falkland Islands Bring & Buy," a forum comparable to Craigslist, where they can make inquiries or post household items such as clothes, books, or toys for sale. Finally, with limited bandwidth, islanders have not had the luxury to stream videos or music online. But most residents have a cable and wireless plan that provides free nighttime internet service from midnight to 6 A.M. Few to no local laws enforce against possession of pirated media, so many islanders set their computers to download torrents of videos during the off-peak hours. They then hold parties or informal gatherings in which attendees exchange multiple hard drives full of entertainment media. While these iterations of the barter system may depart from the idea of a moral economy as a way of addressing food shortages, the spirit of mutual improvement remains intact. Cash might make it easier to order goods from overseas, but within the archipelago, Christmas party invitations stand in for money.

Finally, it bears noting that the islanders' adaptations of a British imperial property ideology contrasts with Argentina's inherited Spanish normative order of state ownership over subsoil. These contradictory principles sharpen the opposition between self-determination and territorial integrity in the sovereignty dispute. The Argentine government asserts national sovereignty over the full South Atlantic continental shelf, as though its symbolism were self-evident. But the history and politics of property within the islands demonstrate that the settlers have used ongoing material practices of enclosure and improvement to stake their claim to British sovereignty through shifting regimes of localized power. If a resolution to the dispute is ever to be reached, it is imperative that we understand these enclosures as manifestations of uneven and inherently contradictory property relations.

IMPERIAL DIASPORA

Deano's, one of Stanley's local bars, holds a weekly "Latin Night" in which Chileans, contractors, and some Falkland Islanders dance to cumbia remixes of pop songs. The party often continues late into the night after the bar closes, when attendees transport cases of Budweiser to the Chilean abattoir workers' smoke-filled portakabins. During one Latin Night, I met Flavio, a slaughterhouse foreman from Chile. Flavio did not speak any English, and he was delighted to converse with someone other than his coworkers who knew Spanish. Since he was only in the islands for five weeks at a time, he never bothered to get a mobile phone. He would check his email in one of Stanley's wireless hot spots, but Flavio's primary method of communication was simply to rush toward me, wave, and whistle whenever he spotted me walking in town.

One day, as I was walking home from my office, Flavio greeted me with a big smile and invited me to have tea in his place: a portakabin in the Murray Heights trailer park, tucked behind the bypass road, up the hill from Stanley. Flavio told me he did not really know his neighbors, but he indicated that they included sex workers and colleagues at the slaughterhouse. Most of the temporary workers in Murray Heights kept to themselves. Flavio's portakabin was comfortable, but, again, there was no phone or internet. He had a collection of thirty to forty mugs, but almost all of his possessions were destroyed or stolen in the aftermath of the devastating 2010 earthquake in Concepción, Chile, where he used to live. This disaster forced Flavio to abandon his career in human resources and travel. First, he went to Peru, where he started to cook for a living. Then, he saw an advertisement for a job as a chef at Mount Pleasant Royal Air Force Base (MPC), and he applied for it. Flavio wanted to learn English, but he worked almost exclusively with Chileans in the Falklands, so he did not acquire any new language skills.

For entertainment, Flavio lifted his bed to find some music or a DVD to play, but the sound did not work on his television set. Flavio prepared *pan amasado* (homemade bread common in Chile), scrambled eggs, ham, and coffee. A talented cook and a gracious host, Flavio had worked in the kitchen at MPC for a year and traveled back and forth between Chile and the islands three times before starting his job at the abattoir. In between these temporary contracts, he worked in tourism in the northern Chilean town of San Pedro de Atacama, and he showed me a slideshow from his travels in the Atacama Desert. Intermixed with photos of geysers and canyons were pictures taken from MPC, including an image of one of the British soldiers in uniform giving him a kiss. At the time, I thought it might have been possible that the soldier was expressing his approval of Flavio's cooking skills. He later told me that he was signaling to me that he was gay.

As we ate, Flavio told me that he had a bad day at the abattoir. He had a confrontation with someone whom he referred to as a *musulmán* (a Muslim Chilean who was visiting the islands to carry out a halal certification inspection). The halal inspector would not let them sell much of the mutton because several of the sheep were pregnant, and this was disappointing. But the real source of Flavio's distress was that the halal inspector targeted him directly. Flavio informed me that the inspector had told him that he would not be able to return for work again because they do not allow homosexuals. Having trained in—and even instructed—human resources, Flavio had a keen sense of his workplace rights, and he felt unfairly discriminated against. Yet, as a temporary migrant, Flavio did not feel entitled to the same rights as Falkland Islanders, even though he felt safe and supported in general. Stanley has a growing LGBTQ community, but the language barrier and his temporary work status prevented Flavio from connecting with permanent residents.

As part of the *matadero* (abattoir) crew, Flavio had been exposed to the seedy underbelly of a surprisingly active nightlife in Stanley, which included alcohol use, sex work, and drugs (usually "legal" highs). The drug use in Stanley pivoted on the consumption habits of Falkland Islanders and British contractors, but action against it was targeted disproportionately on South American and St. Helenian incomers. In addition to the Chilean temporary workers, the meat abattoir employed born-and-bred Falkland Islanders who were themselves deemed "undesirable" or "unemployable," including formerly incarcerated released convicts of sexual abuse.[1] Despite being thrown into this unsavory shadow world, Flavio was also inserted into a supportive network of Chileans, both at

MPC and Stanley, who would get together for barbecues and Latin Night at Deano's. If he had wanted to, Flavio could have endeared himself to one of the Chileans who has permanent "status" and might act as a sponsor. Flavio was trying to earn an honest living, but his sexuality ultimately placed him in the "undesirable" slot, and he argued that this was why his permit was discontinued.

Flavio's disenfranchisement raises critical questions about race, class, gender, sexuality, religion, and nationhood that most writing about society in the Falkland Islands tends to ignore.[2] Who qualifies for Falkland Islands status? Who is included in the "Falklands way of life"? Who is left out, and why? This chapter explores how, even in the Falklands, where referendum results show a strikingly monolithic public opinion, contradictory political strategies of the UK and the FIG have cultivated an ambiguous racial formation among settlers and migrants.

As geographer Klaus Dodds has shown, drawing on the work of Paul Gilroy, political rhetoric underwrote the Falkland Islanders' whiteness as a national cause in the British motherland.[3] Thatcher and other UK leaders spoke of not abandoning "loyal" British subjects, whom they referred to as "kith and kin," "our boys," or a comparable "island race" of "British stock and tradition."[4] In this sense, the UK's support for the Falkland Islanders in the 1982 military campaign stands in stark contrast to its denial of former colonial subjects' rights in other overseas territories.[5] Scholars have begun to analyze these aspects of the war through the lens of imperial history.[6] Nonetheless, by framing the islanders as kith and kin in a benign way as part of "Greater Britain," such studies leave relations of race and dispossession unexamined. Instead, here, I draw on W.E.B. Du Bois's description of whiteness as "a public and psychological wage" and describe how islanders have used the color line to reinforce racial hierarchy and in turn salvage empire.[7]

Responding to Leith Mullings's call for an antiracist anthropology and Stuart Hall's critical theorization of Thatcherism, this chapter interrogates authoritarian dimensions of popular consent that provide the basis for white settler self-determination claims.[8] The racialized power dynamics of authoritarian populism and self-determination are fundamental for understanding what is described here as an *imperial diaspora*.[9] While Falklands residents may claim to be essentially "British to the core," their experiences reveal an emergent, hybrid social process of self-definition at work behind the seemingly rooted and fixed self-determination claim. Building on Immanuel Wallerstein's definition of peoplehood as "not merely a construct but one which, in each particular instance, has constantly changing boundaries," this chapter explores unstable aspects of class, race, ethnicity, and nation in conceptualizations of the Falkland Islanders as a cohesive social unit.[10]

The British colonial government first treated the Falkland Islanders as set-tlers, but they later interacted with them as subjects who have ultimately posi-tioned themselves in the current form of a settler colonial protectorate, as rightful British citizens.[11] In response to the settlers' paradoxical "subjection," a conser-vative, white, working-class movement for representation in government took hold through the establishment of the Falkland Islands Reform Club. As white settlers began to naturalize as "native" islanders, they protested farm managers importing "alien" migrant labor, and then lobbied British colonial administra-tors against any potential sovereignty deal with Argentina.[12] Here, race became articulated as a means of both gaining rights and accumulating capital, as well as what Hall describes as a "modality" through which class is lived.[13] Strug-gles over democratic representation and immigration restrictions indicate how categories of race and class feature in access to jobs and residence, ranging from a white expatriate class of political elites to subaltern groups of so-called unde-sirables like Flavio and his seasonal migrant coworkers in the abattoir.

The prevailing governance principle of the Falkland Islanders exemplifies what Stoler has called degrees of imperial sovereignty.[14] That is, scales of differ-ence and allegiance have allowed racial hierarchies of empire to endure, even after the era of decolonization.[15] As this chapter outlines, the FIG has given such degrees specific values or "points" in the application process for obtaining per-manent residence permits (PRP). Without the necessary qualifications for a PRP, individuals lose the rights to own property, build a career, or access guaranteed health care. But even those with a PRP cannot vote until they obtain the final degree of imperial sovereignty: "Falkland Islands Status." As we shall see, full recognition as a Falkland Islander tends to require that one is locally born. None-theless, legal naturalization is designed to comport new arrivals to the "Falk-lands way of life," a local allegory comprising safety and work ethics that have come to define belonging and propriety in the islands.

Kith and Kelper

In a letter to the *Daily Express* titled "No Civilization Here: A Lonely White Man's Account of a Visit," an anonymous member of the Royal Navy who had been stationed in the islands during the period leading up to the 1914 Battle of the Falklands in World War I described the islands as "a fairly peaceful country, with no savage native tribes." Paradoxically, this same "Empire maker" was anxious to board the HMS *Glasgow* for the "opportunity of seeing civilized white men once more!"[16] Even though he acknowledged the absence of a precolonial Indig-enous population, using racist terms, the sailor implicitly excluded Falkland

Islanders from the white race. Data from periodicals show that at the turn of the twentieth century the islanders were governed similarly to colonial subjects in other parts of the British Empire, but with degrees of white privilege that kept their peoplehood relatively indistinct from imperial administrators.[17] Class stood in for race, as proletarianized settlers, "lacking good manners," clashed with farm managers.[18]

During the same period as the migration of nonlandowning settlers from the Falklands to Patagonia, those who remained in the islands began to build a labor movement, which also served as a vehicle for advancing white working-class conservative values. In addition to the Improvement Society explored in chapter 2, the Falkland Islands Reform Club, founded in 1887, objected to the lack of advancement in labor conditions, which seemed disproportionate to the rising profits of the wool industry. The club's primary objective was political representation through appointment of a working-class member of Legislative Council.[19]

Fallow for decades, new iterations of the group appeared in 1920 and 1933 as the Falkland Islands Reform League.[20] Stagnating wages, unemployment, and precarious seasonal labor in Camp led the Reform League to take a more militant stance to demand legislative representation.[21] In 1940, one member declared: "The day has arrived when [the worker] knows the value of his labour and demands a fair return for it, he lives and has lived since the country was first colonized in a state bordering on subjection, in a mild form no doubt but subjection it undoubtedly is, he is subject to his master's whims and fancies."[22] This social class division thus marked the transition from early settlers referring to the UK as "home" to indigenized subjects criticizing managers for "importing" British laborers. In short, rural settlers came to view themselves as native Falkland Islanders who deserved a right to work and a decent living in their own country.

The term that came to refer to a native of the Falkland Islands was *kelper*. Alluding to kelp, the abundant seaweed surrounding the archipelago (see figure 12), "kelper" usually indexes an individual born in the islands, though there are exceptions to this rule.[23] Currently, this expression is somewhat out of fashion in the Falklands, but repatriated islanders in the UK and mainly elderly people within the islands continue to self-identify as kelpers. Argentine nationalists use it as a derogatory slur. But in the early twentieth century, kelper emerged as a white ethnic community construct through which locally born descendants of British settlers made possessive investments in the future of the colony.[24] This collective awareness of self—as kelpers—motivated islanders to protest against recruitment of labor from Patagonia that might threaten the job security of native islanders.[25] In 1935, a Colonial Ordinance categorized a "prohibited immigrant" as "any person not being native of the Colony" without the required papers.[26]

FIGURE 12.   Giant kelp (*Macrocystis pyrifera*) in the Falkland Islands (Malvinas).

Photo by James J. A. Blair.

Even though the signing of the Atlantic Charter in 1941 had guaranteed the right to self-government, and proposals had been in place to reform Legislative Council, islanders resented the delay in allowing local elections. In 1948, 740 residents of the Falklands sent a petition to the Secretary of State for the Colonies demanding the removal of their appointed governor, Geoffrey Miles Clifford.[27] Significantly, the petitioners pivoted their request for government representation on being a "British community, 100% white, and noted for its loyalty to the crown in the past."[28] As Dodds has argued, the islanders' claims of "kith and kin" and "loyal" UK "stock" effectively positioned the settlers as racially privileged in contrast to other colonial subjects of the British Empire, as well as being differentiated from Argentine citizens.[29] As in other post-abolition, settler colonial contexts of the Americas, the "native" islanders thus chose whiteness, aligning politically by race instead of class.[30] Islanders used their white racial affiliation as a platform for declaring themselves rightful citizens rather than colonial subjects.[31]

While the islanders did gain some unofficial representation in Legislative Council, their racially driven assertions were not always legible to imperial administrators. Entering the era of decolonization, the Colonial Office, which had apparently never sent a minister to the islands as of 1952, compared the "native islanders" to "African or Indian" workers on tropical estates.[32] This inherent contradiction between a British colonial habit of racial classification on the one hand and an assertion of essential sameness as "kith and kin" on the other is representative of a lasting ambivalence between UK administrators and the islanders. This imperial equivocation nearly compromised the islands' British status. In 1965, Argentina pressed the UK on the sovereignty issue, passing Resolution 2065 in the UN's Fourth Committee, which pushed for negotiations that recognized the islanders' "interests" but not their "wishes." There followed a steady bilateral negotiation process between Argentina and the UK that lasted through the 1970s until the breakdown in talks that led to the 1982 conflict.

Meanwhile, the islanders' wishes came to the forefront primarily through the efforts of the emergent Falklands lobby.[33] As they had done in the petition of 1948, when the possibility of a sovereignty transfer arose, the unofficial members of the Falkland Islands Executive Council urged Parliament to reconsider any agreement with Argentina. In a 1968 manifesto, they asserted that the islands have "no racial problem" and lead "a very British way of life."[34] Subsequently, the Falkland Islands Emergency Committee and Falklands lobby mounted a campaign to portray the islanders as a loyal piece of Britain under threat from outside forces. Stanley residents held protests, carrying a banner and signs that read "Keep the Falklands British," as depicted in a famous October 1968 photograph published in the *Daily Express*, which the celebrations following the 2013 referendum would eventually resemble. Synthesizing the authoritarian populist values of the islanders, the accompanying article described it as "a picture not of a crowd—but a people."[35] Later that same year, Lord Alun A.G.J. Chalfont visited the islands—the first time a British minister had done so—but his efforts to persuade islanders to consider potentially beneficial interests of closer ties with Argentina fell on deaf ears. The Falklands lobby, or UK Falkland Islands Committee, continued to emphasize the islanders' white racial composition and British national affiliation as they pushed for political and diplomatic support throughout the 1970s. The islanders ultimately influenced London to defend their wishes for self-determination to remain British, despite the Argentine government's intensifying claim to territorial integrity.

The islanders' self-determination claim contrasted sharply with the way anticolonial nationalists used the same principle to challenge the international racial hierarchy of empire in struggles for decolonization.[36] Nonetheless, the British government wavered in its interpretation of the islanders' racial classification. Lord Shackleton described the islanders as the "locally born indigenous 'kelpers'" or, in relation to population decline, as the "indigenous stock," which he divided from British contractors.[37] At the time, he said that the "indigenous 'kelper' stock" represented 77 percent of the population. He listed their qualities: "honesty, versatility, and hardiness; but there is also an apparent lack of enterprise at individual and community levels, and a degree of acceptance of the status quo which verges on apathy."[38] However, Shackleton also emphasized the kelpers' strong sense of British identity as "an isolated archipelago, remote from Britain, and without any indigenous population, settled by people of British stock; the general 'Englishness' of the culture and way of life."[39] This presents a paradox: how could a people of "indigenous 'kelper' stock" be native without being a precolonial Indigenous population? This ambiguous racialization of British descendants, maintaining an "English" way of life in the global periphery, demonstrates how the settlers' transformation from subjects to citizens remained unresolved in the twilight of the British Empire.

During the period of decolonization, the Falklands lobby established a Falkland Islands office in London in January 1977 (now Falkland House), which pushed for political and diplomatic representation as well as commercial fishing licensing (see chapter 4). In 1980, when islanders perceived Nicholas Ridley, FCO minister under Margaret Thatcher, to be "selling them down the river" through a potential "leaseback" agreement with Argentina, townspeople in Stanley rejected him in a "bellicose" manner.[40] Through furious lobbying, the islanders convinced members of Parliament that a handover of sovereignty to Argentina would amount to race betrayal.[41] As talks with Argentina broke down, leading to the 1982 war, the settlers' cunning assertion of themselves as both indigenized kelper subjects with the right to self-determination and neglected white British citizens ultimately earned the UK's steadfast defense.

The following sections of this chapter discuss new contours of class, status, race, ethnicity and nationalism in the current context of the Falklands, which have seen successive waves of immigration, intensive regimes of temporary labor, and restrictions on permanent residence or citizenship in the postwar period.

Protecting Power

In spite of nativist immigration restrictions, depopulation caused a significant labor shortage in the Falklands during the twentieth century. To supplement the native kelper labor supply, the FIC recruited workers from Chile and Argentina in the 1940s, with a subsequent wave that included Uruguayans in the 1970s.[42] Starting in 1980, the FIG began looking elsewhere in the British South Atlantic Empire for labor, particularly the island of St. Helena. The establishment of the MPC Royal Air Force Base following the 1982 war added not only a permanent garrison of more than 1,000 soldiers but also a wave of St. Helenians who came for service jobs at the military complex. Despite instances of racism and xenophobia, St. Helenians have now arguably integrated into Stanley more prominently than most South Americans and many British expatriates (hereafter, expats), although the Chilean segment of the population is increasing. Groups of migrants continue to arrive for work from other locations around the world, including, for example, the Philippines. Argentine citizens have been allowed to visit the islands as tourists since 1999. However, after a series of incidents—clandestine film projects, child rugby recruitment, and flag waving—a group of islanders gathered 494 signatures in 2015 to petition for stricter limits to free speech, including a ban on Argentine flags.[43] Amid these population shifts, a new division of labor by racial classification has emerged, and the Falkland Islanders have scrambled to protect and preserve the social mores that they refer to as the "Falklands way of life."

The UK's ambiguous racialization of the islanders as indigenous kelper or white British stock did not vanish with the war effort. Rather, the construction of a base, separated by an hourlong drive over a dirt track surrounded by minefields, made the distinction between Stanley residents and UK-born military personnel more apparent. The British soldiers have their own movie theater, bowling alley, bars, and restaurants. Besides visiting a relatively isolated training facility uphill from East Stanley, stationed military personnel (whom islanders refer to as "squaddies") only come into town for nights out on select weekends. Yet, through their limited interactions with locals, the military introduced a lasting epithet that has virtually replaced "kelper" as a vernacular term to index a locally born individual: the "Benny."[44] Deriving from the character Benny Hawkins in the British soap opera *Crossroads*, the term stereotypes local residents as comparable with the witless, backwoods boy who wears a boiler suit and a wool hat. "Benny" refers to a pervasive white working-class conservative aesthetic that blends Scottish and British Commonwealth distinctions of taste with those deriving from the rural US. For example, during Sports Week or two-nighters in Camp, islanders sing along to Lynyrd Skynyrd's "Sweet Home Alabama" while dancing a two-step (also referred to as a "Benny two-step"), a variant of the Scottish cèilidh. While it is still considered derogatory if an outsider employs the term, islanders have now reappropriated "Benny" in an affectionate way. Most locally born islanders who spoke with me rejected the "kelper" label and instead self-identify as a "Benny," as well as "chay," "native," "sourdough" or even "white Negro." These codes not only distinguish fellow Falkland Islanders from so-called military squaddies but also set themselves apart from British expat contract workers

Of course, Benny, squaddy, and expat, much like the prior ethnic community construct kelper, are also indices for whiteness.[45] Santiago, the Chilean Spanish instructor who has lived in the Falklands for more than a decade, has gained an especially keen perception of race and nationality in the islands. He explained, "An expat is a white person that comes from the UK . . . to the Falklands to work, especially in government, and they normally go for two years: contractors." Of course, given their small population, islanders need skilled professionals, and the degrees of imperial sovereignty have allowed for a relatively stable pool of experts and consultants from the UK. Some expats stay beyond their contracts, and a few of them are now elected MLAs.[46] Thus, while most islanders are glad that power has devolved to some extent, they do not necessarily aspire to be fully in charge of all aspects of society, and they benefit significantly from the continued formal links with the UK.

Nevertheless, numerous permanent residents who spoke with me also critiqued how their government administration, including lawyers and judges, consists largely of expat professionals. Irritated at most contractors' lack of commitment to the long-term well-being of the society, some islanders called contractors "imperialist transients" or "breeze-ins." Governor Colin Roberts observed astutely that, unlike before, when islanders used to refer to the UK as "home," now "many Falkland Islanders would say, 'I'm not 8,000 miles away from home. I'm at home, and there is a difference.'" Islanders refer to the Falklands increasingly as a "country," and much governance has devolved to locally elected representatives. At town hall meetings, islanders air grievances about contractors' perks, such as refrigerators ("white goods"). Expats are mocked for not being able to stand the cold or wind, and for preferring the posh chocolate and wine bar over a rusty, understated pub like The Rose.[47] Misrecognition brings islanders' opinions on expats into relief: one Falkland Islander interlocutor was outraged at being labeled "English" in Argentine media—a common mistranslation glossing white Anglo-Saxons as *ingleses*. To avoid mistaken identity, some interviewees introduced themselves to me as islanders because, based on biologized, ethnic absolutist notions of national British culture, they feared being misidentified as English.[48]

The military is not immune to this distanced opposition. Even islanders who participated in the war express outrage at Argentine accusations of "militarization" in the South Atlantic.[49] Despite the degrees of imperial sovereignty, a deepseated mistrust of the UK FCO dates back to the negotiations of Chalfont and Ridley. Some islanders that I interviewed saw the military presence as too secluded from the rest of society and condemned how investment seemed to be decreasing little by little—a process one Falkland Islander veteran of the 1982 war described as "civilianization." Less directly affiliated with whiteness than expats, squaddies are moreover not exempt from racial discrimination in the Falklands. At a footrace competition during Camp Sports Week, one of three British squaddies who happened to be Black took the lead in the one-mile run. From the sideline, I observed a chagrined farmer in the audience joke that one of the younger local boys had been "overtaken by a *n——*." Here, the anti-Black racist implication was that the military officer should have been either impaired by, or given unfair natural advantage according to, the color of his skin. It was unclear how well this soldier knew the farmers at this settlement, because an assortment of spectators gathered from around the islands. Despite their key role in British victories during the two world wars, the military service of Black men has long been made invisible and silenced, as anthropologist Jacqueline N. Brown has observed.[50] Still, it was surprising (and intensely grating) to witness the farmer, a white working-class Falkland Islander ostensibly liberated by British forces, assign such a casually racist epithet to one of the military recruits. As anthropologist Engseng Ho observed in his study of the oceanic ties between British colonialism and diasporic Hadrami Arabs, "a diaspora and an empire were locked in a tight embrace of intimacy and treachery, a relationship of mutual benefit, attraction, and aversion." The envy, affinity, and antagonism expressed in this singular instance of white, rural, working-class conservativism demonstrates how enduring yet contradictory forms of racial hierarchy are inherent to the making and remaking of imperial diaspora.[51]

Meanwhile, when the squaddies flooded into Stanley on weekend nights out, they complained to me about the relative scarcity of local women. Half-hearted, sexist proposals have circulated in Stanley to give the West Island away for "3 forty-foot containers" full of "Argie" women to improve the local population's gender balance.[52] While this particular sex-trafficking scheme was meant to be humorous, Falklands citizens and their British trustees already manage racialized and gendered regimes of labor recruitment. Returning to the predicaments of marginalized temporary migrant workers like Flavio, the following sections offer insights on the everyday afterlife of the British Empire in the South Atlantic.

The "Undesirable" Slot

Bars close at 11 P.M. in Stanley, but once every month or so, a converted shed called The Trough becomes an after-hours venue for live music. During one of my stages of fieldwork, a band called The Pushers consisted of a Filipino singer, a St. Helenian guitarist, a Chilean bassist, and a white British Falkland Islander drummer, who led the band. This group exemplifies the increasing diversity of the local population in Stanley. Nonetheless, the early social division between settlers who were allowed to naturalize (desirables) and aliens who toiled under temporary contracts (undesirables), remains a point of racial tension in the current context of the Falklands.

Most Chileans and St. Helenians—or "Saints," the local shorthand for the group—have become excluded from either the "Falkland Islander" or "contractor" categories. Ethnographic observations of casual racism, precarious working conditions, and diasporic cultural production shed light on some of the predicaments of members of each particular social group, who face different kinds of challenges for gaining permanent residence and Falkland Islands Status.[53] The mechanisms and processes that govern possibilities for peoplehood in the Falklands, however, are generalizable to the historical and political form of authoritarian populism in the settler colonial protectorate.

The Ministry of Defense recruited so-called "incomers" from St. Helena (and later Chile) primarily to support the reconstruction effort after the war. St. Helena, a former British colony (now a BOT in its own right), was a source of cheap labor, and the Saints performed menial tasks such as cleaning toilets at the new Mount Pleasant military base. Once they finished their contracts at Mount Pleasant, many Saints were able to find work in Stanley, but they remained excluded by race. As Santiago, the Chilean Spanish instructor, explained, "Someone from St. Helena who is not completely white but closer to someone from the Caribbean maybe in terms of the way they look, they're not expats; they're 'Saints.' You would never call them an expat; they're 'Saints.' You would probably call an Australian an expat, and we do have Australians."[54] Here, according to Santiago, despite a shared British Commonwealth heritage, Saints would not meet the criteria for the white ethnic community construct of expats in the Falklands that relocated British descendant settlers of Australia may. We can assume that by "closer to someone from the Caribbean in appearance," Santiago meant, "not white"—that is, Black—but what else motivates this typological distinction?

First, besides a heightened British politeness in the Falklands, there is a growing effort to celebrate liberal multiculturalism in contradistinction from Argentina's Eurocentrism.[55] However, myriad examples of race-thinking or casual racism offer critical insights on the assumed norms and values of whiteness, which are often implied in invocations of Falkland Islander peoplehood. For instance, one elderly Falkland Islander portrayed the Saints as extremely nice, insisted on calling their employer "sir," but this interlocutor found the Chileans more "adaptable" as flexible labor wherever needed. This same islander viewed St. Helenians as closer to their families in the "urban" setting of Stanley, and underscored how they have their "boom box on their shoulders." Similarly, one contractor in the process of naturalization described the St. Helenians as the "boys in the hood" of the islands, "riding around playing reggae music." While I did meet St. Helenians that had been living in the islands for more than 15 years and still had never visited any Camp settlements outside of Stanley, these racialized characterizations of urban Black youth blended occasionally with local panic over petty crime in disproportionate and contradictory ways. For example, migrant workers were targeted in Stanley for accusations of littering, even though these are the very same residents who are hired to do the bulk of cleaning and maintenance work.[56]

Breaking down the current division of labor by race in the islands, Larry, the FIC's director explained, "We have sort of rules on who you're going to give [phone] credit to. The general rule is, a born-and-bred islander, no problem. Person with PRP settled from the UK, no problem. St. Helenians, not usually a problem. Chileans, settled here, married here, not usually a problem. Chileans on work permits, a large problem. They tend to come and then leave with the device . . . they buy so much a month and then they disappear." This provides further context for why Flavio may not have gained access to a phone. Moreover, even if they have privileged access, many Falkland Islanders consider it unethical to shop at the FIC's West Store because it generates profit for absentee shareholders. Ironically, shortly after our meeting, white born-and-bred Falkland Islander construction workers were convicted of a series of burglaries.

Among work permit holders, there is a distinct difference in skill level between the expat contractors with specialized postgraduate degrees who direct government departments on the one hand and unskilled laborers migrating predominantly from Chile but also St. Helena, Poland, the Philippines, and the UK or British Commonwealth (and, as discussed later, Zimbabwe) on the other. While there are some high-level Spanish fishing business owners, the crews, comprising primarily Indonesians and Filipinos, barely touch land apart from the Seamen's Mission Centre (see chapter 4). Temporary workers range from qualified carpenters, plumbers, and electricians to a labor reserve maintained to pick up the slack from local school leavers. Some of my interviewees involved in public policy described the population of Saints and Chileans as "backfill." Yet Chileans seeking or having gained permanent residence now manage a number of accommodations and restaurants; they work in retail, in construction, at the market garden, and as hairdressers, and a distinct group of seasonal workers such as Flavio are recruited annually to work at the slaughterhouse.[57] Some migrants, whom islanders I interviewed accused of being "money-oriented," have now entered middle management in shipping agencies or are self-employed. Many of those who have worked both for the Ministry of Defense and in town prefer Stanley, because Mount Pleasant housing offers just a room, not a more conventional house or shared home. Standard Chilean cuisine, including empanadas, *chacareros* (steak sandwiches), *cazuela* (stew), and pisco sours are now basic menu items in town. Still, some prefer working at Mount Pleasant because they find it too stifling to absorb the English language and British culture in Stanley. One Chilean I met who cooks and cleans at one of the several mountaintop military stations told me that she prefers it because she enjoys the solitude and the ability to be virtually "her own boss." Even if it means living in an isolated, militarized environment, she finds the base more accommodating.

Several islanders I spoke with measure class, race, and status based on the kinds of vehicles different groups drive. One multigenerational islander informed me that she had discerned a pattern of Chileans driving slightly worse or older Land Rovers than white local residents, but now that some Chileans are becoming more integrated, it is the St. Helenians who drive the oldest and worst automobiles. To this interlocutor, this confirmed that St. Helenians are currently doing the jobs that Chileans perhaps used to do, particularly service work. However, other interlocutors insisted that St. Helenians have formed a middle-class status group and that Chileans are positioned lower on the socioeconomic ladder (with newly arriving Filipinos even lower). Unlike Chileans, who perhaps arrive with the intention of learning English but may struggle to do so with short-term contracts, the St. Helenians have the advantage that their mother tongue is English. Still, the dialect differs greatly from that of the Falkland Islanders, which, in addition to racial and ethnic identity, has facilitated a comparable enclave effect to that of Chileans.

Settler colonial diglossia described in the previous chapter further separates locals between those who find meaning in their historical South American connections and more xenophobic settlers or white expats seeking a little piece of Britain in the South Atlantic. The most vocal members of the latter group spread fear about the impact of the influx of Peruvians and Chileans recruited to build houses. They seek to limit migrant families' access to education by excluding children who do not speak English. More cosmopolitan islanders learn Spanish, volunteer on Galician fishing vessels, or become romantically involved with Chileans.

On May 6, 2012, Alicia Castro, Argentine ambassador to London under Cristina Fernández de Kirchner, announced that her government intended to begin sending direct flights from Buenos Aires to the Malvinas. The passengers that Castro planned to fly to the islands would not be military personnel, government officials, or even oil company executives, but rather Spanish teachers. This plan to persuade islanders to embrace their neighbors through language education did not come to fruition during the Kirchner presidencies. However, with more Falkland Islanders training in Spanish voluntarily, particularly in Stanley, some of their Anglicized pronunciations are giving way to accents more faithful to formal Chilean Spanish. For example, while the place Laguna Isla used to be pronounced "lagun eyes lee," many islanders now pronounce it as a South American might: "lagoonah eeslah." Despite the prevailing racial divisions, this indicates the possibility of an emergent language shift away from the syncretisms of the high standard English in the islands.

Points for Imperial Sovereignty

Falkland Islanders I interviewed said frequently that they love "their" Chileans or that migration is good for the society. However, there are varying concrete hurdles for temporary workers—even those deemed desirable—that prevent migrants from building savings, gaining citizenship, or owning homes. First, if someone enters the islands with a work permit, they have to pay into their pension but cannot transfer it unless they stay in the islands for at least five years. This could incentivize workers to be less "transient," but most are given two-year terms that are not necessarily open to renewal. Under this system, a work permit usually lasts two years, although this is not set in stone. It is extendable with the right to reside, work, and earn money for two additional years, but an FIG policymaker informed me that around 17 percent of contractors disappear after their contract is up. After another three years, individuals are able to apply for a PRP, which offers more privileges such as the right to own land, the right to operate businesses, medical benefits, and so on. However, they cannot vote until they obtain Falkland Islands Status, which takes a minimum of seven years.[58]

This process may seem relatively straightforward, however former migrants from St. Helena and Chile who had resided in the islands for more than seven years said they were reluctant to apply for status because they did not "feel eligible." One St. Helenian who worked in service and retail came to the Falklands when she was two years old in 1993, two decades before the 2013 referendum. She explained to me, between pensive pauses of uncertainty, that even though she would vote "yes" to remaining British, she did not want to vote in the referendum out of "respect" for the islanders' self-determination. Even though she had had the paperwork for a long time, and most of her peers were aware that some of her family was born in the islands, she had not applied for status. St. Helenians do, however, have an upper hand on Chileans in the sense that they not only speak English but also already have British BOT citizenship. Moreover, the second generation of St. Helenians currently residing in the Falklands benefited from education in Stanley as well as A-levels in the UK. A number of the younger generation St. Helenians had already gained Falkland Islands Status, which their parents arranged, allowing them to vote in the referendum.

Santiago, on the other hand, did not have PRP, let alone status to vote, even though he was fluent in English and one of the most integrated Chilean residents in the Falklands. Born under a strict curfew in the first years of Pinochet's brutal dictatorship in Chile, Santiago was reticent about discussing politics. He espoused strict proprietary standards regarding nationality and language capacity; Santiago critiqued fellow Chileans living in the Falklands who had not learned English, just as he questioned American tourism managers operating entirely in

English with gringo backpackers in Torres del Paine, the Chilean national park where he used to work in Patagonia. He insisted that the Falklands are part of South America (but not Argentina), and he has been motivated to teach local residents Spanish to strengthen regional connections. Santiago first became interested in moving to the islands to practice his own English, and as a translator, he has been especially keen to enhance communicative capacities between and among different people. It is paradoxical, then, that Santiago has held fellow migrants to high standards for adapting their language and culture to fit in yet has maintained that only Falkland Islanders should have the right to vote (in Santiago's view, this should be restricted to residents with birthright). This stubborn principle has made Santiago somewhat of an outcast among Chileans, which became clear to him when he was not invited to the wedding of a prominent Chilean couple.

Nonetheless, Santiago told me that he was quite fond of the "Falklands way of life," its sense of safety, and the people he has met. He has become part and parcel of the generation of young adult Falkland Islanders and British or Australian contractors in Stanley. He was also close with islanders of all generations through his virtual monopoly on Spanish lessons in the archipelago. Even though he fit in well with British Falkland Islanders, Santiago drew great pleasure from "taking the piss out of" people, and he took it in stride in return. However, while he was proud of these tokens of tough love, Santiago was hyperconscious of his cultural difference and occasionally reprimanded peers for veiled or overt forms of racism, particularly racialized accusations of him being Peruvian rather than Chilean, mediated by stereotypical references to llamas and guinea pigs.

Everyday racism was not the reason Santiago had become discouraged from applying for PRP. He doubted that the government would offer it to him, but more generally, Santiago was offended at the permit's emphasis on points for finances and monetary wealth. He asked his Falkland Islander friends to fill in a mock application, and even though their status is secure with birthright, only two out of ten would have qualified for PRP if they had been migrants.[59] Moreover, if he wanted to apply for PRP, Santiago would have to get a letter or certificate of comparability issued by a UK institution to support his translation degree, even though he had already been working as the main translator in the islands for many years. Ironically, the FIG Customs and Immigration Department is one of his primary clients. When they told him his degree would need to be assessed, Santiago asked why, if it were sufficient when they need him for a translation, would it not be adequate when he wanted to become a permanent resident? Even though he has been living in Stanley for more than a decade, Santiago was not sure if he wanted to stay in the islands forever, and this was the ostensible commitment required for such status. He entered the country with a two-year work permit to be a teacher, tutor, or translator for a school that later closed. He then became self-employed, which meant he had no sponsor to vouch for him or cover the airfare if something went wrong. Santiago was happy to stay in the islands, but without PRP he still had to renew his self-employed work permit every two years, which also required him to pay for new medical exams and other fees. Santiago eventually came to refer to the Falklands society in terms of "we." He pays taxes and, unlike Flavio, has felt like he is part of the local community. Far from being considered undesirable, the FIG has relied heavily on Santiago for translation, yet during my fieldwork he was resigned to the fact that he might never be a Falkland Islander.

Owing to such failures, the government, and their voting constituents, have developed what one former MLA described to me as a "schizophrenic" policy on immigration. Do the islanders want to continue the short-term contract system and preserve the aging population of the Falkland Islanders, or do they want to encourage long-term migration of people and their families to integrate into the community? Why? According to the former MLA, "We're schizophrenic because we want to protect our community and protect our history and our culture and what are we as a 'Falkland Islander' going to become if we grow? And how are we going to grow?" Here, repetitions of the word "protect" offer a possible explanation: the sociopolitical entity of the Falkland Islanders as a white ethnic community construct has emerged (with limited growth) as a settler colonial protectorate that preserves a selective British history and culture through degrees of imperial sovereignty, which are assigned specific values in the point system for permanent residence permits. The FIG has gone through many iterations of immigration reviews to tinker with the point system, yet the policy has changed relatively little. As the settler colonial protectorate became geared toward offshore oil production, fears of new waves of people streaming into the islands for work in the industry, who might outnumber "Falkland Islanders," caused some FIG councilors to express the authoritarian populist need to "get a grip on unfettered immigration" so that the local community is not "eroded."

The next section explores how, amid a temporary workforce that threatens to outnumber the native population, Falkland Islanders are defining themselves and their distinctive lifeways as "a people."

Defining the Falkland Islander

In law, the FIG defines a "Falkland Islander" as "a person who has Falkland Islands Status."[60] However, as Santiago demonstrates, a Falkland Islander is generally considered to be a white person who was born in the islands and/or who possesses a lineage of British settlers. The "locally born" classification, in particular, allows native islanders to define and separate themselves from the racial, ethnic, and national categories discussed above.[61] However, degrees of imperial sovereignty also allow some who may or may not have been born in the islands rights to permanent residence through Falkland Islands Status. This reveals uneven access to the "Falklands way of life," a local allegory of imagined Britishness that promotes safety, work ethic, and kinship.

The primacy of birthright became clear when I listened to Santiago's interpretation of the local order of people in the Falklands. He told me, "You're only a Falkland Islander if you were born here, and that's how they see it. And I will never be a Falkland Islander . . . a Benny is [also] only someone who was born here." One concrete example Santiago pointed out is that there is a prize awarded annually in the marathon for the first Falkland Islander to cross the line. To win it, one has to have been born in the Falklands, irrespective of residence status. Similarly, when two Falklands-born individuals get married, the celebration is magnified. Islanders and expats described such couples to me as the "hope" of the islanders' longevity as a people.

However, residents do not always adhere strictly to the categories that Santiago mastered. Friendly family rivalries exist over who is "more Falkland Islander" when, for example the parent of a child was not born in the Falklands but the child was, or if the child of a parent with a long lineage of Falkland Islanders was born outside of the islands. Santiago viewed a European-descendant naturalized citizen as still "not a Benny, not a Falkland Islander," even with PRP or Falkland Islands Status, unless they were born in the islands. He asserted that they would still be referred to properly as expats if they were of British or Commonwealth origin, or by their nationality if from elsewhere. Yet islanders use degrees of categorization for such individuals, including terms like "half-Benny" or "full-Benny," to index their status.

Take, for example, my interlocutor Gus, who migrated to the Falklands from Spain when he was eight years old because his father worked in the fishing industry. While he had moved several times within Spain, Gus eagerly latched onto the resolutely local Falkland Islander identity while trying to maintain Spanish nationality and heritage. Unlike his older brother, who retained his Spanish accent and returned to Spain, Gus acquired speech habits like most native English-speaking Falkland Islanders, including Anglicized Spanish syncretisms and British humor. (Gus also speaks Spanish, having concentrated in Latin American studies and traveled extensively throughout South America.) His closest friends growing up were all Falklands-born islanders or British expats.

Gus might have qualified for PRP as early as 1996, but he went away to college in 1997, interrupting the naturalization process. The FIG covered the costs of his education because, at that time, individuals did not need to be residents to have rights to overseas education as long as they paid taxes (they did need to be residents at the time of my fieldwork). Gus moved to London after university, which meant that he stopped being a resident, and his tenure as an aspiring "Falkland Islander" was reset to zero. After living three years in London, Gus wanted to go back "home" to the Falklands, but even after spending his formative years in the islands, he was told he would need a work permit. Just three years out of university, with limited financial resources, Gus had relatively little work experience. He does not complain, though, because when he applied, the regulations were less stringent. Gus was allowed to apply for PRP from abroad, which he obtained relatively quickly because, he told me, it was a more "subjective" process before the elaborate point system that is in place currently, and everyone knew him. After all, they had paid for his education, he identified with the Falklands, and he still had a living link with his father residing there. Even after receiving PRP status, Gus had to wait six more years to become naturalized with BOT citizenship, which is basically the equivalent of British citizenship but without the right to vote in Parliament. Being naturalized allowed Gus to apply for Falkland Islands Status one year later, after a total of seven years of being back in the islands. Since returning, Gus has worked in commercial fishing, government, and journalism. During my research, he was a senior member of the Mineral Resources Department and the individual primarily responsible for crafting new legislation.

Related to these racial/ethnic classifications is a local allegory referred to as the Falklands way of life. Usually what people mean by this is (1) knowing each other (i.e., engaging in everyday practices of familiarity, such as greeting each other on the street or at least raising a finger from the steering wheel to acknowledge someone driving on the other side of the road); (2) having very little crime, which is usually demonstrated by the freedom to leave one's keys in the ignition with the vehicle unlocked or to leave home with the front door open; and (3) work ethic (islanders pride themselves on having virtually no unemployment, apart from a small portion of so-called unemployables or undesirables who have trouble holding a job).[62] Finally, the Falklands Way of Life embodies a nativist fear that the islands will be overrun by outsiders who may not continue the traditions of horticulture, collecting penguin eggs, or cutting peat.

These values of sociality, safety, and work ethic intersect with the prevalent "British to the core" homogeneity of national identity. Linked to this is a certain imperialist nostalgia, which manifests in discouragement of dissent on regional and global politics, even if islanders enjoy a spirited divergence of views on local, intra-island matters.[63] "In some ways we're all incomers . . . and people are welcomed here, as long as they don't go around saying what a terrible place it is," one Falkland Islander told me in Stanley. Another islander in Camp asked me if I knew who the three who voted "no" were, so that "they could shoot them and tidy up the place." The violent sarcasm from this last interlocutor came as quite a surprise because it was out of character with his generally cosmopolitan, if secular, outlook. He made frequent statements about being an international man, praising Chilean employees of his tourist lodge and decrying violence he viewed as being caused by religion in unidentified geographical and social contexts around the world. Despite this authoritarian populist remark, he thought of the islands as a place to forget about politics.

These contradictory notions of community compel us to question whether Falklands society is truly becoming diverse after rounds of migration, or if practices of multicultural tolerance are symptomatic of degrees of imperial sovereignty. Take, for instance, a birthday party I attended in Stanley. It was held for the expat spouse of a contractor who was in the process of applying for Falkland Islander status. The spouse had spent the previous six years teaching elementary school in Thailand and the Amazon region of Peru. Their home was decorated with paintings from Peru and Bolivia, depicting colorful shantytowns and Andean peasant women working with baskets. We consumed an array of appetizers from a variety of global cuisines with a Spanish Mediterranean theme: squid, tortilla, chickpeas, chorizos, and so on. Two Falkland Islanders, one working in the fishing industry and his contractor partner, a traveling teacher, were present, as were Gus and his Antipodean date, a contractor teaching high school, another British contractor working in construction, two Chileans who own and manage a hotel and restaurant, another Chilean woman who has a high-level job in a shipping agency, a St. Helenian couple working in communications and flooring, and a Thai woman employed at Government House. The latter individual and our host sat on the floor and spoke in Thai for much of the evening.

A cynical observer might view this multicultural gathering simply as settler colonialism dressed up as a kumbaya performance, but it also served as a safe space for candid discussion of struggles with the immigration process. The Chileans who were present all had footholds in the emergent oil industry through investments in accommodations or supply chain services, but at least one of them decried the "hassle" to get PRP, naturalization, and status. Again, most of their Chilean peers did not bother applying because it would require a rigorous background check, including a letter from Chile, employment and bank account criteria, age and number of years in the islands, and more. The forty-five points to which these qualifications must amount are impossible to accumulate for some. Even Chileans who had given birth in the islands were unable to obtain status for themselves or their children. Still, having been in the islands for ten years, a Chilean who spoke out at the gathering had sought Falkland Islands Status in order to gain a political voice and vote, as well as using the Falklands as a possible stepping stone to relocating again in the UK or Europe (potentially a less viable plan after Brexit). Meanwhile, the St. Helenians in the room already had British citizenship. They had been living in the islands for nearly twenty-five years and were recruited originally to "clean shitbowls" at MPC. The couple has continued to have trouble getting their son and newborn grandson a work permit to return, even though they are proud to call the Falklands home, and they raised their son there for most of his life.

While white British contractors are already members of government council, when I asked residents of the Falklands if they thought Chileans or St. Helenians with status could get elected and serve in office, the common response was, "It's not quite there yet." However, since my fieldwork ended, the islands' first Chilean-born resident with Falkland Islands Status was elected as an MLA, offering inspiration for others to follow in their path.

Securing the Falklands Way of Life

A grim reminder of the 1982 war, apart from the British and Argentine cemeteries, is the series of 117 minefields embedded in the Falklands' rolling landscape. In fact, a cemetery on West Falkland Island was itself mined, in addition to a water source, a corral that doubled as a public barbecue area, and sandy beaches that penguins inhabit. To fortify their occupation of Stanley and Camp settlements, the Argentine military planted some 15,000–20,000 antipersonnel and antitank mines in 1982 (some estimate up to 25,000).[64] As a precaution, vast swaths of land are fenced off with red signs warning "Danger Mines" with a white skull and crossbones symbol; these signs have now become an iconic merchandise item sold to tourists in Stanley's gift shops.[65] Even though Falkland Islanders have grown accustomed to being surrounded by live explosives, the UK has been responsible for removing the mines in accordance with the 1997 Ottawa Treaty, the Anti-Personnel Mine Ban Convention. Every few years, the British government has issued a tender for contractors, who recruit cheap labor from around the world with the necessary certifications to do this unenviable job. For the last several seasons, the teams of workers, known as "de-miners," have come from Zimbabwe.

My field research in the Falklands coincided with periods of mine clearance. I could hear the explosions detonate in the evening, and I met several of the forty-two de-miners at public events in town like a flea market in the Town Hall. I kept up with one of them through email, and after their contracts expired, I got to know the former de-miners who decided to stay in the islands permanently.

Shingai became friendly with Falkland Islanders by playing football (soccer). One of his teammates, who runs a local flooring company, offered him a job as a carpet layer; he also works evenings as a waiter at the fine dining restaurant in Stanley's swanky hotel, the Malvina. He decided that he liked being in the Falklands, so he sent for his family, including a wife and three children. Shingai's best mate, Joseph, who was also part of the team of de-miners, decided to stay, too. He works in the FIC warehouse and is arranging for his family to relocate to the South Atlantic, as well. They volunteer in Stanley's fire brigade and have even represented the Falklands as athletes in the Island Games and Commonwealth Games. Islanders point to Shingai and Joseph as model minority immigrants: proof that anyone, regardless of race, class, or ethnicity, who works hard will be valued and accepted in the Falklands.

De-mining in the Falklands is especially difficult because, even if the mines were laid out on a perfect grid, the peaty land or sandy beaches allow the location of the mines to shift considerably over time. Moreover, the explosives left in the Falklands are minimal-metal mines, making them challenging, if not impossible, to detect. Shingai and Joseph are therefore extremely methodical in following a specific process.[66] Shingai walked me through what goes through his mind when he is digging for explosives:

> So, it's normal, but it's a cautiousness. It's building it up. OK, I'm conscious of what I'm doing; I know what I'm doing. I know what I have to do. This is it. First things first, I have to be safe, and the other person will be safe, and everyone will be safe. And that's how we do, just go like that. Simultaneous. Simultaneous, like it just happens. You just know the moment that, you just focus on your job and you just get on with it. And you know you want to do it, and you actually do it if you want to. But if you feel you're not ready that day, you have to say: I'm not ready to go into it.

As they dig and clear one-meter lanes, the de-miners lay chains and rope to mark where they have been. With their hands in front of them, they slowly sift through rocks and dirt. If they feel something, they place sticks around the edges, and when they get to the mine, wherever it is, they actually hit it from the side (not the top!). "The moment you hit it, you actually forget about it," Shingai said. After chipping away enough soil to feel resistance, the de-miners place a ribbon on it and continue digging with a trowel to excavate the mine. Finally, they use an electro-jet to detonate the explosive with what is essentially another explosive and retrace their steps with the mechanical method.

Shingai and Joseph entered into de-mining through careers in security. After three months of training and another three months on the ground, they served in rapid response teams, clearing mined war zones and border areas. Zimbabwean de-miners are deployed in Mozambique, the Central African Republic, and elsewhere. Despite the high risk of physical harm and low pay, these jobs are highly selective and sought after among Zimbabwe's surplus population of unemployed workers.

After returning from Afghanistan in September 2010, Shingai was tired and had nearly had enough. He wanted a steady job that could support him to be with his family. At this conjuncture, he got a three-month contract to do de-mining in the Falklands in early 2011 for £450 per month. He said to himself, "OK, I'll just go there for three months, get my money and . . . stay with my kids [in Zimbabwe]." However, there was no steady job for Shingai there. The 2011 contract was his second trip to the Falklands and, having already met islanders through pickup games of recreational football, he inquired about job opportunities and became a floor layer. Shingai's skills transfer seamlessly: who could be more precise and meticulous working on the ground than a mine-removal specialist?

Shingai and Joseph miss Zimbabwe. "We have people starving . . . but it's still one of the most beautiful places," Shingai said. On a typical weekend night, Shingai will finish his shift at the Malvina restaurant and chat with his extended family over Skype video from midnight until 6 A.M., while it is both daytime in Zimbabwe and off-peak hours for free internet in Stanley. He sleeps during his days off.

Shingai and his family live in one of the early settlers' original houses, a historic building on Pioneer Row. They rent it from the daughter of their sponsor, an FIG MLA whom Shingai considers a father figure. Shingai invited me to have dinner at his house on Pioneer Row with Joseph and his family. The two de-miners picked me up with Shingai's two sons, while his daughter and wife, Gwen, finished cooking what I did not realize was going to be a feast: trout caught by Joseph's coworker, fried squid, beef vertebrae, and maize meal (grits). After dinner, the men took what was left of the savory squid into the living room, where the news was on TV. Paralympian Oscar Pistorious was on trial at the time for the murder of his wife. Shingai and Joseph were convinced he was guilty. They said that the South African judge, who was Black, should see through the defense's case about an intruder, even if it is true that there are break-ins in South Africa. Moreover, they said that the security of the house was so high that it should not have been possible. As trained security professionals, they spoke with authority about their trade.

Shingai and Joseph told me that they are quite happy in Stanley. Echoing the values of the Falklands way of life, they described it as quiet, clean, tidy, safe, and suitable for raising kids. Despite having put themselves into harm's way in some of the most intensely disputed areas in the world, Shingai and Joseph told me that they do not follow the political drama of the sovereignty dispute. I asked them what it is like coming from Zimbabwe, another former colony where white European settlers tried to claim self-determination. In contrast to the Falkland Islanders, white Zimbabweans lost power and 33 percent of the country's land when leader Robert Mugabe's "fast-track agrarian reform" began around 2000.[67] Shingai and Joseph regretted elements of greed but highlighted common understandings with white Zimbabweans who, they said, "grew up there and have got the right to say what they want." This emphasis on local belonging and equal rights to possession through continuous occupation is consistent with the Falklands way of life, and by clearing mines, Shingai and Joseph have offered an additional safety valve for the settler colonial protectorate.

One night, when I met Shingai and Joseph at a pub, there was only one other patron: Norris, a Northern Irish veteran of the 1982 war who returned to the islands and works as a painter. I had met Norris before, and he had shared horrific memories of hand-to-hand combat during the conflict. I invited him to join us. Unbeknownst to me, Norris and the de-miners knew each other quite well because they work together: Norris paints the houses where Shingai puts in the flooring, and Joseph coordinates shipment of the building materials. Interestingly, despite their unanimous agreement about how grateful they are to be living in the Falklands, and how nice and clean of a place it was, all three complain that locals are underskilled, chronically hungover, and lacking in work ethic (the exact opposite of the islanders' historical values of improvement and self-sufficiency). Norris, Shingai, and Joseph also bonded over extensive experience with war and violence. Norris said that he "enjoys" war: he cherishes the thrill and immediacy of life or death. Here, the de-miners drew a significant distinction between what Norris has done and what they have done; they said that Norris has been in *danger*, whereas what they do is *risky*. They downplayed their own precarity and expressed profound respect for Norris's ability to stay out of harm's way.

In sum, by securing the landscape, the de-miners are enhancing the Falklands way of life. In exchange, the islanders are opening their settler colonial protectorate to the de-miners. Falkland Islanders have offered Shingai and Joseph new lifeways and stability. They miss their extended kin but no longer need to risk their lives in the ruins of war. They can provide for their families in an agreeable environment where they have become local heroes in their own right.

This chapter analyzes how the sliding scales of race, ethnicity, and nationalism have shifted over time, as represented by the predicaments of different groups of individuals struggling with migration, citizenship, and assertions of peoplehood in the Falkland Islands. Reacting to their subjection by British colonial administrators, settlers formed a rural, conservative, white working-class struggle for government reforms that would give them a democratic voice. Settlers naturalized as "native" islanders, identifying themselves as a separate white ethnic community formation, apart from though still aligned with the British, that is, as kelpers. Influential circles of nativists pressured farm managers to halt labor importation. Building momentum for authoritarian populism from the periphery, the islanders lobbied British Parliament against a potential sovereignty agreement with Argentina. This backstory is often overlooked in analyses of the 1982 war, yet it remains crucial to the current British position on refusing to negotiate on the paramountcy of the islanders' wishes.

The chapter also discusses contemporary categories of race and immigration status that impact what kinds of individuals may or may not claim to be a Falkland Islander, as well as how this settler colonial logic differs in practice among residents. In turn, this distinction delimits access to social programs and voting rights. If class stood in for race during the subjection of settlers in the earlier colonial period, then race has become a key criterion for defining Falkland Islanders as a people at the current conjuncture. Phrases like "kith and kin" and "British stock" may have marked Falkland Islanders as white by essentialized British descent in defense of British sovereignty. Yet, during and after the war, military personnel labeled islanders "Bennys" to signify backward hinterland rurality, and islanders have reclaimed this epithet to represent pride in white, rural, working-class conservativism. The current settler colonial milieu includes a patchwork cohabitation of residence with a range of racial, ethnic, and national identity formations. The playing field is far from even in the enduring racial hierarchy of imperial diaspora. Besides race and ethnicity, there are certain qualities, including gender, sexuality, language, and class, that divide and protect elite "desirables" from dispossessed "undesirables." Nonetheless, new connections through migration, affinity, and multigenerational residence are transforming the racial formation of the Falklands.

While race and diaspora may be relatively fluid in the Falklands, the islanders' self-determination claim requires an assertion of self that remains firmly fixed in a resolutely British territory with degrees of imperial sovereignty.[68] This privilege is unevenly distributed in the Falklands through a point system for permanent residence. Falkland Islanders have preserved popular consent to British sovereignty through forms of racial, ethnic, and national exclusion that limit access to political power. Nevertheless, the international community remains reluctant to accept the islanders' claim to peoplehood. While the islanders' formation of whiteness may appear to give them ethnic evidence of British kith and kin, the rub is that it also prevents decolonized nations from recognizing them as a distinct people from the UK with an inalienable right to self-determination. As one Falkland Islander told me, "Not all the minefields are fenced in."

Part 2

WRECKAGE

4

DOES THE SEA LION ROAR?

As a remote island chain, the Falklands have often been portrayed as insular.[1] However, before the Panama Canal's emergence as a "global infrastructure," the vast majority of trade and travel between the Atlantic and Pacific Oceans passed by the archipelago.[2] With firm footing on dry land, settlers were seemingly indifferent to the existential angst of distressed seafarers clinging to planks of wood amid common occurrences of catastrophe due to the harsh South Atlantic conditions. Beyond the mere curiosity of a spectator, Falkland Islanders desired wreckage as a strategic enterprise for gaining profit.[3] They seized the opportunity to swindle vessels struggling to journey past Cape Horn as they took refuge in the islands for provisions or ship repair. As Atlantic historians have examined in other frontier areas, shipwreck salvage was an extractive economy in its own right.[4] Elsewhere in the Atlantic World, salvors recruited Indigenous people for diving, and wrecks stimulated trade with coastal Native societies receiving them. As we saw in chapter 1, the colonization of Tierra del Fuego and Patagonia included a series of conflicts punctuated by intercultural exchanges involving wreckage. But in the Falklands, salvage extraction became a business exclusive to settlers. Initially, settlers repaired and refilled damaged vessels at a lower rate than what might have been charged in Montevideo or other ports of South America, but they eventually used their position of advantage to charge several times the cost of repairing a ship in San Francisco.[5] The Falklands' development as an extractive frontier for marine resources—from sealing and whaling to squid fishing and oil exploration—was thus built on salvaging the value of wrecks and extorting captains of damaged ships.[6]

While part 1 of this book uncovered the initial crises of colonial dispossession, land enclosure, and racial formation in Falklands society, by focusing here on marine resources, part 2 offers a deeper immersion in the archipelago's "shipwreck ecology," drawing on literature in environmental humanities and social sciences that connects the history of shipwrecks with the development of globalization as well as critical thought in Black diaspora and Indigenous studies.[7] According to Black queer theorist Christina Sharpe, the ecology of the ship, the wake, and weather formed a climate of anti-blackness and capitalism, extending from the slave trade in the Middle Passage to current transportation of goods and property in container vessels.[8] Similarly, Tiffany Lethabo King has analyzed how offshore geological and oceanic formations that presented hazards for shipwrecks—shoals—may offer a disruptive vantage point from which to unmoor the overlapping layers of genocide and slavery in the Atlantic World.[9] In what follows, we consider how South Atlantic settlers have regenerated resource wealth from the environmental wreckage of such shipwreck ecologies that haunt the imperial present.

Building on notions of imperial "ruination," this chapter takes shipwreck salvage as a point of departure for understanding the global sedimentation of rubble and debris that Walter Benjamin envisioned as the "wreckage upon wreckage" of modern progress.[10] Benjamin famously described progress from the perspective of the "angel of history" as a mighty storm: an immense succession of catastrophic events, leaving behind a pile of debris. Here, I follow postcolonial theorists Edouard Glissant and Ann Stoler, who analyze how the "uneven sedimentation of debris" reveals more than just relics or ruins of empire that memorialize the past, but rather an ongoing process of "ruination."[11] Specifically, I connect the intensification of marine resource exploitation for commercial fishing and oil exploration to successive modes of opportunistic wealth extraction in the South Atlantic. Ultimately, this chapter arrives at how the islanders have attempted to craft a new offshore oil regime to futureproof their settler society from a perceived resource curse, building on past extractive economies dating back to shipwreck salvage.

According to Anna Tsing, "salvage accumulation" describes the process through which "raw" nonhuman materials and human labor become understood as capitalist resources and gather value as commodities.[12] In the case of the Falklands, salvage accumulation was first literally just that: generating wealth by treating shipwrecks as resources. Stranded wrecks were gifts of the built marine environment, products of "second nature" that settlers could coopt to accumulate surplus value.[13] Islanders developed a thriving business out of seizing and beaching wrecks, auctioning off their cargo, repairing them, and reusing them as hulks for storage. Masts and ballasts were even re-purposed as gate posts, blending marine resource extraction with practices of land enclosure we explored in chapter 2.[14] In 1855, eleven vessels wrecked on the shores of the Falklands: two on fire, eight others in extreme distress, several with sick crews or without food or water. Cargo warehoused and reshipped exceeded £21,000, including several hundred tons of expensive machinery, manufactured goods, and copper ore.[15] In more violent acts, settlers-turned-pirates would "make a wreck" through more coercive means, by exercising their local authority to detain vessels.[16]

Extending from this historical precedent of shipwreck salvage, this chapter will outline Falkland Islanders' strategies for capturing wealth through marine resource exploitation and explore their preparations for producing oil at the Sea Lion well of the North Falkland Basin. During my fieldwork, the Falklands ideal of climbing social ranks through autonomous personhood was perceived as under threat by the potential catastrophe of mismanaging resource wealth.[17] The "resource curse" is the label for a set of dynamics that economists and political scientists argue is associated with high dependence on natural resource rents.[18] These dynamics include nondemocratic political orders, stagnant economies, and environmental disasters. Resource dependence has distorted institutions in oil-producing nations such as Nigeria and Venezuela.[19] As Fernando Coronil analyzed astutely in the context of Venezuela, rent dependence empowered the landlord state to "sow the oil" in democracy and perpetuate a myth of progress.[20]

By rethinking the resource curse hypothesis through wreckage, I aim to consider how the Falkland Islanders attempt to transcend or circumvent political dependence by asserting themselves as exclusive owners of a bustling oceanic extractive frontier on the edges of empire. While salvaging wrecks was an industry based on dispossessing others' catastrophes, embracing an economic future based on production and exchange of oil and other marine resources has placed the islanders at the mercy of speculative investors in an unstable commodity market that demands complicity in the shared future wreckage of global climate change and mass extinction. This raises the question of how to maintain the Falklands ideal of self-sufficiency through increased dependence on resource rents with acute social and ecological impacts.

The following sections analyze how the Falkland Islanders try to avoid a resource curse and preserve personhood in the formation of various marine resource licensing regimes. I describe tensions between the new fiscal scheme and a growing international network of technical, financial, and diplomatic oil trustees interested in developing the Sea Lion well of the North Falkland Basin. Ultimately, ethnographic analysis of visits from oil partners shows how attempts to salvage settler autonomy become partially eclipsed by a new charismatic face of corporate personhood. But first, I examine how the offshore oil projects, which Argentina's government has considered illegal acts of piracy, are the outcome of past strategies by islanders to capture value on the margins of empire in addition to salvaging shipwrecks, particularly sealing, whaling, penguining, and commercial fishing.

Sealing, Whaling, and Penguining

There is a saying in the islands that "the only difference between a sealer and a pirate is opportunity." Both occupations were based on salvaging resources and gaining profit from spectacular acts of violence. Irrespective of the sovereign status of the islands, transient crews, originating primarily in North America and Britain, had long used the many sheltered harbors, coves, and inlets of the archipelago as bases for exploiting the natural abundance of marine mammals. While the agrarian colonies of Spanish and Ríoplatense settlers were concentrated on East Falkland Island in the eighteenth and early nineteenth centuries, sealers established bases primarily in the northwest outer islands of the archipelago. At West Point Island, a stone's throw from the northwest corner of West Falkland Island, a hut remains intact (see figure 13), marking a key harbor for up to eight American sealing schooners at a time, including gardens, introduced pigs and rabbits, and a blacksmith shop.[21] Focused primarily on the exploitation of fur

FIGURE 13.   Sealers' hut, West Point Island.

Photo by James J. A. Blair.

seals for pelts sold in China, sealers clubbed and lanced hundreds of creatures in a single swoop. In smaller satellite islands, they burned through the thick tussac grass covering the littoral areas to gain access to seal colonies of the interior.[22] Having exploited most of the colonies onshore, by the turn of the twentieth century, they took their hunt to the open sea.[23]

The Spanish governors tried to enforce regulations against the unauthorized destruction of seals in West Point. As discussed in chapter 1, Luis Vernet, political and military commander of the Malvinas for Buenos Aires, also enforced restrictions on sealing, which led to his colony's demise.[24] Yet Vernet was not the only target of American intimidation in protection of sealing interests: when British colonial governor Rennie arrested and tried Americans who admitted killing sea lions "for fun," a US warship arrived in Stanley Harbor, compelling their swift release.[25] Sealing thus inhabited a gray zone at the boundaries of imperial powers.

Marine mammal exploitation was not exclusively a business for foreigners. Besides the Spanish, who had tried primarily to restrict rather than encourage marine mammal exploitation, British (and French) settlers took seals and whales, as well as penguin and albatross eggs, exporting hides and oil for profit.[26] Prices varied for sealskins, depending on whether they were fur seals, sea lions, or elephant seals.[27] Whale oil was more profitable than seal oil.[28] In the mid-nineteenth century, the FIC, as well as rival merchant John Markham Dean—in collusion with resident sealers—seized the opportunity for buying and transshipment of skins, oil, and whalebone, as well as supplying sealers with provisions.[29] There was no season for sealing, so schooners and cutters sailed into Stanley every three to four months to sell their cargo of skins and oil for reshipment, and to cash in their receipts on drinks, merchandise, and ship repair. This system kept the sealers indebted to Dean and the FIC year-round.[30]

While some settlers gained individual autonomy and profited off the sealing industry, others saw its presence as a hindrance to moral progress in the nascent British colony. Similar to the way the British settlers (and Argentine *criollos* on the mainland) regarded gauchos as degenerates (see chapter 1), the conduct and language of the sealers and whalers were considered "relics of barbarism."[31] The local church-sponsored *Falkland Islands Magazine* attributed a murder, which led to the assailant's execution, to the social decay caused by the moral standards of sealers and whalers. Quality of character aside, sealers and whalers did operate at the limits of the rule of law; poaching on the far reaches of the archipelago, such as the Jason Islands, was common.[32] The capture of marine mammals therefore presented settlers with prospects for both commercial blessing and social affliction.

In the 1860s, settlers attempted to develop their own branch of the business by boiling down penguins in cauldrons, some of which were parts taken from ships' boilers.[33] "Penguinners" marched the seabirds into the large casks that became known as "try pots," with slits cut down their sides for oil to ooze through when pressed.[34] The FIC even tested the market for exporting penguin skins as "ladies wear."[35] (Fortunately, this never became fashionable.) The government eventually passed ordinances to prevent the FIC from performing commercial killing of penguins on Crown land. This legislation was meant not to preserve the seabirds but rather to allow settlers to continue collecting eggs for food.[36]

Like hydrocarbons today, sealing and whaling inspired speculative investments that were just as likely to sustain a loss as a profit. The demand for oil fluctuated, collapsing in the 1870s, and marine mammals became overexploited in the Falklands. This pushed the industry's frontier farther out to South Georgia and the South Shetland Islands. Yet, as these islands were "Dependencies" of the Falklands, Stanley benefited by regulating their whaling stations through a licensing regime beginning in 1906.[37] This kept the economy afloat as maritime traffic decreased due to the construction of the Panama Canal. Local residents speculated about a possible fad diet of canned whale and panicked about their possible extinction. Whalers sang songs about the "thrilling toil" of producing "eight thousand tons of oil," and shareholders celebrated full cargoes of up to 40,000 barrels produced per ship.[38]

Despite numerous efforts to revive it, the whaling industry remained less profitable within the Falklands. In 1908, the Norwegian whaling company Salvesens, with bases in Scotland and Leith Harbour on South Georgia Island, constructed a whaling station at New Island. Governor Allardyce (1904–1915) was hopeful that the new station would discourage poaching and deliver wealth to the Falklands. But the whaling operation at New Island did not last long, as oil production paled in comparison to the factories on South Georgia.[39] Sealing picked up again momentarily in the Falklands when a temporary halt to the business expired in the early 1920s. John Hamilton, settler of the western Weddell Island, invested in a short-lived sea lion oil factory.[40] Islanders attempted to develop a sealing industry, unsuccessfully, twice more, from 1928–1940 and 1949–1952 at Albemarle Station, West Falkland Island.[41]

Other potential marine resource industries, such as guano from penguins and kelp seaweed for iodine and potash, never became commercial.[42] As discussed in chapter 2, sheep farming would long define the Falklands' economy, making wool the islands' principal commodity export until the 1982 war. Nonetheless, the ambiguous boom and bust legacy of these early modern marine resources industries endures in islanders' personal aspirations for salvaging wealth and progress in present regimes of fishing and oil exploration.

The Post-1982 Commercial Fishing Boom

In 1982, the same year as the military conflict between the UK and Argentina, the UN held its Law of the Sea Convention, which defined the international maritime regulations for national territory in ocean waters. The convention set the spatial limit of an Exclusive Economic Zone (EEZ) to extend 200 miles from national coasts. Even after surrendering in the war, Argentina's government began regulating its natural resources according to such an EEZ, without relinquishing its claim to the islands as part of the nation's official territory.[43] Nevertheless, Falkland Islanders lobbied the UK FCO to create two different fishing zones in the waters surrounding the islands in 1986 (the FICZ, originally meant for military protection) and 1990 (the FOCZ).[44] The revenue islanders accumulated from East Asian and Spanish squid fishing activities subsequently lifted the sheep-based colonial backwater into a stable economy with a new enclave of wealthy local license brokers.[45]

Since the establishment of the commercial fishing zones, the FIG has issued licenses to at least fourteen different companies, accumulating revenue of between £10 and £29 million each year.[46] While finfish such as Patagonian toothfish (*Dissostichus eleginoides*, marketed as "Chilean sea bass") are licensed increasingly in the Falklands and South Georgia, squid is the most frequent catch. The *Doryteuthis gahi* (hereafter, *Loligo*) squid species, marketed as "Falklands calamari," is consumed primarily in Spain, and the larger *Illex argentinus* (hereafter, *Illex*) species is consumed mostly in East Asia (see chapter 6 for more on this distinction and scientific monitoring of fish stock). With a total annual catch of more than 250,000 tons, the squid fisheries make up 5–10 percent of the global supply and around 50 percent of Southern Europe's *Loligo* calamari. The trawlers fishing for *Loligo* are primarily joint ventures between license-owning businesses founded by islanders and Spanish fishing fleets, while the "jiggers" (figure 14) fishing for *Illex* tend to be Taiwanese or Korean fleets holding licenses directly from government. No matter what flag a given vessel flies, the nationalities of crews range from Indonesian and Chinese to Filipino and Peruvian.

Without a doubt, fishing has become the most thriving commercial sector of the islands' economy, representing 45–60 percent of GDP on a given year. For nearly two decades, the business consisted of little more than reselling annual licenses, but in 2005, the local fishing companies became more involved in actively managing quota through the establishment of the Individual Transferable Quota (ITQ) system for all fishing except *Illex*. Under this system, the FIG allocates quota to companies based on their previous track records, and individual firms may borrow money or trade out of their quota if they desire. Only companies registered in the Falklands can hold ITQ, and the system allows for

FIGURE 14.   Jigger in Stanley Harbour for licensing.

Photo by James J. A. Blair. Reused with permission from NACLA.

longer-term licenses starting at twenty-five years. Islanders working in fishing use a variation of global value chain analysis to explain that their regime is just "first value" because the Falklands companies are not involved in processing (second value), storing (third value), or selling (fourth value) products to market.[47] Some islanders now "own steel" (i.e., a fleet of trawlers), but they still primarily act as agents distributing quota. To the chagrin of some fisheries managers, who wish the islanders were more committed to the industry, fresh fish is not even sold locally. One can purchase locally caught fish at either of the two local supermarkets, but only in frozen packages. Other than the few weeks per year when dozens of jiggers crowd into Stanley Harbour to renew their licenses, islanders rarely see the fishing crews, who are ironically excluded from popular depictions of Falklands society as prosperous and multicultural, despite producing the bulk of the area's wealth.[48]

The inhumane working conditions of the jiggers have been the subject of heightened controversy. Recalling the distressed seafarers or sealers who abandoned their vessels to seek refuge in the Falklands for survival, each year, when the jiggers come in to harbor to collect their licenses, cases of crewmembers jumping overboard and drowning are reported. Innocuous stories circulate about homesick crewmembers swimming ashore to call their families and then returning to their ship or their home countries. Some crewmembers jump to escape the harsh conditions or physical abuse. Even before Argentine media called attention to the brutal and unfair labor practices of the jiggers, the local *Penguin News* investigated the problem of crew welfare.[49] According to the report, from 1985–2014 there were thirty-five reports of "man overboard," with eighty-nine persons lost at sea, and 122 deaths on board ships caused by murder, fires, or illness. Of course, sea slavery and servitude is a worldwide problem.[50] The flag state of the vessel, and not the licensing government, is responsible for operation on board, and the FIG fisheries focuses primarily on monitoring stock, not labor. The Seamen's Mission Centre, a charitable organization, offers refuge, advocacy, and pastoral care, including telephone access, basic provisions, and accommodation in Stanley.

The unevenness of suffering is especially stark in Stanley Harbour when luxury cruise ships escort tourists past the rusty jiggers collecting licenses. Ironically, the brutal conditions of the factory fish vessels yield far more local wealth, albeit more unequally distributed, than tourism does. In short, while the FIG and individual islanders have become more autonomous persons as tax collectors and quota holders, they have done so through dependence on overexploitation of crew labor. Despite a gap of several decades, these dynamics show continuity with the construction of personhood through the salvage accumulation of ship repair, sealing, and whaling, as the islanders transition into broader scales of resource licensing through oil drilling.

A Company Named Desire

Since the early twentieth century, Falkland Islanders have envisioned a rosy future of modern progress built on oil development.[51] Yet while British and North American geological studies date back to the 1920s—with mineral licenses and seismic surveys beginning in the late 1960s and early 1970s—production has long remained a pipe dream.[52] The British government considered joint oil exploration as part of a potential leaseback agreement with Argentina in the 1970s, but the sovereignty dispute prevented these plans from materializing.

One British petroleum geologist devoted much of his life to making this dream a reality. In 1977, Colin Phipps, a former Shell employee turned Labour Party politician, published a controversial Fabian Society pamphlet on oil prospects titled *What Future for the Falklands?* Militant nationalist Argentine writers have vilified Phipps colorfully as an opportunistic British imperial "pirate," profiteering from potential environmental wreckage in the contested waters of the South Atlantic, much like the islands' early settlers.[53] But it is important to note that in this original proposal, what Phipps actually outlined was a strategy for resolving the sovereignty dispute with Argentina through the stated principles of mutual understanding, cooperation, and joint development of hydrocarbons.[54] While he defended the right of Falkland Islanders to self-determination, Phipps proposed that an "accommodation" with Argentina would be in their best interests; he viewed offshore oil development as the catalyst for breaking the geopolitical deadlock.[55]

Of course, Phipps did not foresee the Argentine invasion and subsequent war.[56] The UK government, particularly the Department of Energy and secretary of state for energy David Howell, had already crafted policy to defend British access to oil in the South Atlantic seabed and subsoil.[57] After the British victory, the FIG sought unilaterally to model an offshore oil bonanza on the fishing boom of the late 1980s; this alarmed British Treasury officials, who demanded that potential tax revenues from oil should go to the UK rather than the FIG.[58] Nonetheless, the FIG licensed a round of seismic surveys in 1992, and the oil company licensees, Geco-Prakla (a subsidiary of Schlumberger) and Spectrum, received "letters of comfort" from the UK and "letters of discomfort" from Argentina. Unfazed, the FIG's new Oil Management Team began crafting mineral resource legislation, including a fiscal regime of 9 percent royalties on potential profit.[59]

At this conjuncture, Argentine foreign minister Guido di Tella, serving under the neoliberal-Peronist President Carlos Menem (1989–1999), launched a "charm offensive" (*estrategia de seducción*). Exiled in Oxford during the Dirty War, di Tella had developed a sophisticated theory of the frontier, which tied technological change to economic expansion.[60] He proposed that Argentina—a country built on one big wave of agricultural production—might benefit from expanding its territorial frontier to capture newly discovered rent through spurts of mineral resource extraction. Given the opportunity to apply these insights concretely as foreign minister, di Tella considered the autonomous personhood of the islanders in unprecedented and creative ways. In addition to agreeing to cooperate on oil interests with UK foreign secretary Douglas Hurd, di Tella made regular phone calls to FIG Legislative Assembly members and sent residents hand-signed Christmas cards and Winnie-the-Pooh teddy bears. As relations warmed, the UK and Argentina formed a Joint Declaration on Cooperation over Offshore Activities in the South West Atlantic in 1995, a nonbinding umbrella arrangement over natural resources, not territory or sovereignty.[61]

Reenter Colin Phipps. Having built Clyde Petroleum into a £130 million firm since its founding in 1977, Phipps had become an authority on offshore oil in the North Sea.[62] In 1995, Phipps retired from Clyde, and in 1996, he promptly founded Desire Petroleum, which partnered with the British firm Lasmo. Desire took its name from the Falkland Islanders' motto, "Desire the Right."[63] In total, the new joint licensing regime attracted fourteen oil firms, including majors and supermajors, such as Hess and Shell.[64] Investing £100 million, the licensees drilled six exploration wells in the North Falkland Basin in 1998, finding subcommercial amounts of hydrocarbons in five. None of the wells drilled in the Argentine part of the Malvinas Basin was considered commercial.

When the price of oil fell in 1998, most of the licensees opted not to renew their licenses in the risky frontier area, with notable exceptions: Argos Resources (a company also connected to fishing) and Phipps's firm Desire Petroleum. Unlike other licensees, a majority of Desire's shareholders were Falkland Islanders and British veterans of the 1982 war.[65] Rex Hunt, former Falkland Islands governor during the invasion, sat on the company's board of directors. The Falkland Islander shareholders are not necessarily elites; one informant described them as "everyday Joe Blows" who have invested anywhere from £1,000 to £15,000—often significant chunks of their savings. Some of them just buy and sell shares when they know a rig is contracted and operating or not. These practices of incremental salvage accumulation indicate that, unlike sealing and whaling or fishing, oil was perceived as a condition of possibility for a wider distribution of personhood.

Desire kept the Falklands flame burning long enough for the oil price to rise again, and in 2004, a new cohort of small risk takers—Falkland Oil and Gas (FOGL), Rockhopper Exploration, and Borders and Southern—collaborated to gather further seismic data and drill again. During the Kirchner administrations, relations with Argentina soured, but this did not prevent Desire and the other small firms from going forward with their exploration plans. Having found no commercial hydrocarbons in the South West Atlantic, Argentina's government backed out of the Joint Agreement in 2007. Argentine congress passed laws declaring the exploration projects illegal, leading to lawsuits and threats of significant fines and up to fifteen years in prison, but this legislation is domestic and thus difficult to enforce. Moreover, none of the licensees had assets in Argentina. On the contrary, Desire and the other firms courted tens of thousands of private London investors by branding themselves patriotically as Falklands companies. When the major firms abandoned the area, these smaller ones acquired vast license areas with high equity participation and 100 percent interest. The business model was set up to fund relatively minor seismic surveys, then float their potential on the stock exchanges to raise cash for exploration and attract partners to pay for further drilling in exchange for equity.

Assuming that the natural resources in its conservation zones were entirely the property of the British Overseas Territory, the FIG continued to license new rounds of drilling unilaterally. Colin Phipps died in 2009, before the presence of commercial oil was ever confirmed in the South Atlantic, but Desire continued to operate under his son's direction. Finally, in 2010, Desire announced that it had discovered oil in the North Falkland Basin. However, this statement was overly hasty: far from turning water into wine, Desire revealed that this oil "find" was actually just *water*.[66]

During that same campaign, one of the new companies, Rockhopper (named after a local penguin species, *Eudyptes chrysocome*), actually did discover commercial amounts of oil (the equivalent of 400 million barrels) in the adjacent Sea Lion well of the North Falkland Basin.[67] While companies have drilled elsewhere, the focus since then has remained primarily on the Sea Lion complex and adjacent northern wells.[68] Through a series of mergers and acquisitions, as well as subsequent rounds of exploration and appraisal drilling in 2012 and 2015, Desire combined with FOGL and "farmed out" its acreage to Rockhopper and the larger independent firm Premier Oil. A more "grown up oil and gas company" with experience producing oil in other frontier areas, Premier's involvement, along with that of Houston-based Noble Energy, raised expectations once more for oil production in the South Atlantic.[69]

The Norwegian Model

In January 2014, Argentina's foreign minister Héctor Timerman challenged British prime minister David Cameron over the sovereignty of the Falklands/Malvinas. Provocatively, he reasoned, "The United Kingdom's refusal to sit down at the negotiating table is compelling evidence that, in the question of the Malvinas Islands, the lion roars but does not inspire fear any more." Presumably, Mr. Timerman was waxing poetic about British imperial nostalgia in general, and not the Sea Lion well in particular. But if he meant the latter, in retrospect, he would have seemed clairvoyant. Despite mounting pressure from Argentina, development plans for offshore oil production instilled a sense of importance among the islanders and their licensees.

The Sea Lion discovery was not a typical "gusher" moment. The results of the wireline test, a radioactive measure of the type of rock and potential liquids, were initially unclear. After a second look, news broke of the first official oil discovery, and the FIG Mineral Resources Department was abuzz with excitement. Government administrators celebrated with champagne and bacon rolls, and individuals investing in the emergent energy industry felt as though their patience had finally paid off.

Even after discovering oil, Premier and Rockhopper's decision on whether to push the final button to develop the Sea Lion well would depend on multiple rounds of appraisal, as well as the price of oil and the profitability of production.

But facing a sudden potential upsurge in the economy, FIG councilors and directors sought immediate advice for managing the future. The FIG's fiscal regime had thus far mirrored much of the UK's legislation on North Sea oil, and as a BOT, any potential field development plan had to be approved by not only locally elected FIG officials but also, ultimately, the secretary of state for the FCO.[70] One of the first steps legislators took therefore was to travel to Aberdeen, Shetland, and Norway to gain insight on crafting appropriate policy.

Economic advisors around the world praise the Norwegian model for avoiding the problems associated with the resource curse by accumulating a sovereign wealth fund of more than $1 trillion. In the whole panorama of "resource-cursed" oil-producing states, Norway is considered to be the one outlier. Explanations generally emphasize sequencing, that is, Norway's development of strong democratic institutions, oversight and monitoring agencies, social safety nets, and so on *prior* to the North Sea bonanza. To make the most of the nonrenewable resource wealth, only 4 percent of Norway's sovereign wealth fund entered the national budget. In addition to a corporate tax rate of 28 percent, Norway's fiscal regime featured a special tax rate of 50 percent. This may seem relatively high, but Norway incentivized companies in multiple ways, for example, by covering up to 50 percent of equity for exploration, and unlike most other oil producers, the state did not collect bonuses or royalties. The Norwegian government also owned two-thirds of the firm Statoil, holding a majority of licenses.

While they intend to emulate the prudent Scandinavians, the FIG's fiscal regime differed from Norway's significantly (besides the obvious demographic disparities). While Norway offered much of the front-end costs and charges no royalties, FIG licensees must self-fund their exploration activities and pay a royalty of 9 percent. But to encourage business in the frontier regime, the FIG would not collect a Norwegian-style special tax on top of the islands' relatively modest 26 percent corporation tax. Far from the extortionist legacy of shipwreck salvage, or even the more dominant posture of settlers toward sealing, whaling, and fishing licenses, islanders sought to create a welcoming atmosphere for oil companies. Furthermore, the FIG did not plan to create a formal sovereign wealth fund. The basic principle borrowed from Norway, then, was simply to keep its fiscal regime from oil static and separate from other funds.

Still, the specifics of setting requirements for distributing funds from oil differently from other revenue streams have been somewhat inconsistent. At first, the government proposed an Oil Development Reserve, which would be allocated toward advancing the energy industry before any other uses. Then, the FIG switched to a more holistic approach to managing the cash it holds on the stock exchange. Treasury separated financial reporting practices on nonrecurring revenue, such as tax revenue from oil. This allowed the government to form a budget based on the state of its consolidated fund without having to account for the spikes in tax revenues during drilling campaigns.

Of course, this process requires that licensees report their taxes accurately. Yet, when Rockhopper farmed out 60 percent of its license for Sea Lion to Premier in 2012, conflicting assessments on capital gains led to a serious disagreement. Rockhopper submitted a $122 million tax bill to the FIG and took steps to distribute reserves in the UK, but the FIG calculated corporation taxes tens of millions of dollars greater. The expected proportion due at the time of the joint agreement versus later (after "first oil") was also subject to debate, and, after an agreement in principle, it took eighteen months before Rockhopper and the FIG reached a binding agreement on $146 million ($42 million due immediately). The international business media picked up on this negotiation, portraying the FIG as greedy and overly conservative.[71] In turn, the tax debacle did irreparable damage to both the perception of the Falklands within the oil industry and the way ordinary, individual islanders view their government. Even though the FIG saw itself as emulating Norway by legislating rather than negotiating, the disagreement gave some islanders the impression that the government had botched the regulations. During a dinner party, one Stanley resident compared the tax deal to him waiting until I got on my bicycle, ready to ride home, and then giving me a bill for the meal. This put the government in the embarrassing catch-22 scenario of either appearing difficult or rude, if they enforce regulations, or coming off as a passive victim of the bullying fossil fuel juggernaut, if they are more accommodating.

To avoid the massive taxes on capital gains, other companies in Rockhopper's position might have just sold the company after finding oil. But the way Rockhopper structured the sale of interest in Sea Lion indicated that the firm had aspirations to grow into a competitive London-listed company.[72] The transaction was the first of its kind in the Falklands, but it would not have been unusual in the North Sea. Rockhopper assumed that for "to be determined" items in the fiscal regime, the FIG would simply follow the UK precedent of twenty to thirty years ago. Instead, the FIG took a hardline approach, insisting on credit protection with more discretion than the industry anticipated. Rockhopper and other oil executives view this fiscal conservatism as symptomatic of a broader state of nervousness within the FIG not to jeopardize islanders' personal fortunes. While Norway's guiding principle of long-run consistency might have influenced the FIG, oil executives interpreted the controversy as a short-run failure to look at the bigger picture of encouraging development and progress.

During the 2012 drilling campaign, I had the opportunity to discuss the uneven local adaptation of the North Sea model with a crew of Norwegians who were laid over in Stanley waiting for provisions so that they could return to their search and rescue support vessel. Kaj, a twenty-five-year-old crewmember training to be captain, was proud and almost evangelical about his nation's oil successes. He is usually at sea four weeks at a time, working six-hour alternating shifts. But Kaj only has to work half of the year and has an annual starting salary of $70,000. In addition to the comfortable wages for relatively few hours, all members of the crew receive electronic tablets and enjoy various other perks. Even though they were staying in the two highest-rated hotels in town, Kaj and his mates were struck and disappointed with how unprepared the islanders seemed to be when one of two supermarkets ran out of necessary supplies. Unsympathetic to the islanders on the issue of sovereignty, and wary of wasting the oil company's money, he proposed that they get supplies from Chile, Brazil, or even Argentina if the retailers didn't move faster. Kaj felt that the whole town was "waiting around for money to flow."

At a FIG Legislative Assembly meeting, one councilor addressed this sense of uneasiness. Calling for a public debate, the MLA warned, "Even though oil hasn't started to flow, money [from licenses and third-party service agreements] has flowed through the community," adding, "it is not making people happy, but rather anxious." Frustrated with the unpredictable cycles of exploration that bring in revenue one year and bring in nothing the next, individual islanders wondered what they should be doing now that they were nearing the trigger point of production. Still, this trepidation was mixed with a gleeful excitement that the dreams of autonomous personhood and salvage accumulation of resource wealth could soon be fulfilled. Another councilor confided in me, "In a year or two, we will have enough oil money flowing to bring tears to your eyes."

The Missing "Falklink" in the Oil Supply Chain

In Stanley, the aftermath of the Sea Lion discovery brought concern over managing money, but it also made residents scramble to make sure all the pieces were in place for the industry to carry out its work. Uncertainty, curiosity, and suspicion colored local residents' perceptions of whether the oil companies would use local services, skills, and accommodations, as opposed to importing labor and facilities. An extraordinary majority of islanders identified as pro-oil, but a patchwork of different sectors of the society and economy produced a range of personal and public anxieties: Will the fishing industry be held to the same high health and safety standards as oil? Will oil installations damage the image of the Falklands as a pristine environment for tourists? Will revenue from taxes go to developing roads in the Camp? What kinds of public works are necessary to make oil happen?

Oil managers ultimately gave credit to the FIG for following through with legislation, but some industry interlocutors doubted if ordinary islanders understood just how much money sat on the horizon. They drew parallels between Stanley and the fictional Northern Scottish village of Ferness depicted in the film *Local Hero*. In this 1983 comedy, an ambitious Houston oil company representative is tasked with convincing the whole town to sell their land in order to develop a refinery. The picture represents the locals as naïve and accommodating, apart from one beach dweller claiming rights to a key coastal area. The oil business elites ultimately become seduced by Ferness's quaint lifestyle. They embrace conviviality, concede to the demands of the lone dissenter, and opt instead to develop their refinery offshore in the North Sea. The head executive, who forms a cosmological bond with the eccentric beach activist, even offers to donate an astronomical observatory and ecological research facility to the people of Ferness.

As in Ferness, the question remained of how to make oil work for the Falklands rather than the other way around. As discussed in a different environmental context in chapter 6, islanders of various stripes described a need for "gap analysis" for oil readiness. To advance the skilling process and prepare the islands' economy for smooth integration with the oil industry, an entrepreneurial former civil servant associated with Stanley's Chamber of Commerce summoned the expertise of Terry, a veteran of the North Sea oil industry living in Aberdeen, where I met him in 2015. A chemical engineer by training, Terry had worked for Schlumberger, BP, Chevron, and Amerada Hess; he spent much of his life trotting around globe wherever exploration-drilling rigs required geophysical or electrical engineering, particularly in Southeast Asia.

When Hess employed Terry in the 1990s, the company emerged as a leading partner of the joint venture for wildcat drilling in the South Atlantic. But when the oil price fell (leading the company to discontinue its license from the FIG), Hess "diversified" Terry, transferring him from engineering to business services and logistics. Terry was tasked with developing a cross-industry "cost reduction initiative" to navigate a way past the accusations that the petroleum industry was undertaxed. To rescue the industry from being "taxed to hell," Terry compiled all the costs involved with a full cycle of producing a barrel of oil and mapped the entire hydrocarbon value chain.[73]

With the insights gained from the supply chain mapping project, Terry was well positioned to team with islanders in 2012 and create "Falklink." The aptly named project is a skills gap analysis designed to align training services from the North Sea and developing local businesses oriented toward the needs of the oil industry in Stanley. With funding from the FIG, the Falkland Islands Development Corporation (FIDC), and Scottish Enterprise, Falklink also held workshops in Stanley and in Aberdeen. Once sufficient data were collected, Terry dusted off the supply chain map he had created for Hess in the 1990s and compared it with Stanley's to identify where the differences lay. This helped Falklink identify gaps to fill through partnerships with existing businesses, such as food, or by enlisting migrating businesses that may not be likely to establish a permanent presence in Stanley, like steel mills, manufacturers of gas turbines, production chemistry labs, and so on. Terry had become Stanley's "guy in Aberdeen," and as an intermediary, he helped to set up a series of joint ventures to certify and train islanders to inspect containers, test pipes, and certify rigs.[74]

From the initial seismic surveys to the production stage, Terry projects that for every job offshore, there are at least three jobs created onshore: from government resource managers, technicians, roustabouts, managers, and control room operators to maintenance planners, production engineers, warehouse and pipe yard operators, truck drivers, helicopter technicians, pilots, and more. There is already a semifluid labor reserve of casual workers who hold full-time jobs but gain additional income as tour guides, but they do not have the skills necessary for these occupations. Moreover, these drivers may be even more needed for permanent crew changes if oil development occurs. Falklink began to address some of these "gaps," but questions lingered: How could these career opportunities be maximized for local individuals without distorting other businesses and institutions in the Falklands? How should investment in building housing and other support facilities be timed if the licensees were not yet fully committed to building production installation? How should the islands "open up the doors" to migration for the workforce needed to support one or more well developments?

Engaging in Oil Diplomacy

Islanders' uncertainty about the future impact of the oil industry became compounded by economic pressure from Argentina under the Kirchner administrations. The sovereignty dispute has stunted the growth of the energy industry in two ways.

First, in addition to impacting the traffic of cruise lines and other ships through the Gaucho Rivero bill (see chapter 1), containerization was significantly impacted. Leaning on all large container companies to ban Falklands containers, Argentina's government was successful in influencing the main commercial carrier Hamburg Sud to end business with the Falklands. Buenos Aires blocked ships found to have containers in the Falklands; these ships are typically charter vessels charging rates of $35,000 to $40,000 per day, so the diminished rewards made the link uneconomical. To circumvent this problem, the islanders began using their own leased containers, so-called grey boxes not tied to any shipping line (see figure 15). The Falklands' only container ship (apart from the Ministry of Defense's cargo vessel) flies the neutral flag of Antigua and Barbuda in order to gain access to South American ports.

Second, the political controversy has made the Falklands a difficult place for the FIG or its licensees to attract new partners with the required capital to commercialize. In addition to the added risks of the North Falkland being a new basin with complicated supply chain logistics in its remote location (i.e., a "frontier area"; see chapter 5), the geopolitical deadlock has limited the bidding process for mergers and acquisitions.[75] Banks that function as arms of government, such as many of the firms associated with the oil industry in Asia, are closed for business. And while British or international banks with no business in Argentina have been supportive of oil in the Falklands, the big UK banks with branches in Argentina cannot finance the oil licensees for fear of the Argentine government freezing their assets.[76]

In spite of these difficulties, the FIG has altered its geopolitical and diplomatic strategies since the watershed moment of the Sea Lion discovery. Courting opposition lawmakers throughout the Americas and beyond, councilors use the promise of oil in the South Atlantic to portray allying with Argentina as tanta-

FIGURE 15.   Container yard, Stanley.

Photo by James J. A. Blair.

mount to foolishly missing out on business opportunities. UK FCO minister Hugo Swire went so far as to say, "I think ironically that the Falklands can play a part in the economic recovery of Argentina, but that requires Argentina to recognize the sovereignty of the Falkland Islands, and to recognize the fact that there are commercial opportunities here in the interest of both parties."[77] The islanders also point to licensees Premier Oil, which has licenses in Brazil and Mexico, and Noble Energy, which holds exploration licenses from Nicaragua, to demonstrate that being party to the Falklands' energy industry is not mutually exclusive with holding concessions in Latin America.[78]

During my fieldwork, delegations visited the islands from Uruguay, Brazil, and Trinidad and Tobago, and a potential oil partnership was a major theme for each. The Uruguayans (figure 16), representing the minority moderate to conservative opposition parties, emphasized a history of cooperation in shipping, education, and medical services and expressed hopes for new partnerships to enhance common objectives for business, tourism, and culture. They took pride in the fact that gauchos from Uruguay had a presence in the islands dating back to the nineteenth century. Uruguay, much like the Falklands, is poised to develop an offshore oil and gas industry and is also planning development of a deepwater port. A modest start would be a direct air link. This is Deputy Trobo of Uruguay's dream, as he put it: "to board a plane in Montevideo, and after a

FIGURE 16. Uruguayan delegation.

Photo by James J. A. Blair.

snack, in two and a half hours we would be walking along the waterfront of Stanley. And that [the islanders] could do the same in Montevideo." Islanders that have become involved with the oil industry told me that the only reason Uruguay, Chile, or other neighboring countries continue to support Argentina's sovereignty claim is because they have something to lose by severing ties. But introduce a couple of producing oil fields in the South Atlantic, and all of a sudden there is something to be gained in a relationship with the Falklands. From this perspective, oil becomes a tipping point for potential partners on the continent to bypass Argentina's sanctions and collaborate with the islanders.

The Brazilian group consisted of scholars interested in geopolitics and international law, while the Trinidadians represented the commercial oil and gas industry without addressing the sovereignty question. The Brazilians were split between supporting the islanders' wishes and giving lip service to Argentina's claim, but all were impressed at the comfortable living standards achieved in Stanley. However, they judged that the population of around 3,000 residents (2021) would be too small for having the minimal scale for being viable, so they urged "more traditional" islanders fearful of being overrun by new national and ethnic groups to consider a more inclusive immigration policy. If there were a way out of the political quagmire, Brazil could eventually supply pipe or other materials or services, but Stanley Chamber of Commerce members remain suspicious of the "tap being turned on and off by our nearest and dearest neighbors [Argentina]."[79]

Like the evangelical Norwegians, the Trinidadians presented themselves as authorities on how to make oil work for the producing country, not the other way around. The Trinidadian trade mission showcased similar skilling and training as the Falklink project: the group suggested that they would provide the same expertise as Aberdeen, only closer and cheaper. Based on their visit, they envisioned townspeople as future mechanics, engineers, drillers, and health and safety officers. They invited islanders to Trinidad so that they could actually "touch and feel" platforms on live facilities or simulators before venturing onto the real thing. Many Falkland Islanders saw the Trinidadians as relatable for being from a comparable small-island state.[80] By connecting with Trinidad and Tobago, as well as other Commonwealth countries modeled on the North Sea (particularly the provinces of eastern Canada), Falkland Islanders have become enmeshed in a technopolitical network of not just diplomats, politicians, and foreign ministers but also engineers, business groups, economists, and energy experts in the Americas.

Throughout their oil diplomacy efforts, the message islanders have sought to make clear, which was obvious to nearly all of them already, is that they undoubtedly want to be British. Trade missions unabashedly leveraged their BOT status as an advantage. Chamber of Commerce members described it as "quite the franchise," a "dependent partnership," or a "modern-day protectorate." The good governance clause enforcing UK standards in the FIG constitution may limit autonomy of individual personhood, but islanders conceded that it provides oil companies dealing with a small country like the Falklands the reassurance that the government is going to uphold its agreements.

The Charisma of Corporate Personhood

While the Latin American delegations met with select representatives in government and the private sector, another special visit from one of US oil firm Noble Energy's most senior executives, Cara, drew a full crowd (seventy-five to one hundred people) to Stanley's Narrows Bar. The Falklands' weekly newspaper, the *Penguin News*, hailed Cara as an "inspirational Noble lady" for being the highest paid woman among a 2009 list of 500 Houston-based oil executives.[81]

A loud "Hey-yaw!" from Noble's Texan country manager hushed the eager crowd, and the presentation began. Cara described how, as an impressionable middle child of five from Canada, she had done an internship in the Arctic Circle and felt much nostalgia arriving here in the sub-Antarctic. For Cara, the settler-colonial, circumpolar, "energy frontier" milieu provided the pretext for Noble's compatibility with the Falklands.[82] Cara later commented on this with me, connecting her experience prospecting for uranium mining companies in First Nation territories to exploring for oil in the Falklands. Spending a summer "in an Inuit community that reminded [her] a lot of the Falklands," Cara would "map, tent, and fly out." Cara explained that Noble takes explorations seriously, particularly in a "frontier area" like the Falklands; she defined a frontier area as one where no oil industry is yet established with production and infrastructure.

Cara quoted Noble's slogan, which, to her, sets it apart from her previous employers in the oil industry: "Energizing the World, Bettering People's Lives." "We are all about this," she affirmed. "We are a bunch of people in a company, not a corporate body." Similar to anthropologist Elana Shever's description of a Shell executive building corporate community partnerships in a shantytown bordering a refinery in Buenos Aires, Cara portrayed Noble not just as a corporation or a legal "person" but also as a collectivity of people.[83] Emphasizing how the company, like all human beings, wants to "make a difference," she universalized the company's drive for revenue to an essentialized notion of anthropogenic progress. In doing so, she eliminated unequal forms of personhood and represented its employees', shareholders', and partners' interests as reciprocal to the islanders' desires.

While she impressed islanders with Noble's overall success rate, in an assertion of corporate "transparency," Cara informed her audience of the low probability of oil discovery. She gave some preliminary figures from the over 40,000 square kilometers of seismic surveys conducted to date. There was a potential for a 14 billion barrel gross resource equivalent, out of which they might be able to recover 20 to 30 percent at the most. Here the presentation got technical, despite the stated principle of transparency, which Andrew Barry argues makes knowledge not only available but also marked by significant absences.[84] Nevertheless, Cara was very careful to apologize for using jargon, and to describe the "play types" of offshore minerals, she referenced recognizable features of the onshore landscape. For example, she compared the depositional fan to the iconic "stone runs" of the Falklands that had mystified Darwin when he visited the islands during his journey of the *Beagle*.

Returning to the self-defining question of "Who are we?" Cara paused and theorized: "A company isn't just an entity, it's actually people." Here, Cara employed a depoliticizing discourse that decomposes the notion of corporate personhood: she suggested that the corporation is a group of well-meaning, human individuals. She said that Noble is "true to its values" and that integrity is one of its differentiating aspects. To drive home the point, Cara's tone became almost menacing: "We take this very seriously. We do not accept limits." True to Marx's theory that capital does not abide limits but rather turns them into barriers to overcome, Cara asserted that Noble believes "in limitless possibilities."[85]

During the reception, I spoke at length with Zach, one of Noble's exploration managers. Young for his rank, Zach reminded me of a Texas card shark, despite his Midwest origin. He even had a habit of using gambling metaphors. In describing cost assessment, he differentiated between a "rig rate" and a "spud rate." It costs millions of dollars each day just to rent a rig vessel, he explained, let alone to actually begin drilling. Zach pointed out that companies like Exxon or Shell are seen as reaping huge amounts of money but actually only get 6 to 7 percent on return: "It is a numbers game." Zach explained that there is only a 20 percent chance of commercial discovery in the Falklands, and it's likely that something will go awry—even when Cara and Zach are willing to "bet their shirts on success."

More modest than Cara in his expectations, Zach conceded that the situation with Argentina was something they had to consider. He was among those who had received letters threatening legal action, and it had in fact influenced Noble to pull employees off a plane before landing in Argentina. Still, Zach said that the geopolitics would not impact development potential. He gleaned from the FIG that there is an "official political policy" on the one hand, and a "business reality" on the other. Immigration and labor recruitment would not be issues either, as Noble would import workers from the United States or elsewhere. Contrary to Cara's speech about Noble's commitment to the community, Zach indicated that they would offer very few permanent jobs.

Even if direct benefits might be minimal and short-lived, Falkland Islanders informed me that they were inspired by Noble's arrival. On that same visit, Cara gave another presentation to the business community in Stanley's Chamber of Commerce. Derek, one of the attendees, told me that he was especially impressed with the feeling of informed confidence that Noble brings. As he put it, "Noble means that the American dollar is here." One of the only Falkland Islanders with a high-ranking position with one of the oil licensees, Derek is far from naïve. He pays attention to the seismic survey results and recalibrates his expectations when firms scale back engineering designs or downgrade an exploration well. He came to understand how shareholder money factors into the companies' strategies for risking and de-risking. Nonetheless, Derek saw lasting benefit for the livelihood of the islanders as a "people" from Noble and its shareholders "driving everything." For islanders like Derek, Noble's corporate personhood had become a panacea for both the peril of a potential resource curse and problem of contested peoplehood in the sovereignty dispute. The settler colonial protectorate thus acquired new local meaning as a potential oil producer.

Hydrocarbon Principles

The FIG and its licensees have undertaken a series of socioeconomic impact assessments since the Sea Lion discovery to address some of the issues discussed above, but the question of state capacity remains a significant concern. During my research, the FIG invited me to apply for a tender to conduct a social impact monitoring study for an upcoming round of oil exploration. I was flattered and intrigued at this opportunity to apply my work concretely and to contribute something of practical value to my host community. However, because the FIG as well as the oil companies would fund the study, taking the position might have risked me being banned from research and travel in Argentina. After considering the ethical and political consequences, I ultimately declined to apply, but I observed closely the formation of the below "hydrocarbon principles" that resulted from these studies.

First, to manage what the industry glosses as "nontechnical risk," in 2011/2012, Rockhopper hired a group of consultants from Plexus Energy Ltd. to conduct the first report that was specific to the potential development of Sea Lion. From Sea Lion alone, Plexus estimated that revenues could more than double the islands' GDP (£116.9 million). (Of course, this would depend on the prevailing oil price.) However, remoteness and other logistical issues, such as restrictions on migration and transportation, a relatively small local workforce, limited housing, and a lack of infrastructure (as discussed in chapter 5) presented limitations. The report saw upgrading, expansion of port facilities, investment in housing, and immigration reform as the most serious risks.[86]

Subsequent to the Rockhopper-Plexus study, the FIG initiated its own Oil Readiness Programme, including another socioeconomic impact study and monitoring program, which consultants from a company called Regeneris carried out in 2012/2013 and 2015. Unlike the Plexus report, this Regeneris study was not limited to Sea Lion; the consultants continued to base their predictions around the one Sea Lion oil field, but they expanded their figures for alternate scenarios of various combinations of one to three oil fields as well as a possible gas field, LNG plant, and ongoing exploration. The consultants recommended that the FIG build up to forty homes per year and/or build more portakabins for the surging population. As discussed elsewhere, the FIG was also encouraged to actively recruit migrant laborers and develop various forms of infrastructure through a capital investment plan centered on a proposed new port.

Regeneris found that the primary source of distress leading into potential oil development has been over how to preserve the Falklands way of life. To most respondents, this meant, again, the sense of safety to leave one's house unlocked and car keys in the ignition, and the community and cohesion stemming from a small population where everyone recognizes each other. But with an estimated peak of around 550 extra onshore workers (or up to 800 if an onshore liquefaction plant is built), including support for 170 jobs in the islands and 125 offshore, the economic effects multiply and accelerate. This demographic change threatened to distort the degrees of imperial sovereignty that constitute status in the islands. The transformation would likely lead to more expansive urbanization of Stanley and further depopulation of Camp, as well as the need for fewer constraints for migrant workers. Those consulted, however, highlighted "a preference for English-speaking immigrants and for avoiding too great an influx from any one nation/community."[87] Still, the consultants acknowledged that migrants should be able to get access to permanent resident status and property ownership in a more efficient way. Regeneris recommended that the FIG consider adopting a "proactive" recruitment policy geared toward families from "certain countries," as well as reforms to open borders through existing legislation.

While Regeneris's social impact monitoring program enlisted local partners, some Falklands residents viewed the consultations as a frivolous check-box exercise. Regeneris faced some resentment, especially when the consultants assumed that townspeople had already experienced a greater surge in population when the Mount Pleasant military complex was built. This revealed that the researchers did not understand how removed the military base is from Stanley and how little interaction there usually is between ordinary Stanley residents and military personnel. Other concerns were the availability of accommodations not just for oil but also for tourism. There were only about 200 respondents to the Regeneris survey, so while it might have established a baseline, it was not comprehensive. Ultimately, it revealed few surprises.

To recenter its role in the new commodity economy, FIG's Executive Council published a list of "Hydrocarbon Principles," setting out the islanders' regulatory powers:

1. Hydrocarbons in Falkland Islands waters belong to the people of the Falkland Islands and their exploitation must be to the benefit of the people of the Falkland Islands, both those of *today and future* generations.
2. The Falkland Islands Government will maintain *constant* supervision and control over all hydrocarbon activities within the Falkland Islands Designated Area.
3. Petroleum discoveries must be *efficiently* managed and exploited to maximise economic recovery and to ensure the development of a *long-term* industry presence that will benefit the Islands *for decades to come.*
4. Development of the hydrocarbons industry must ensure the protection and *conservation* of the Falkland Islands' environment and biodiversity.
5. Development of the hydrocarbons industry must take into consideration *existing* commercial activity and promote the development of local business capacity.
6. The exploitation of *finite* natural resources will be used to develop *lasting* benefits to society across the whole of the Falkland Islands.
7. Transparency and accountability *must be present* throughout the hydrocarbon development process from all parties involved.
8. The Falkland Islands will only consider onshore hydrocarbon facilities if they are considered to be in the best interests of the Falkland Islands, and can be proven to satisfy all of the above policy goals.[88]

The above emphasis on temporal security and insecurity is mine. Here, the FIG articulates islanders' visions of prosperity while simultaneously claiming authority as a people. As discussed further in chapter 5, seizing the moment to control resources thus became a way in which to reassert political stability.

FIG's Legislative Assembly issued a similar statement, rejecting lawsuits filed against the oil licensees in Argentina. Citing the 2013 referendum, the elected councilors defend the "right to develop our economy, including the hydrocarbons industry" as well as the "inalienable right to determine our own future."[89]

Political theorist Timothy Mitchell's suggestion that energy from fossil fuels materializes democratic claims thus takes on new meaning in the current context of the Falklands.[90] In this particular hydrocarbon protectorate, settler peoplehood combined with corporate personhood, and the universality of self-determination served to eliminate fetters for the global commodity market.

Focusing on marine resources, this chapter traces how settlers of the Falklands/Malvinas have crafted personhood through social incorporation into global commodity markets. Treating shipwrecks as resources, settlers took advantage of their geographical position at the limits of the modern Atlantic and Pacific Worlds to extract value through "salvage accumulation."[91] Settlers captured wealth from sealing and whaling by colluding to offer transshipment and supply services that forced transient crews into their debt; speculative investments in industrial manufacture of animal oil yielded mixed results. After the 1982 war, islanders lobbied for the rights to establish a commercial fishing license regime, which has swelled government coffers as well as the wallets of a new enclave of elite license brokers.

Seeking to establish an offshore oil industry on the heels of the commercial fishing boom, the islanders' oil licensees began drilling for oil in the 1990s, in cooperation with Argentina. Oil findings in the South Atlantic were not declared "commercial" until the Sea Lion discovery of 2010, but by this time, Argentina's government had rescinded their agreement on joint development of hydrocarbons. Nevertheless, bracing for commercialization, the islanders have begun to prepare for oil production. Attempting to model a stable fiscal regime on Norway's sovereign wealth scheme, the FIG experienced difficulty enforcing it, as disagreements over capital gains taxes followed the Sea Lion discovery. Entrepreneurial islanders enlisted North Sea experts to identify missing links in the oil and gas supply chain. And to counter the Argentine government's challenge to the oil industry, they have engaged in new forms of oil diplomacy, spreading the message that oil concessions in the South Atlantic are protected by popular consent to British sovereignty.

Multiple islanders described the time between the Sea Lion discovery and Premier's unknown decision on developing the well as a "nervous period." With poised trepidation and heightened interest reminiscent of the early receivers of wrecks, local residents deliberated about what kinds of changes their society might undergo, and what they could do to (1) profit from it as individuals in ways that will last beyond the depletion of reserves, and (2) do so in a way that will not risk losing their identity as "a people." Oil executives like Cara, exploring other areas in Falklands waters, dazzled islanders with a charismatic face of corporate personhood, characterized by swashbuckling confidence and the power of the US dollar.

However, from the oil industry's "subsurface perspective," the question was not how but when the stuff will ultimately be extracted, monetized, and sold to the highest bidder. Oil company representatives were quite candid that if the islands were part of Argentina, the regulations and logistics might actually be easier. One confidential interlocutor went so far as to say that the whole operation would be smoother for the oil companies if the islanders would just go back to the UK. Nonetheless, the islands' designation as a BOT, through the islanders' assertion of self-determination, conditioned the possibility for the energy industry to emerge in the South Atlantic.

Some islanders have viewed resource wealth as a way to "buy" full decolonization. UK government administrators have hesitated to link oil with independence, reflecting on doubts over the saga of Scotland; however, local residents envisioned possibilities not only for self-government but also for reimbursing the UK for the costs of defense. Discussing the desire for independence, one Camp resident said, "If you've got the money, and you've got the commodity that the world wants, then you're in it. . . . Nothing in this world today is about the ethics of human rights. It's nothing to do with human rights. It's all about trade." Here, rather than asserting how self-determination is enshrined in the UN Charter—the FIG and UK's usual argument for peoplehood—the islander emulates the swindler of shipwrecks, building self-sufficiency through self-interest.

This formation of personhood based on the willful dependence on commercial markets for marine resources parallels the slow devolution of political power in decolonization. Here, settlers appear to exchange "the right to rule for the right to make money."[92] This sense of dependent personhood conjured previously unimaginable possibilities of new state independence. However, British imperialist nostalgia and authoritarian populism, which the Falklands referendum conjured, also motivated UK citizens to vote to leave the European Union in 2016. Excluding the Falklands from Brexit would ironically impose costly tariffs on Falklands fishing exports, thus raising the stakes for a successful oil industry in the islands.[93]

GROUNDING OFFSHORE OIL

Early in the morning on March 27, 2014, a motorcade of sport utility vehicles descended on the sprawling oil yards of Stanley. Parked along the seafront, spectators watched as the *Pacific Hickory* tugged the Noble Frontier temporary dock facility through the harbor's narrows (figure 17). The local weekly *Penguin News* held a photo contest for the best shot of the floating dock's arrival, and the winner was awarded £100. This eagerly anticipated event represented an advanced stage in the protracted development of oil exploration in the Falklands. The barge was designed to accommodate new rounds of drilling by Noble Energy and Premier Oil in the North Falklands Basin. Yet local residents conveyed to me that the aptly named Noble Frontier was also an object of keen interest because the hope for oil was propelled by a desire for self-determination in the British-controlled archipelago, which Argentina's government continued to dispute as a violation of national territorial integrity. This raises the question: What is the role of infrastructure in the making of a geopolitical frontier? More specifically, how does offshore oil infrastructure condition the possibility for imperial salvage in a UN-designated non-self-governing territory like the Falklands/Malvinas?

This chapter provides an ethnographic account of the operations of oil companies in the frontier phase of exploration drilling in this remote geopolitical hot spot. While the previous chapters focus more closely on the history of settlement and the contested peoplehood of local residents through changes in land tenure and resource regimes in the islands, here I demonstrate how

FIGURE 17.   Noble Frontier temporary dock facility upon arrival in Stanley Harbour.

Photo by James J. A. Blair.

petro-capitalist infrastructure projects perpetuate a frontier condition of temporal instability in disputed imperial formations like the Falklands.[1]

During the 2013 referendum on self-determination, I spent much of the voting day on one of the unpaved mobile polling routes outside of Stanley in Camp. Farmers welcomed me into their settlement houses and served me tea and cookies. Again, none of my interviewees deviated from the "Yes" decision to remain British, but some residents of younger generations raised the possibility of greater independence, specifically by redistributing future oil revenue to reimburse the UK for defense costs.[2] Kim, a young parent, told me, "If Argentina got sovereignty—God forbid—and we had to move there, it would be like moving to a foreign country." Falkland Islanders like Kim are eager to reject Argentine accusations of the British "militarization" of the South Atlantic. Yet, rather than ending Royal Air Force exercises or banning nuclear submarines, they prefer to pay for their own protection with oil money, thereby demonstrating that the islanders are an economically self-sufficient, if not politically independent, population.

On the fringe of both an ongoing sovereignty dispute and an emergent energy industry, the Falkland Islanders find themselves compelled to ask two related questions: "Are we a people?" and "Are we commercial?" These key questions have contrasting temporal registers (ideas relating to time). As discussed in part 1, peoplehood, on the one hand, is a concept in constant flux yet built on the common notion of a shared past.[3] While the Argentine government has dismissed the islanders as a "non-people," Falkland Islanders back their claim to self-determination as a people with up to nine generations of occupation.[4] Commerciality, on the other hand, is future-oriented, as revealed in the islanders' Hydrocarbon Principles, discussed in chapter 4. Here, commerciality refers to the capacity of oil firms to produce a profit, a fraction of which the local government may recuperate through taxes and royalties. While Brexit may pose a threat to the islanders' lucrative commercial fishing industry, Argentine sanctions have limited the energy sector's growth, leading Falkland Islanders to accuse their neighbors of "economic colonialism." The islanders and their oil licensees have thus sought to craft a subnational petro-state at the cusp of contested imperial frontiers: a settler colonial frontier on the one hand, and an oil frontier on the other.[5]

Amid this dilemma, residents and oil companies grappled with how to salvage empire through resource infrastructure.[6] How might oil infrastructure ground the possibility of a commercial future? This chapter focuses primarily on research conducted with residents, government planners, oil managers, and engineers to consider how infrastructure contributes to what I call the South Atlantic's general condition as a *perpetual frontier.*[7] Building on studies of oil in anthropology and related disciplines, this chapter focuses on the "how" of oil production, paying particular attention to the construction of the Noble Frontier temporary dock, designed to support the proposed development of the Sea Lion oil field in the North Falklands Basin—the first "commercial" discovery in the region.[8] I track how the dock's temporary nature forecloses possibilities of asserting settler colonial authority through a preferred permanent deepwater port, requiring islanders to retrofit the Noble Frontier for extended use. This is followed by a closer look at local debates over proposed oil rig infrastructures, through which islanders seek to further materialize stability in different extraction apparatus proposals presented by the industry. In analyzing how the Falkland Islanders and their oil licensees try to secure their place in contradictory ways, this chapter attends to the unintended consequences of infrastructure that allow for a breakdown between technologies and social meaning.[9] That is, ethnographic analysis of the construction of the barge and the broader extraction system reveals not only how infrastructure's unruliness limits the profitability of oil but also how it perpetuates instability in the frontier.

Situating the Perpetual Frontier

Geographer Michael Watts has defined the *oil frontier* as "the spatial, temporal, and, of course, vertical dynamics in which fields within a province are discovered, developed, and recovered—from primary reserve creation to tertiary recovery from existing 'matured' reservoirs."[10] In spite of family resemblances, the South Atlantic differs from other oil frontiers in multiple ways. First, unlike onshore oil frontiers in the Niger Delta or the Amazon, no Indigenous population inhabits the extraction site, and, apart from the 1982 conflict, political violence and confrontation are rare.[11] Second, while the North Falkland Basin saw twenty-two wells explored by 2013, seafloor of comparable area might have 11,000 wells in the North Sea.[12] Unlike the Gulf of Mexico, wells explored in the North Falkland Basin have not met the technical criteria of "ultra-deep" drilling.[13] Yet, similar to the Arctic, gusting winds, swelling waves, and frigid temperatures make working in the South Atlantic hazardous.[14] The Falklands are also physically remote and artificially secluded by polar-national borders and Argentine sanctions.[15] According to Watts, "oil frontiers have no simple life cycle; they are dynamic and unstable, they are discovered and rediscovered—opened and reopened—rather than simply open, closed, and depleted. In short, the oil frontier appears as a permanent prospect."[16] In the South Atlantic, this propensity of private capital toward flexibility and short-term mobility threatens to eclipse settler colonial plans for permanent infrastructure.

The purpose of a frontier in the classic (albeit loaded) sense of the concept—derived from Frederick Jackson Turner—is to pave the way for a civilization or industry that will ultimately replace it.[17] For Turner, the frontier was both a romantic territory of freedom and a dangerous space of "savagery." In this framing, the frontier risks degenerating into a primitive state of uncontrolled violence, as settlers occupy Indigenous peoples' lands in an extrajuridical "space of exception."[18] As discussed in part 1, as a late colonial frontier, the Falklands/Malvinas served initially as a staging ground for the conquest and settlement of Patagonia. Similar to their status in other cattle frontiers of the Americas, gaucho cattle herders were subjugated in debt peonage and marginalized as barbarous vagrants, yet held some forms of empowerment beyond state control.[19] Vessels struggling to journey past Cape Horn took refuge in the islands for provisions or ship repair, and even though agriculture would become central to the islanders' economy, ship chandlery and wreck salvage helped to define the archipelago as an extractive frontier. As maritime traffic declined, the Falklands underwent successive modalities of resource extraction: from a cattle frontier and sealing haven to a sheep station and commercial fishing export zone. Throughout these transformations, the Falklands has remained a "salvage frontier," which Anna Tsing describes as not "a place or even a process but an imaginative project capable of molding both places and processes."[20] Chapter 4 describes how Argentine government functionaries like Foreign Minister Guido di Tella envisioned the South Atlantic as a frontier for economic expansion of Argentina and sought to charm the islanders into comanagement of resources. Nonetheless, inroads toward cooperative relations with Argentina deteriorated, and the islanders attempted to accumulate capital and salvage empire by developing an exclusionary oil frontier.

With the prospect of increased tariffs on commercial fishing due to Brexit, the islanders' current vision is to model an oil bonanza on their fishing boom. However, oil infrastructure remains largely provisional. In place of the inexorable modernization process and immediate fierceness of Turner's frontier, we find a twenty-first-century British protectorate, where curtailed capital projects embody the settler society's unfinished ethnogenesis, explored in part 1. The sovereignty controversy, price fluctuations, and the temporality of oil infrastructure have contributed to a constant state of deferral in this perpetual frontier.[21] Instead of Turner's notion of westward expansion and regression into barbarism, successive modes of speculative resource exploration have formed a perpetual imperial frontier in the Falklands, trapped between the countervailing desires of peoplehood and commerciality. The oil industry's commercial imperative for high-risk, low-cost exploration calls for temporary infrastructure, even though the islanders aspire for permanent infrastructure to undergird their presence as a people.

The Long-Term Interim

By envisioning oil wealth as a means of buying British sovereignty, islanders render it into a concrete object that can be acquired with money, rather than a lofty international legal principle like self-determination. The Falkland Islanders are, for the most part, an elite enclave of license brokers with ironclad control over resources. However, an abundance of temporary infrastructure, in incremental stages of development, maintains an enduring sense of temporal instability. Even though Falkland Islanders are adamant about their rights to exploit resources and continue residing in the islands, the manner in which they establish their dominion as a people appears materially transient. Residents prop up "kit schools" and "kit churches" out of containers, and portakabins threaten to outnumber houses. The vast roads network outside of Stanley is unpaved, and all rural dwellings are still referred to as "settlements" in Camp. This raises the question of whether the chronic temporariness of infrastructure is a colonial leftover or something inherent to extractive industries. I argue that the chronic temporariness of infrastructure in the Falklands not only engages future uses for oil but also suspends the islands' frontier condition in a liminal stage of settler colonial nostalgia as a disputed territory.[22] Here, I examine this ethnographically by taking a closer look at how the islanders attempt to ground their offshore aspirations in murky terrain, between land and water, through visions of a future permanent port.

Since the 1982 war, Falkland Islanders have relied on a temporary floating dock in Stanley Harbour called the Falklands Interim Port and Storage System (FIPASS). The British military installed FIPASS after the war, and the FIG inherited it as a stopgap before a hypothetical permanent port. Having been what a shipping company director described to me as a "long-term interim" infrastructure for decades, it is only a matter of time before the rusty dock falls apart.[23] Previous oil exploration campaigns utilized FIPASS, but when Noble Energy started scoping for a 2015 exploration campaign, the firm's inspectors told the FIG that the risk of a potential catastrophe was too high. Thinking contingency, the FIG supported the oil companies' proposal to build the Noble Frontier, or what some islanders referred to as a mini-FIPASS.

Before the FIG approved the new temporary dock, the government considered a number of proposals to either transform FIPASS or build a new permanent port. Susan, port missioner at the Seamans Mission Centre near FIPASS, reasoned, "I think that the islands' economy has got to continue to progress, and I think that we need that deepwater port." Development of a port was thus enticing for some Falkland Islanders seeking to address chronic temporariness through a conventional development discourse. Yet, in the course of my fieldwork, I found that islanders offered contrasting visions, leading me to question what value a port actually had locally.

This inquiry goes to the foundations of British Falklands society explored in part 1. There is an ongoing debate about whether the archipelago's first British settlement, Port Egmont, was situated on Saunders Island or if it was actually the name used for the sheltered harbor of water. Similarly, the harbor around which settlers built the capital, Stanley, was once called Port Jackson. The harbor eventually became known as Port Stanley, which still later led many to refer to the town itself as Port Stanley.[24] Some historical maps point to an offshore location, but prevailing contemporary perceptions of a port as grounded have made the onshore version a more salient "spatial history" among Falkland Islanders.[25] The commonsense notion of a port as dry rather than wet has inspired an industrial waterfront image of proliferating wide-berth docks, crowded wharfs, towering steel cranes, and rows of stacked containers. In their promotion of a permanent deepwater port just beyond Stanley at a location called Port William, FIG planners and residents like Susan latched onto this vision of onshore infrastructures that appear to materialize modern progress and commerce.

However, residents employed in tourism were skeptical about this grand vision. An expat FIG planner predicted that the new port, "if it's ever built," will accommodate the 30–40 percent of tourists who are unable to visit by tender (a small ferry) each year due to inclement weather: "We're hoping that we'll trap more tourists, and we'll have fewer of them to turn away." Jenny, a Falkland Islander shipping agent who coordinates much of the Falklands' cruise tourism industry, anticipated that "when oil comes, [she] will be able to get stuff done," but she would prefer if the government invested in paving roads to shuttle visitors instead of a port.[26] A local engineer who has worked in tourism and fishing told me:

> I think this dream about an oil port is sort of . . . my gut instinct is that that temporary port will turn into a permanent port. Look at FIPASS: it's been interim there since 1983. *Interim.* And it's doing the job well . . . I think there is a psyche that we need to have a port. But what do you expect that port to be doing in ten years' time? This is not for fishing. You're telling me it's for oil. Tourism is a separate issue.

Contractors and shipping agents across various sectors predicted that the three-year temporary use of the Noble Frontier would become permanent because the lay-down areas for pipe bundles and exploration equipment were already located by the harbor. Premier Oil had proposed extending FIPASS with a permanent structure, and a few local businesses invested in design plans. However, FIG planners and legislators did not allow a permanent port to be located in Stanley's harbor due to concerns over noise and light pollution. Moreover, the government expected the oil companies to contribute an estimated £100 million toward the permanent port. A shipping company director remarked that this was "a hell of a lot of money to spend on a port with a population of 3,000 people." Oil company representatives maintained that they were in the energy industry, not the port construction business.

Acting autonomously from the FIG, a group of elite shipping agents who had partial ownership of Desire Petroleum developed plans for an alternate location for the permanent port on the opposite side of East Falkland Island. Many residents interpreted this as a satirical stunt based on its name, Port Smyley. The shipping agents explained to me that the name was chosen for the historical significance of the place, Smyley Village at Port San Carlos. William Horton Smyley (1792–1868) was an American sealer and swindler known for rescuing stranded seafarers in his schooner and salvaging cargo from their shipwrecks.[27] Hiding his ship in what would become known as Pirate Creek at Port Howard,

Smyley was described by contemporaries as a pirate and charlatan for dressing as a British authority to seize American vessels and vice versa. Smyley became American consul in the islands and formed powerful alliances with settlers in disputes with absentee landowners and the colonial government over rights to cattle, land tenure, and labor recruitment. Smyley's resourceful buccaneering offered a storied legacy of a past colonial utopia that the entrepreneurial shipping agents sought to salvage by locating the port in Smyley Village.[28]

The FIG refused to take the Port Smyley concept seriously and scheduled any decision to build a new port to be coterminous with the golden goose of Premier's final decision on whether to produce oil from the Sea Lion discovery. This plan left nearly all parties unsatisfied. Amid the chess match of speculative land acquisition and investment in planning, contract workers hired to build roads for the new port were forced to stop working until "first oil." My interlocutors in fishing and tourism industries felt undervalued because they would not be allowed to use the hypothetical port until at least five years after installation. From the perspective of the oil companies, shipping capacity would be most necessary during the initial surge of development, not during the gradual decline of exploitation over decades. Ultimately, between FIG policymakers' frustrations with mollifying "3,000 professional opinions" on the one hand and the islanders' seclusion in what one resident described as "two islands surrounded by a sea of advice," contending interests failed to harness the desire for a deepwater port that settlers had long viewed as central to their establishment as a people. FIG directors I interviewed blamed the failure on the high prices of a "greedy local private sector," while business owners viewed the bureaucratic meddling of "expat" FIG contract workers as relics of British "imperialism," relating to tensions between the government and private sector discussed in the previous chapters. This nonstarter created the conditions for the Noble Frontier, an interim infrastructure that holds in tension the archipelago's chronic temporariness. The next section discusses how the Noble Frontier, designed to be portable, became an unruly temporary fix that, like FIPASS, would become relatively permanent by default.

Not All Frontiers Are Made Alike

Even though the Noble Frontier was meant as a temporary fix, it augured commerciality. Backed by the dollar, the barge came to materialize hope for an offshore oil boom in this new oil frontier of Port Stanley. Following is an analysis of the dock's construction and discussion of how the barge came to represent the offshore oil industry's commitment to the long-term wishes of Falkland Islanders, even though it was designed to be temporary.

Noble's outdoorsy, Texan manager, William, explained how the floating dock got its name. "The Falklands is part of the International Frontier Business Unit, being frontier exploration, so the Noble Frontier seemed to make sense. It's the marriage of Noble and one of our frontier interests."[29] The "covenant" of this marriage was that the Noble Frontier would have a ten-year life expectancy, but only a three-year "use life" for Noble. Noble's head engineer, Alex, described the dock as "purely temporary in its purest form." It boasted no fuel, water, or chemical supply—not even a toilet. The Noble Frontier thus follows the licit capitalist logic of what anthropologist Hannah Appel calls "modularity."[30] Describing a trip to an oil rig off the coast of Equatorial Guinea, Appel analyzes how the industry works to disentangle its global profit-making motive from place-based particularities through "self-contained" infrastructures that make offshore drilling occur in a similar fashion across oceans. This intermodal quality of offshore infrastructure follows the dominant global trend of containerization in shipping since the 1950s, geared toward lowering freight bills by maximizing transport efficiencies.[31] Like rigs or containers, floating docks like the Noble Frontier are temporary in the sense that they are movable from port to port or from port to land. Yet, unlike offshore drilling platforms or shipping vessels, the Noble Frontier was designed to extend from the coast, rendering it part of the Falklands' landscape.

Once plans were approved, the Noble Frontier completed its forty-seven-day journey from the Gulf of Mexico. Despite being absentee-owned, the FIC satisfied local content restrictions for Noble but recruited labor from another UK-based company, together forming SATCO (South Atlantic Construction Company). Apart from the stepson of an FIC manager, no Falkland Islanders were part of SATCO's construction team.[32] Alex wished more islanders were directly involved in the construction process, but instead of redistributing risk to "seep into" lives of ordinary residents, health and safety regulations mandated that the oil firms hire certified welders.[33]

As construction neared completion, Alex invited me to tour the dock. Upon entering the causeway's new walkway, a noisy steel screech signaled the pile adjusting to the changing tide. A few of Noble's American welders, peppered into the British SATCO construction crew, wore face shields painted with the Union Jack (see figure 18). "They didn't have any stars and stripes at Lifestyles [the local hardware store]," one of them told me. A worker approached Alex and asked if he had anymore galvanized eyebolts for the walkway. Rust had been a significant problem during construction, but the desired hardware was missing, so Alex instructed the worker to use nongalvanized ones as a temporary solution. As I

FIGURE 18. US welder wearing a shield sporting the Union Jack working on the Noble Frontier.

Photo by James J. A. Blair.

backed away from a moving crane, I noticed a wooden pallet with the word "Pemex" (Mexico's national oil company) stamped on its side. Alex reflected that the Noble Frontier was built to Gulf of Mexico standards, which proved incompatible with the harsh South Atlantic climate for welding, and the schedule failed to account for the loss of productivity. Moreover, the SATCO subcontractors refused to accept extra hours offered because they were paid in a lump sum. At the time of my visit they were running two and a half times behind schedule.

Alex spotted a truck approaching and told me that this was the big moment, the "first run." I happened to be there on the first day that a tractor trailer would attempt to make it over the causeway. It was unclear whether the ramps would allow safe passage of a fully loaded truck, depending on the tide. This was a problem Noble anticipated, yet the vehicle's bolts still slammed against the metal mat on the first attempt. Eventually, Noble modified the level of lead onto the ramp—with temporary wooden materials. However, the engineers had severely miscalculated bathymetry (underwater depth), for when Premier took over operations, they found that the dock was too shallow for most of their equipment vessels, which delayed drilling. Rather than securing a self-contained platform to accommodate drilling, this modular infrastructure was unstable due to labor asymmetries, a failed Gulf of Mexico template, and a depth problem. The Noble

Frontier demonstrates how fossil fuel industries require temporary materials to minimize "frictions" across frontiers, as what Anna Tsing calls "zones of not yet."[34] Mishaps reveal problems inherent to global infrastructure supply chains, suggesting that not all oil frontiers are made alike.

In the absence of a permanent port, and with "first oil" revenue years away, the FIG was ultimately left with no option but to plan on buying the underperforming Noble Frontier. The dock may have been temporary by design, but once piles were drilled, causeways put in place and fenders installed, the infrastructure was fixed in Stanley Harbour. Even Alex, who had insisted that the Noble Frontier would be purely temporary, eventually conceded that it was built to be permanent. One Noble employee observed suspiciously, "all that concrete—for a temporary facility." Though it was meant to be portable, the concrete components of the Noble Frontier solidified public desires for permanence, as cement has done elsewhere.[35] In this sense, the Noble Frontier became indigenized as another symptom of the Falklands' chronic temporariness. Infrastructure built out of ancillary materials may be modified for an extended use-life, but its temporary nature challenges the myth of permanence in teleological framings, such as Turner's frontier trope.[36]

Playing with Legos

Despite the constant deferral of a permanent port, the material presence of the Noble Frontier instilled a proximate sense of importance for development plans of the Sea Lion oil well among the islanders and their licensees. This brought another layer of local controversy surrounding oil infrastructure for offshore drilling. As the main operator of exploration at Sea Lion, Premier Oil had initially engineered for an extraction system centered on an FPSO (floating production and storage offloading) vessel. The FPSO construction, which oil company representatives described as "putting together Legos," is a low-budget apparatus common to frontier areas. It usually consists of a converted tanker and temporary flow and injection lines. The FPSO's grid of multiple far-reaching subsea wells would require vast piping, thus creating a greater onshore presence of pipe bundle facilities than the FIG had warranted. Furthermore, Premier envisioned that an FPSO should be bespoke, brand new, and designed for a twenty-five-year fatigue life, making costs spiral. Premier therefore shifted plans in 2014 to instead construct a tension leg platform (TLP), a sturdier and more compact alternative design that might support a more permanent drilling rig.

To inform about a dozen curious townspeople of what this change in design would entail, Premier's country manager, Les, presented a slideshow he had shown previously to encourage London investors to commit to the project despite the lack of infrastructure, remoteness, and "unfriendly neighbor" (referring to Argentina). Les described himself as a "big fish in a small pond." While most other oil sites would have an industry already present with a preexisting social network, the frontier aspect of the Falklands prompted him to spend his leisure time with local residents. He was a guarded yet energetic, unseasonably tan gentleman who had built rapport with local business elites by organizing weekly poker games, cooking reindeer meat, and inviting them to stay in his Italian villa.

Outlining the technical specs of the new TLP idea, Les explained that its main advantage is that, unlike the FPSO, the TLP actually has a drilling rig on it, so there would be no need for a separate rig or subsea wells. To calm audience members concerned about safety, Les stated that the TLP would be installed and fixed with tethers reaching 450 meters below the surface for the full twenty-five years of expected production at Sea Lion; it was what he called a "closed system," with no need for flaring gas.

However, far from a totally compact unit, the TLP setup would require an additional FPSO, as well as one other tanker. Audience members asked well-informed questions about the linkages between the TLP and FPSO:

"How does oil get from the TLP to the FPSO tanker?"

"How does the manifold compare to the TLP in that both drill laterally?"

Les explained that they had already drilled twelve wells, so they should be able to install the TLP quickly. Joan, the local historian involved in heritage projects, then nudged Les to differentiate between the construction phases of the FPSO and TLP.

"Where does this fit in the twenty-five-year prediction of time you will drill actively?"

Hesitant to make any firm estimates, Les replied, "It is hard to predict the price of oil later, costs, and so on."

To avoid dwelling on the touchy issue of finances, Les bolted into the mechanics of the TLP. He suggested that it would be "made to order," causing an impressed resident in the audience to remark that it appeared to be "a real pioneer design, isn't it?" Les, however, rejected this frontier imagery. He insisted that Premier's emphasis was on using existing, proven technology and compared the design to Shell's analog Olympus TLP at the Mars B field in the Gulf of Mexico (ironically, an oil frontier notorious for disaster). Then Les showed an image of the ship required to deliver the TLP, which looked almost like a truck carrying an oversized house on its bed. In spite of the tremendous, unwieldy load, Les made it sound easy to just "pop it on the back of one of those boats." The crowd laughed at the tall task, and Les accepted that it might be more challenging than he had claimed—to which a local newspaper reporter added, "to say the least!" Here, Les described the TLP as an innocuous toy—like "playing with Legos"—and in a self-effacing manner, attributed his cavalier demeanor to the natural predisposition of "boys and their toys."

The journalist, a woman born and raised in Stanley, voiced her observation that "as boys get older, their toys get more expensive." While Les sought to convey that the rig would be efficient to assemble, the islanders were concerned about long-term costs.

After conveying the technical details, Les went through the timeline to "first oil," or as he liked to call it, the "water under the bridge." Premier would need to finish the engineering concept, submit a development plan for FIG approval, find a partner, contract a rig, explore further appraisal wells, and if all went according to plan, "first oil" could occur in four to five years. Les mentioned the cost figures, which were really only included for investors in the City of London. But local residents wanted to know what this all meant in terms of local oil rent collection:

"What would be the estimated time to payback?"

"What is the percentage of market value, and what are the royalties on it?"

"How much money is fixed?"

Here, Les curiously drew a blank. He tried to calculate the rate of return but came up empty and asked the audience members to give him a call, and he would get back to them. He confirmed that there was a 9 percent royalty of total oil and added sarcastically, "Your government will make sure they are fairly compensated."

While it seemed to be a superior design, the TLP blueprint assumed that there would be a pool of capital available to fund its construction. Most local residents fully grasped this weakness. While public consultations about the TLP plan focused primarily on the technical aspects of the front end engineering design, Premier downplayed a key detail: to "share risk," finance equity, and debt for the TLP, the company would likely need to sell 30 percent of stake in the project to a potential partner. Les attempted to gloss this, saying, "Premier can't *quite* afford it," and attributed the need to "farm down"—he stopped short of calling it "selling"—to the company's costs in the North Sea. Here, a sharp audience member asked:

"Would [such a transaction] be understood as an ongoing investment? Would Premier repay their potential partner?"

Through such savvy questions, these local residents tried to figure out not only the role of Premier in the petro-capitalist joint venture but also whether Premier would maintain and expand their presence as part of the wider settler colonial frontier milieu of the islands. His hand forced, Les answered, "No, it is a sell." Another participant asked if the "farm down" were truly critical:

"If Premier does not find a partner, would they not develop at all?"

Here, Les responded circuitously, accepting that they would need some kind of financing to manage Premier's "exposure" for shareholders. Even though the "farm down" (also known as a "carry") had yet to occur at that time, when the oil price fell in 2014–2015, Premier scaled back its engineering plan, retreating to a slower-phased, less challenging FPSO development design to quell investor concerns over mounting costs. Premier's initial forecast did not match the volatility of the global price of oil and its impact on the privatization of risk in speculative capital.[37] One oil executive told me that "commerciality moves around," because what might be considered commercial in a $100/barrel "price environment" is not necessarily commercial at $50/barrel. Service costs, such as the rate of contracting a rig, might decrease in tandem with the oil price, but a low price squeezes the margin for "viability." Provisional infrastructure plans materialized such temporal instabilities.

After the public meeting, Stanley residents and colleagues at the research institute where I was based told me that they began to see Premier as "fickle." A satirical website covering news related to the Falklands published a parody of Premier's "more simplified approach that will allow for clarity of operations as well as flexibility should things change in the future." Mocking the tone of Les and his Lego, the false story reported, "The initial copy reads as follows: 1. Drill hole. 2. Insert straw. 3. Suck up oil. 4. Fill up a tanker. 5. Sell oil. 6. (Roll around in a bed full of money?)"[38] Other cynical local interlocutors began to suspect that the company would back out of the Sea Lion project altogether. While the industry considered Sea Lion commercial upon discovery, the wavering decisions on the port proposal and the rig engineering design made it apparent that the well's profitability was in constant motion.

Through ethnographic analysis of oil infrastructure development, this chapter considers how extractive capitalist projects suspend disputed territories such as the Falklands in a perpetual frontier. Provisional infrastructures play a significant role in making the frontier temporally unstable, calling into question who and what are permanent in this unusual settler colonial situation. Government planners in the non-self-governing territory vied with business elites and the broader public over ways to assert self-determination and become commercial through plans for a desired deepwater port. However, these plans were abandoned when oil firms won permission to install a new interim floating dock, the

Noble Frontier. After a series of unintended setbacks, the construction of this temporary barge thinly satisfied interests in low-cost drilling without addressing the islanders' wishes to ground a more substantiated sense of belonging in a permanent port. Instead, the FIG was forced to retool the temporary barge as a semipermanent stopgap. While oil companies put the temporary infrastructure into place—paradoxically through promises of corporate commitment—a boomerang decision on the proposed rig system for Sea Lion prolonged a sense of instability.

Noble Energy has since stranded its assets in the South Atlantic, and oil supermajor Chevron acquired the company amid the global coronavirus pandemic and economic crisis in 2020.[39] FIPASS will likely be dismantled, and designs have been in the works for a new port in Stanley Harbour.[40] Premier continued with plans for the FPSO engineering design for oil production at the Sea Lion well, but the oil frontier has been held in suspension due to a global glut in supply. This has left the oil frontier's commerciality uncertain.

How does oil infrastructure ground a commercial future and ultimately salvage empire? As Elizabeth Ferry and Mandana Limbert point out, resources "inscribe teleologies; and they are imbued with affects of time, such as nostalgia, hope, dread, and spontaneity."[41] One well-traveled teleology is the theory that overdependence on resource rents has "cursed" modernizing nations with economic stagnation and authoritarian regimes.[42] Anthropologist Gisa Weszkalnys has shown how social actors take up this "resource curse" hypothesis as a black box to explain the unexplained.[43] The boom and bust cycles associated with extracting nonrenewables presuppose noncontemporaneous presents and pasts as well as futures.[44] Holding these regimes of time in tension, malleable infrastructures like the Noble Frontier are designed to mitigate present problems, such as a perceived lack of shipping capacity. However, such infrastructure carries future risks of ruin if it is not maintained to address contingencies.[45] These ethnographic observations unveil the social relations involved with construction and maintenance that make a "finished" product appear unstable despite being imagined as permanent.[46] The Noble Frontier is thus a technology that seems at first glance to manage the problem of sovereignty and the promise of oil.[47] Nonetheless, the unruliness of the temporary dock's modular model shows us that what might be appropriate in one oil frontier may not meet the conditions of another.

This chapter's ethnographic focus on the temporal politics of infrastructure has broader implications for our understanding of not only petro-capitalism but also settler colonialism in a perpetual frontier. Departing from Turner's romantic ideal of the frontier, scholars have theorized settler colonialism as a structure rather than an event, yet one that is challenged by Indigenous sovereignty.[48]

As a contested settler society with no present Indigenous population, the Falklands departs further from the prevailing narrative. Far from a totalizing structure, the arrival of the Noble Frontier in the Falklands may be understood as a mere infrastructural event.[49] It has become part of the Falklands' partially built environment: "a vanishing frontier" that "refuses to vanish."[50] Rather than paving the way for modern statehood by "sowing the oil," the Falklands remains unpaved and nonindependent.[51] As the frontier advances and recedes, this disputed territory lingers in a suspended state of development. The FIG's salvaging of the Noble Frontier for other uses shows how infrastructure sinks into the structure of settler colonialism as yet another piece of wreckage floating in the South Atlantic.

Oil workers I met appreciated the stripped-down surroundings of this former colonial outpost with a fond sense of nostalgia. Alex told me that, unlike other oil frontiers, "it isn't the Third World. Everybody is friendly, and it almost feels like you've gone back thirty years." He said that what he liked most about the place was that "the entire country does not have one single traffic light." William stole his thunder when he said he had just seen one. Of course, the traffic light was, itself, *temporary* (see figure 19). The colonial nostalgia that temporary infrastructure invokes in the Falklands does not confirm Turner's notion of

FIGURE 19. Temporary traffic light, Stanley.

Photo courtesy of Miguel Barrientos.

"regression" of the frontier to primitive violence. Rather, it suggests that globally circulating ideas of the frontier and practices of frontier making on colonial peripheries may not always converge.[52]

If infrastructure both generates and materializes projects of capitalism and settler colonialism, temporary infrastructure can only hold these projects together in a dynamic and precarious tension. Here, the frontier goes toward neither civilization nor barbarism. Instead, unruly infrastructures perpetuate a temporal crisis of colonial nostalgia. Focusing on the temporal politics of infrastructure allows us to make visible the precarious links between carbon commodity chains and settler colonial situations, enabling a greater understanding of the frontier as an unstable limit of power.

Part 3
SURVIVAL

THE GEOPOLITICS OF
MARINE ECOLOGY

In his self-published exposé *The Falklands Regime*, independent British penguin biologist Mike Bingham tells the impassioned story of his fight to save the Falklands' penguin population from decline.[1] The "regime" in the book's title is an ironic allusion to the kind of antidemocratic dictatorships that the 1982 war was ostensibly supposed to have ousted when Thatcher ordered her military to recover the islands from the Argentine junta that had seized the archipelago. Again, the British victory in the South Atlantic established Thatcher as the formative figure responsible for the conjuncture of state domination with popular consent discourse that Stuart Hall called "authoritarian populism."[2] The sensational cover photo of Bingham's book displays a beach covered in corpses. This was also reminiscent of the widely circulated images of human carnage left on the battlefields in 1982.[3] However, on Bingham's cover, instead of soldiers, the dead bodies are a gruesome array of penguin carcasses that washed ashore in the 1990s.[4] Bingham placed the blame for this apparent penguin decimation squarely on the emergence of a prosperous commercial fishing industry, which he claims had removed the seabirds' main food source.[5] Moreover, Bingham's early resistance not just to commercial fishing but also offshore oil drilling remains one of the only local acts of defiance against local conservation groups with entrenched interests in the Falklands' booming extractive industries. Whistleblowing earned Bingham personal threats from political and business elites in the islands. He accused the government of a litany of coercive acts, including attempted bribery; false accusations of data theft, burglary, and fraud; framing with possession of contraband pornographic film; planting firearms; delivering death threats and malicious phone calls; committing perjury; fabricating documents; and sabotaging a vehicle. Bingham ultimately went into exile in Argentina and later Chile, but only after taking his freedom of speech case all the way to the British Supreme Court in 2003—and winning.[6]

Mike Bingham's controversial case throws light on emergent entanglements between biodiversity conservation and authoritarian populism.[7] Part 3 examines how technoscientific aspects of petro-state formation have helped empire to survive in the South Atlantic in the form of new environmental governance systems. When it comes to petroleum, we tend to associate seabirds with shocking images of oil-soaked feathers, contaminated in the wake of a toxic spill or disastrous blowout. However, seabirds and the data they generate have also become key actors in the crafting of environmental governance systems for new oil drilling projects. This chapter examines how penguin science has come to play a significant role in the development of data infrastructure for the emergent offshore oil frontier in the South Atlantic. It does so through ethnographic description of scientific initiatives to fill data gaps, a term that scientists use to label apparent absences in published research, specifically used here as they pertain to biodiverse species in an underanalyzed geographical location: ocean waters surrounding the Falklands. In addition to the lack of infrastructure discussed in chapter 5, a dearth of scientific data has made the South Atlantic a "frontier" of knowledge. This chapter describes how scientific research has intensified in the islands since the emergence of commercial fishing and oil exploration. Similar to how the research of Antarctic explorers was financed through revenue accumulated from whale, seal, and even penguin oil, much of the current scientific research in the Falklands goes hand in glove with natural resource exploration and exploitation. Taking an ethnographic approach to examine emergent scientific programs in the South Atlantic, I contend that, by performing transparency, new scientific data infrastructures may authorize future pollution of the marine environment in resource frontiers.

This chapter takes an anthropological approach to uncover the geopolitics of marine ecology.[8] The Argentine government and the FIG have expanded their respective scientific research programs centering on the disputed maritime territory of the South Atlantic: the Pampa Azul (Blue Pampa) and the South Atlantic Environmental Research Institute (SAERI). A 2016 UN ruling expanded the limits of the continental shelf, and the Argentine Foreign Ministry has maintained that the islands are within its waters.[9] The Pampa Azul program has allowed the Argentine state to assert the country's geopolitical sovereignty claim over maritime territory in the ostensibly benign name of knowledge. In this framing, the Argentine government seeks to leverage science in the South Atlantic for socioeconomic development.[10] In turn, the Argentine scientific community hopes to fulfill its long-standing national goal of transcending technical dependence on North Atlantic nations and industries.[11]

To examine the islands' own scientific vision, this chapter draws mainly on participant observation from within SAERI's office, as well as co-research in the field. Analyzing the institute's Data Gap Project, which is financed jointly by the FIG and the oil companies, I describe how marine ecologists tag penguins with tracking sensors as part of a new "spatial data infrastructure" of the South Atlantic.[12] This digital information platform has supported the deployment not only of remote sensing devices but also models, standards, and institutions that serve collectively to fortify existing public and private geopolitical arrangements. Data representing the penguins' foraging patterns and other curated ecological facts are intended to contribute new degrees of confidence to environmental impact assessments, influencing oil drilling. By filling data gaps, scientists reinforce corporate-national control over territory and natural resources in a popular-democratic framing of good environmental governance. These contrasting uses of science to reassert sovereignty claims are situated in historical and ethnographic analysis of past and present environmental knowledge production.[13]

South Atlantic Universals

Over time, the South Atlantic has become what Andrew Barry calls a "technological zone" that does not cleave neatly to terrestrial state borders.[14] Using the comparative tools of science, technology, and society (or science and technology studies, STS) to reconceptualize this disputed area, this chapter seeks to address two questions. First, what is the relationship between independent researchers seeking to advance objective scientific knowledge on the one hand and their state or corporate sponsors interested in asserting contested claims to sovereignty, self-determination, and resources on the other? And second, how does such scientific research materialize these contending geopolitical claims through competing representations of the South Atlantic as either a possible future Argentine national oceanic territory or home to an archipelagic people rooted in an imperial past?

Beyond geopolitical saber-rattling, a historical dearth of scientific data on the South Atlantic has influenced the British and Argentine governments to imagine the islands and their surrounding waters as a knowledge frontier. During a temporary Spanish settlement of the late eighteenth century, Alejandro Malaspina delighted in analyzing wild celery and charismatic marine fauna.[15] In his short-lived colony, Luis Vernet made detailed maps. FitzRoy complemented this work with further surveys on the *Beagle* with Darwin—the same vessel used to transport O'rundel'lico and other Indigenous youth to endure a "civilizing" process—an imperial scientific encounter revisited in chapter 7. But during the so-called Heroic Age of Antarctic research, scientists used the Falklands primarily as a launching pad to support voyages farther south, notably Sir Ernest Shackleton's Imperial Trans-Antarctic Expedition on the *Endurance*.[16] Colonial powers have long used archipelagos opportunistically as stepping stones for land acquisition.[17] At the height of the British Empire, the Falklands served as the administrative center for research in the expansive Falkland Islands Dependencies, now British Overseas Territories, which included South Georgia, the South Orkneys, the South Shetlands, the Sandwich Islands, and the territory of Antarctica known as Graham Land. A series of research stations were established for the Falkland Islands Dependencies Survey, now the British Antarctic Survey.[18] The South Pole's gravitational pull for imperial science left the Falklands under-researched by comparison.

The emergence of marine resource economies like commercial fishing laid the foundation for scientific research focused more squarely on the South Atlantic, but the sovereignty dispute has posed a challenge to stock management and preservation. Since the end of the 1982 war and the subsequent emergence of new marine resource regimes, the ecological niche of squid, which straddles the border of disputed economic zones, may be understood as what STS scholars call a boundary object, which maintains its identity across diverse public and private networks.[19] From 1990 to 2004, a boundary organization called the South Atlantic Fisheries Commission conducted research surveys and collaborated on scientific observation and monitoring in both the Falklands and the Argentine EEZ. When the FIG introduced the ITQ system, establishing longer-term fishing quota rights for Falklands companies (see chapter 4), the Argentine government's scientists stopped participating in the commission. This fallout over quota, in conjunction with a broader geopolitical strategy of the Argentine government to isolate the UK with stronger regional alliances, built momentum for the launch of the Pampa Azul program.

By juxtaposing how Argentine government scientists and Falklands-based British researchers not only carry out scientific procedure in data collection but also perform geopolitical strategies through representations of their own actions, this chapter shows how these material practices support what STS scholars call contrasting "sociotechnical imaginaries." Sheila Jasanoff and Sang-Hyun Kim define sociotechnical imaginaries as "collectively imagined forms of social life and social order reflected in the design and fulfillment of nation-specific scientific and/or technological projects."[20] Jasanoff and Kim developed this concept to analyze how the governments of the United States and South Korea promoted nuclear power to reimagine apparently similar national values in different ways. The role of the liberal modern state became central for narrating and defining risk, benefit, the public good, and nationhood as naturalized political imperatives presented in a supposedly unbiased scientific framing. This chapter shows how geopolitical machinations of contending sociotechnical imaginaries illustrate the ways Argentina's oceanic assertion of territorial integrity contradicts, rather than complements, British channeling of the islanders' self-determination claim.

My employment of sociotechnical imaginaries to consider new dimensions of this geopolitical conflict also builds on Michel-Rolph Trouillot's reconceptualization of modernity as a "North Atlantic universal," a fictionalized utopia that prescribes a seductive vision of the world through projections of transhistorical power.[21] I propose that Pampa Azul and SAERI mobilize particular scientific practices to promote either sovereignty or self-determination as *South Atlantic universals.* These universals are examples of sociotechnical imaginaries that present contending geopolitical aspirations, yet they are framed as objective facts to shape popular understandings of the maritime area. Here, South Atlantic universals may be either (1) the expansive oceanic reach of littoral Argentina or (2) the surrounding archipelagic environment of the non-independent society of Falkland Islanders.[22] First, Pampa Azul serves as a government-led scientific tool to regain *epistemological* freedom, if not lost national territory itself. Second, to counter Argentina's national oceanic territorial imaginary, the UK promotes Falklands-based science through a different modern vision based on the universal rights of island societies.[23] SAERI's sociotechnical imaginary makes this possible by linking the overseas territory to Great Britain in a continuous Atlantic island chain of data infrastructures.

Argentina's sovereignty assertion and Falkland Islanders' self-determination claim represent South Atlantic universals in their own right, insofar as respective state actors have sought not only to replicate North Atlantic models but also to insert their subjectivity from the global periphery into dominant regimes of historicity through science.[24] Yet, despite clear historical imbalances, neither set of actors mobilizes these South Atlantic universals from a purely subaltern position.[25] In the South Atlantic, the intensified commercial shift toward marine resources has pitted Argentina—a stagnating export economy, seeking to regain its modern promise as the "Europe of Latin America"—against a wealthy assemblage of predominantly white British settlers pursuing a different utopian island dream of the "not yet."[26] Imperial science provides a line of continuity between the violent colonial past and the extractive geopolitical present of the Atlantic World.[27]

Science in the Service of National Sovereignty

In 2014, the Argentine Ministry of Science, Technology, and Productive Innovation invented a new way of imagining the South Atlantic: as the Pampa Azul. The Pampa is the name for the vast breadbasket of the so-called interior of Argentina: the cattle frontier and agricultural legacy upon which the Argentine nation was built through violent dispossession of Indigenous peoples.[28] By reconfiguring the Atlantic Ocean as the Pampa Azul, or Pampa Sumergida (Submerged Pampa), the Argentine state has strategically asserted the country's sovereignty claim over maritime territory through an oceanic sociotechnical imaginary. The Pampa Azul's main objectives are exploration, conservation, and technological innovation expressly geared toward the "productive sectors related to the sea," as well as supporting the country's sovereignty in the South Atlantic with "scientific information and presence."[29]

This "blue economy" supported by sustainability science in the South Atlantic may be understood as a variation on what environmental anthropologist Marcos Mendoza calls the Kirchner governments' "green economy" in Patagonia.[30] Yet it is significant that the government chose the ecological image of the Pampa rather than Patagonia, even though the South Atlantic has more shared regional history with the latter. While Mendoza describes how Patagonia has been interpreted as a remote "sublime" ecotourist destination, the Pampa evokes a conquered and exploited territory central to national identity and in closer proximity to the capital of Buenos Aires.[31] The Pampa Azul initiative allows the Argentine government to argue that it manages marine resources that the country claims through the objective scientific distance of "trained judgment," even though the South Atlantic, unlike the actual Pampa, is—for the time being—out of reach or control.[32]

During my research in Buenos Aires, I attended a conference where leaders of the social movement Pueblos por Malvinas (People for Malvinas) assisted the government in launching the Pampa Azul campaign at the Argentine National Congress Circle of Legislators. "Malvinas: Final Frontier of the Planet" was the name of the event, which featured international relations scholars and politicians. The talk was held in an ornate gallery. Photographs of the founders of national political parties lined the walls, centering on influential populist former president Juan Domingo Perón. The presentation offered a South Atlantic universal vision of the country's present and future as a modern nation: "Argentina is a country of the South that is Oceanic. It is South American, but also European and developed. A more positive national brand could give Argentina the opportunity to join the community of emerging economies."[33]

Argentina was on the verge of another default at the time, so it was proposed that the nation should take a pragmatic approach to producing and consuming resources. In this context, the Pampa Azul campaign framed Malvinas as "the global frontier of natural resources." The speaker argued that ordinary Argentines are "madly in love" with Malvinas, but when asked why they think the islands are Argentina's, they downplay its importance as a minor battle over a couple small pieces of land. Instead, the presenter emphasized that, if listeners broaden their perspectives on the South Atlantic, considerable reserves of energy, fish, and possible precious metals come into view. Yet, as long as it is under British rule, the speaker argued that the South Atlantic serves as a "militarized highway" for exclusive UK exploration leading to Antarctica. As the staging ground for the broader South Atlantic and Antarctica, the Malvinas, then, are not an insular phenomenon but rather a massive opportunity: "Malvinas is a question of the future." Rethinking the South Atlantic as Pampa Azul recalls place-based universals of political memory, justice, and democracy, rooted in specific experiences of losing an "absurd war."[34] With Pampa Azul, instead of invading the Malvinas again with armed forces, Argentina would occupy the South Atlantic with scientific vessels. As the speaker put it, "The greatest triumph of the UK was not the victory of 1982. It was the colonization of our resources." The talk ended with the determined statement: "We will return to Malvinas. Latin America with science."[35]

The Circle of Legislators event provided a vivid image of the strategic purpose and sociotechnical imaginary of Pampa Azul, but did the perspectives of Argentine scientists also support the sovereignty cause? Edith, a principal marine ecologist of the campaign—who is also a resident of Patagonia with five generations of descent from Welsh settlers—told me that individual researchers in Argentina had maintained long-established relationships with UK scientists to share resources, software, and data without any political issues. However, since ties between Argentine scientists and their counterparts in the islands have broken on joint management of marine resources like squid, they have not been permitted to interact with Falklands-based scientists in an official institutional capacity. Edith and her colleagues had even been invited to collaborate in workshops in the islands, but the Foreign Ministry restricted them from participating. While Edith is not representative of all Pampa Azul scientists, she emphasized that her team seeks to transcend these geopolitical impasses and identify aspects of the ecology impacted by potential exploitation of resources. They follow British scientific advancements with admiration and are not actively involved in crafting the Argentine government's geopolitical strategy. However, they echo their Foreign Ministry in questioning potential conflicts of interest between British scientists and private extractive industries in the islands.

An appointed governor of the Falklands dismissed Pampa Azul in an interview with me as "simply a sovereignty argument." The Circle of Legislators presentation shows some truth in this, but because Pampa Azul is a long-term project that is publicly funded, Edith argued, "epistemologically, the researcher can think more freely." She described Pampa Azul as "a tool that can provide a lot for future generations . . . to generate knowledge and logistics in order to know the sea, which is almost as big as our [landed] part of the continent."

The next section explores the UK's contending sociotechnical imaginary, SAERI, which serves as an archipelagic counterpoint to Pampa Azul's oceanic perspective.

Archipelagic Imaginary

Parallel to Argentina's Pampa Azul campaign is SAERI, formed in Stanley. SAERI was not created as a reaction to Pampa Azul (SAERI predated it slightly); rather, it was born out of a separate FIG economic development strategy to grow a "knowledge economy" in the Falklands. With increasing numbers of scientists based at FIG Fisheries and Falklands Conservation, a quasigovernmental affiliate of Birdlife International, local knowledge production had expanded in the 1990s and 2000s. However, research at FIG Fisheries was relatively confined to stock management, and since Bingham's work, Falklands Conservation had been more focused on its watch group and penguin rehabilitation center. It was therefore proposed that a research institute would recenter more wide-ranging scientific work throughout the various island British Overseas Territories in the new knowledge frontier of the South Atlantic.

With funding from the FIG, the Government of South Georgia, and the UK, the British Antarctic Survey conducted a feasibility study for the potential research institute in 2010. Considered a "pristine environment," various ecological aspects of the South Atlantic invited further research on elements of air, earth, and water: (1) the atmosphere offers a clear, unpolluted sky beneath the ozone hole; (2) an emergent oil industry and peaty soil imply prospects for petroleum geology and climate change studies; and (3) the thriving fisheries, underanalyzed benthic ecosystems (seafloor habitats), and oceanography present opportunities for marine ecology. The British Antarctic Survey and the FIG found ample reason to create an institute, and in December 2011, the FIG Executive Council approved SAERI's formation with three years of seed money.

SAERI's birth coincided conveniently with my own multiphased research in the Falklands, so I was able to carry out in-situ fieldwork that I had not had the opportunity to do at Pampa Azul stations of Patagonia and Tierra del Fuego.

Renting office space throughout my fieldwork in Stanley gave me an inside perspective on SAERI's activities, procedures, and operations. Being in such close quarters with these researchers allowed me to take part in their fieldwork, supporting shallow marine dive surveys and tagging penguins with locational trackers. I was welcomed as part of the budding SAERI community: I answered phone calls, participated in office meetings, and planned events. SAERI included me in their events and appreciated my critical and applied social scientific perspective.[36] I participated in meetings of the FIOHEF advisory group for the Data Gap Project.[37] I also attended public meetings and provided comments on a new marine spatial plan for the South Atlantic. Toward the end of my fieldwork, SAERI hosted a public talk that I gave for local residents. And after departing the islands, SAERI invited me to participate in a workshop on data analysis and marine spatial planning at the University of Cambridge. The FIG also invited me to apply to lead a socio-ecological impact monitoring study, but since the oil licensees (considered to be operating illegally according to Argentine law) were partial funders, this activity may have risked fines and imprisonment of up to fifteen years in Argentina, so I politely declined to apply. However, as a result of the workshop in Cambridge, I served as an external reviewer and provided comments for the Gap Project's Phase 1 Report. SAERI welcomed my suggestions and has made efforts to address them. Nevertheless, these scientific studies have generally advanced rather than deterred oil development in the South Atlantic.

Projects at SAERI that overlapped with my fieldwork included the creation of an information management system and GIS data center, an inshore fisheries research project, marine spatial planning, and a data gap analysis project jointly funded by the FIG and its oil licensees. The latter project is the primary focus of the following sections, as it related most closely to the preparations for environmental management of potential offshore oil.[38]

SAERI's office is located in Stanley Cottage, a historic building with low ceilings where the altruistic Mrs. Orissa Dean hosted lavish dinner parties for the colony's elite merchants, transients, and swindlers in the late nineteenth century.[39] Stanley Cottage is centered in the town's main seafront thoroughfare, Ross Road, which made it an expedient base for my research. The office has three rooms, with several tables and chairs to accommodate a maximum of ten to fifteen people. Canvas prints of field photographs capturing albatross, sea lions, and lobster krill, as well as posters stating SAERI's mission statement and research aims and abstracts of published articles, decorated the otherwise sterile, white interior. Scattered among SAERI's open plan tables were microscopes, stacks of plastic slides with biological samples, maps, and bills. My time at SAERI predated the hire of a business and development manager, so much of the shared day-to-day work consisted of securing external funding for the fledgling institute, tweeting and blogging to increase its web presence, negotiating how to allocate money to different projects, and monitoring the costs of researchers' airfare or internet usage.

While SAERI may be less explicit than Pampa Azul about the political interests of its sociotechnical imaginary, the sovereignty issue inevitably leaks into the institute's affairs. The governor and the FIG have insisted that there not be "crossover" between politics and science; however, on visits to the islands, UK FCO ministers and directors asked SAERI researchers how they might be able to use SAERI's research to spoil Argentina's claim. SAERI's managers explicitly held science to be apolitical. Yet, when Daniel Filmus, Argentina's then-secretary of Malvinas, critiqued the oil exploration activities for risking a potential "eco-disaster," SAERI used the remarks to leverage funding.[40] This was not a knee-jerk response. Rather, it was a strategic move to push back against the accusations by asserting that there is already an objective research institute taking measures to identify data gaps and assess environmental impact. Redirected this way, Pampa Azul might have actually helped SAERI's case for more funding from the UK.

To showcase their role as environmental stewards, SAERI researchers participate in FIG diplomacy trips. They try to brand the institute as having a wide, island-to-island scope, ranging "from the tropics to the ice," with research in Ascension Island, Antarctica, and the other British Overseas Territories in the South Atlantic, as well as the Caribbean and Southern Africa. Finally, seeking to become incorporated and grow independent from government, SAERI has swiveled into the field of energy consultancy. SAERI became registered as a Charitable Incorporated Organization and is a 100 percent shareholder in the SAERI (Falklands) Limited trading subsidiary. Current and former government directors continue to serve as advisors and trustees, but oil consultancy contracts have become a key source of private capital for expansion of the institute's scope of analysis and governance. As shown below, rather than serving as the local regulator that reviews environmental impact statements (EIS), SAERI has begun to work directly for oil companies. This contract work includes original environmental research as well as expert assessment of third-party reports. The motivations of the institute's knowledge production have thus shifted from that of a public service initiative to a private enterprise. This has significant implications for the nature of data infrastructure when scientific facts have become imbued with value beyond, or even contrary to, the environmental cause of sustainability or biodiversity conservation. Moreover, this broader transition toward deregulation has accelerated resource exploitation and capital accumulation through new strategies to generate information about the environment.[41] With its ever-expanding archipelagic sociotechnical imaginary, SAERI has strived to connect the dots between self-determination, ecology, and natural resource management in the South Atlantic.[42]

Tracking Penguins, Sensing Petroleum

One of the ways SAERI became an instrument for propelling the Falklands' emergent offshore oil industry has been through the Gap Project, an initiative to highlight and address lacunae in environmental research data on the South Atlantic. With joint funding from the FIG and the Falkland Islands Petroleum Licensees Association, the Gap Project was a product of FIOHEF. Once the forum had identified some of the "priority gaps," SAERI recruited two researchers to carry out the project on a two-year contract. The intention was to complete the gap analysis by "first oil," an indeterminate starting point of oil production.[43] Both of the researchers came to the islands from Australia: Keith specialized in benthic systems as well as oceanic and fisheries data, while Wendi focused on higher predators, specifically penguins.[44] A third independent scientist later contributed supplemental research on pinnipeds (seals and sea lions).[45] The proposed concept was that, together, these data would fill potential gaps in upcoming environmental impact assessments (EIAs) and environmental management plans. Nonetheless, unlike most cases of misassessment, in which third-party consultants typically disregard local or Indigenous communities out of ignorance of social dynamics, Falklands-based experts used their own rootedness and objectivity as responsible scientific authorities on the region to instill public confidence in what they considered to be minimal environmental risks of a potential oil spill.[46] They have done so through particular data practices that transformed the EIA into a political-ecological process that minimizes risks through particular transparency measures and selective or ambiguous information regarding transboundary impact and local pollution.

Most of Keith's research took place on his computer or outside of the islands. The seafloor covering the oil wells is too deep for human divers to reach, so Keith intended to hire an ROV (remotely operated underwater vehicle): an amphibious drone equipped with a camera and arms for collecting samples. This had not yet taken place during my fieldwork, so Keith's main remit was to build a foundation of data infrastructure by collating, reviewing, and analyzing extant evidence collected during previous research. In some senses, this made Keith's task less onerous, because the industry provided access to most of the benthic data that it held. In addition to meeting with consultants who have done surveys in the past, Keith established affiliations with the British Geological Survey (BGS) and the Natural History Museum in London, making his objective not only to collate or store the data but to literally curate the data, that is, to build up ecological and biological records needed to underpin environmental data sets that would enable searchability and recoverability.

However, developing a strategy for forming a unified taxonomy out of the extant benthic data proved a challenge. First, it turned out that BGS had little to no oversight of sampling in the Falklands. There was a complete disconnect between research undertaken by professional survey companies in the field and remote, subcontracted storage and analysis in London. As Keith put it, "It's all a much bigger mess than we thought, but it does highlight the need to be centralizing this in the Falkland Islands, especially if we're going to have an oil and gas industry which is going to require environmental data." Second, Keith found that the voucher specimens collected thus far were largely missing, unlabeled, or described incorrectly, making them useless as references. Moreover, if any were stored at all, they were contained in a freezer that failed, causing samples to perish. According to Keith, the industry consultants who collected the samples had been concerned more with the company's bottom line than with science, a tension that he had come to understand through his previous consulting work. Design flaws underscored how the surveys had been done as cheaply as possible in order to check off boxes; for example, if the requirement was to collect X number of samples, they did what Keith referred to as a "double grab," which is simply collecting a larger section of ocean floor and cutting it in half, rather than collecting two independent samples. This technique amounted to pseudo-replication.

Having worked with Chevron, BHP Billiton, and other companies operating offshore in Western Australia, Keith praised the collaborative nature of the Falklands' nascent energy industry, which seemed less guarded than other extraction sites. There is extraordinary access to both government and industry, largely because of the multistakeholder environmental forum. "This is what we should have been doing twenty years ago [in Australia]," Keith said. "It showed a lot of foresight and a lot of vision." Elsewhere, Keith said that environmental science "tends to scare" oil companies because, as he put it, "they always think it's a stick that they're going to be beaten with."[47] Environmental research related to oil in the Falklands had been considerably less adversarial, in part because, before the emergence of SAERI, hired consultants had conducted much of it. The drawback was that, according to Keith, such third-party consultancy studies cut corners. Here, Keith had encountered a structural problem that political ecologists such as Heather Bedi have identified: extractive industry consultants tend to produce cursory or misleading EIAs, often at the expense of local or Indigenous communities and their environments near production sites.[48] SAERI had positioned itself to end this common tendency of poorly executed environmental governance from a distance, through locally rooted data infrastructure and more proactive research programs like the Gap Project, with considerable support from the FIG and the oil industry.

Wendi, the second lead Gap Project scientist, complemented Keith's vision. Her initial research focused on Antarctica and the Southern Ocean, where she developed ways to use particular species in these environments as indicators of broader ecological health. She used species at the top of the food chain or food web, such as elephant seals and penguins, to make assumptions about the rest of the system. Her research asked what kinds of parameters could be measured in these species to provide information about the ecosystem writ large, and how that information might be used for environmental management. Based out of Cambridge, UK, Wendi then moved into science policy related to spatial monitoring and decision making, and also worked as a tour guide on cruise expeditions back "down South."

For the Gap Project, Wendi was also tasked with collating and reviewing data from previous researchers. In addition to baseline surveys and coastal mapping, scientists have long conducted seabird monitoring using remote sensing technology. According to one researcher, the practice can be traced all the way back to the Romans, who tied strings to birds in 200 B.C.E. Throughout the twentieth century, Antarctic expeditions adapted military homing devices as tools for measuring penguins' migration patterns in the region. Devices were modified over the years to ensure that seabirds "behave naturally," and since the 1960s, trackers have gone from analog to digital, enhanced by the emergence of the global positioning system (GPS) for military purposes and wildlife management.[49] Once tracking devices earned researchers' trust as reliable, nonhuman "actants," they tinkered with perspective: for bathymetric and temporal data, scientists attach watches to birds that stop and start when diving underwater.[50] For data at the scale of the body, internal and external sensors detect when birds feed, and light sensors or magnetic sensors record day length periods to calculate distance. For data on a longer and wider scale, scientists employ video cameras, camera traps, or satellite images to observe remote populations or new colonies.[51]

While previous data were collected on penguin foraging areas in the Falklands by biologists and wildlife conservationists such as Bingham, they primarily served the purposes of counting for demography and lacked consistency in terms of the type of species and time of year. For example, they might focus on the chick-rearing period but not incubation or the winter season. Wendi's time and energy therefore went into creating a new penguin-tracking program to understand how feeding synchronizes and interacts with the market for oil. She focused primarily on the Northeast and South of the Falklands, areas hosting penguin colonies whose foraging patterns might overlap with drilling areas. For each colony, Wendi was interested in patterns of foraging, such as differences in age, sex, and consistencies in the annual cycle that might provide a more comprehensive understanding of their interaction with oil work.

To track the penguins, Wendi primarily used global location sensing (GLS) tags, which were attached to the birds' ankles to record light levels twice per day, and GPS satellite tracker tags glued onto their backs. The latter are far more precise, but the former are relatively inexpensive, so most of the tags used were GLS. There are certain efficiencies that make the GLS tags less expensive than GPS trackers. First, their spatial resolution is poor, so they are only accurate beyond 100 km. If penguins travel within that distance from their colony, the tags are not particularly useful due to the large margin of error. The recorded light levels allow researchers to "develop a feeling for error" by working out latitude and longitude using an algorithm, but multiple sunlight readings per day cause aberrations.[52] Extensive data cleaning during analysis makes statistics difficult to run. Second, these tags archive data, which means that data are not detectable remotely and the sensors need to be physically retrieved for analysis. While penguins tend to return to the same colony each season, recouping tags can prove difficult, especially for Magellanic penguins (*Spheniscus magellanicus* or jackass penguins in local vernacular), which burrow into the ground and are often infested with fleas. Thankfully, the penguin tagging work for which Wendi enlisted me involved Rockhopper penguins (*Eudyptes chrysocome*) rather than Magellanics.

Together with Wendi and two other researchers, I volunteered to remove location sensors that had been attached to Rockhopper penguins at Cape Bougainville on East Falkland Island, a breathtaking coastal area (figure 20). Wendi selected this site, among others, because she expected penguin colonies there to be most affected by a potential uncontrolled oil spill at Premier's Sea Lion well. She referred to them as "priority gaps." We spent most of the day staring at the Rockhoppers' pink ankles, searching for any sign of the black band and green disk of the GLS tag. Gently, we took turns prodding at the Rockhoppers' ankles with poles and, if we found a sensor present, we isolated tagged birds with a fishing net. We then brought the selected birds to Wendi, who held them in her lap for me to cut the tag from around their ankles. I recorded the number-letter label on the device as well as the apparent sex of the penguin. I then plucked four feathers as samples for isotope tests, which would be used to analyze diet and other biological features (figure 21). Next, I ran a small blue bungee cord under the bird's belly and attached it to a hanging scale to measure the bird's weight. Finally, Wendi reintroduced the bird to the rest of the colony, where it hopped

FIGURE 20. Rockhopper penguin colony at Cape Bougainville.

Photo by James J. A. Blair.

FIGURE 21. Penguin tagging process.

Photo courtesy of Amélie Augé.

around (hence the name) to avoid defensive couples guarding their eggs. We carried on with this method with steady success, locating and processing nine out of the twenty previously tagged birds.

At that point, we hit a wall. Hours went by without finding further tags. Our eyes glazing over in the maze of penguin ankles, we decided to try and thin out the colony to get past the perimeter of tagless birds and allow others room to creep closer in. The method we came up with could not have been more apt in the Falklands: of course, a corral! As discussed in previous chapters, before the advent of conservation in the islands, gauchos corralled feral cattle, while sealers and settlers used the same technique on penguins to funnel them into try pots to be boiled down for their oil. Ironically, we were now corralling the birds so we could protect their colony with scientific knowledge and close the data gap on new potential oil riches, which might, in turn, pose a threat to their livelihoods if an uncontrolled spill were to occur. As illustrated by the continued practice of the corral, the data gap thus functioned as yet another form of enclosure.

Pulling the Land Rover closer to the colony to shield us from the gusting wind, we wrapped chicken wire around plastic stakes and reinforced them with ties and metal stakes. While one researcher captured random birds with the net, I picked them up by the ankles, carried them over (avoiding their beaks), and laid them down gently inside the corral. This method was effective in keeping the penguins temporarily separated, thus allowing Wendi an opening to the core of the colony. Wendi waded through the inner circles of the colony on all fours, but no further tags were to be found. Eventually, we accepted our limitation and let the group become whole once more. We were not quite satisfied with nine out of twenty tags. We had even less luck at other locations, but Wendi had retrieved twice as many, eighteen out of twenty, the day before at Diamond Cove. Wendi and her collaborators planned to return to Cape Bougainville for the remaining tags during the incubating period.

Data from the Gap Project would ideally inform new strategies to minimize and mitigate impact in environmental management plans, but Wendi acknowledged, "In all honesty, I don't think there's any way they'll stop going ahead with the drilling." Tracking penguins may offer insights on the birds' foraging patterns, which will likely interact with oil exploration activities, but the GLS tags did not yield significant data at the time of my research. Further tracking, combined with new forms of modeling, has helped to fill in the blanks, but new transparency measures do not fully address transboundary impacts or forms of local pollution that may arise from offshore oil development.

Conflict and Confidence

The Gap Project was supposed to feed into the crafting of environmental regulations for offshore oil exploration in the Falklands. As part of a broader EIA structure and process that ranges from baseline data analysis to public consultations, an EIS precedes each round of drilling licenses. Social scientific studies of EIA and social impact assessment procedures in mining and energy development projects highlight a common tendency to emphasize risks of damage that the industry considers technically feasible to mitigate.[53] According to Fabiana Li, this stems from a logic of equivalence between "pollution" and "impacts," identified as manageable in EIAs. Li argues that "practices of accountability prioritize mining interests and enable corporations to define the standards of performance that governments will use to establish compliance."[54] In this sense, EIAs have become neoliberal tools that harness legitimacy to prioritize the growth of commodity markets in a liberal democratic framing of public participation.[55] Many of these studies identify a structural problem in the use of underqualified independent consultants rather than local experts. This has led to disingenuous or poorly prepared EIA evaluations and, ultimately, the marginalization, exclusion, or displacement of Indigenous or local communities and ecosystems in proximity to extraction sites.[56] Indeed, British oil consultants I met in the Falklands described the EIS to me as a "box-checking" exercise: in some cases, consultants had literally copied and pasted sections from previous statements based on Northern Hemisphere standards.

However, as Keith noted, SAERI's unique role as a locally rooted quasigovernmental research center, located in a non-independent settler society without the counterhegemonic presence of Indigenous peoples, conditioned the possibility for new experiments with "community-controlled impact assessment."[57] During SAERI's initial period as a government-funded center, its lead scientists would review the EIS and pass comments on to the regulator, FIG Mineral Resources. With its pioneering role in the Gap Project, SAERI was well positioned to contribute new local knowledge to the FIG's vetting process. Accustomed to rigorous science that advances understanding rather than simply meeting accepted standards, SAERI and its independent collaborators attempted to transform the quality and structure of the EIS by tweaking its methodology. For instance, they questioned the qualitative nature of oil spill fate modeling that had long been based on incomplete data. Initially, the common assumption was that the Sea Lion discovery was far enough away from the islands—around 100 miles north—that if anything were to go wrong, prevailing currents would carry a blowout or spill away from the islands (this obviously did little to address Argentine concerns). However, a British engineer who worked with British

Petroleum following the Macondo disaster adapted a computerized "plume 3-D" model that suggested otherwise. The model, developed by Norwegian experts using twenty years of "hind-cast" data incorporating winds, currents, and waves, suggested that 2 percent of the hypothetical blowout, equivalent to about 250–300 tons, *would* arrive on the northern coast of the Falklands in a worst-case scenario. The model showed that the especially waxy Sea Lion crude would solidify and form an emulsion of individual "waxlets," which would rise to the surface, making wind a stronger factor than currents.

This model caused alarm among Falkland Islanders worried about the pristine coastal image, but local scientists from SAERI and FIG Fisheries were less concerned than the oil company representatives. They questioned the conservative manner of modeling and the data sources, arguing that they neglected oceanographic research on the region's unique system of geostrophic currents: eddies flowing in a western-northwestern direction.[58] This marked a significant difference from the common scenario scholars have identified of marginalization through technocratic procedure in the EIA process. Here, instead of the typical case of independent consultants downplaying potential impacts due to ignorance of community voices, local experts from SAERI and FIG Fisheries drew on their own place-based scientific authority to dismiss the alarm generated by contracted consultants' modeling of a potential oil spill.

Once it began to distance itself from government and become incorporated as an independent organization, SAERI not only reviewed statements as the regulator but also accepted contracts from the oil companies to produce them. This shift from the use of knowledge for environmental management to industry-affiliated research signals what Jenny Goldstein calls "divergent expertise," which has a tendency of fomenting "green grabbing."[59] Conflicts of interest may be inevitable in such a small, remote island community. In this case, SAERI found itself in an awkward position because it remained partially government funded. To circumvent a potential conflict, the FIG issued a letter stating that the government would not interfere in the process and that SAERI was essentially an independent organization with regard to the consulting work. Of course, SAERI did not review its own EIS, and the authors of the EIS indicated that if they produce them regularly, they will not continue reviewing others because SAERI might be perceived as more damning to competitors. If SAERI both reviews and produces EISs consistently, then it would need to split into separate legal entities (a separate subsidiary was later formed). Lead scientists were quite candid about the decision to begin producing EISs: reviewing them pays peanuts or nothing at all by comparison, and why not utilize the local skills and expertise SAERI has to generate an income that may support other forms of academic knowledge production? Similar to the petroleum industry's role in enhancing biodiversity conservation science in Ecuador's Amazon region, SAERI researchers thus viewed oil partnerships as potential paths toward more sophisticated research technologies and stronger environmental governance.[60] Oil company managers who spoke with me focused on what they were trying to achieve and asserted that SAERI likely possessed the best possible information on the subject, providing the added value of local knowledge that UK-based consultants were not capable of contributing.

To make their own presuppositions more transparent, SAERI introduced "degrees of confidence" into the EIS. Armed with a vulnerability score or index parameter, the authors could judge that the less confident one is in an assessment, the more conservative one would need to be in estimating potential risk. In public consultations, Premier's local representatives and environmental officers highlighted that the data gaps were included in these degrees of confidence, saying that they were "proud of the world class product of the Falkland Islands."[61] SAERI built on this public-private commitment to global excellence through transparency in science to recruit and publicize teams of visiting scientists, including climate change experts from throughout the Americas and Europe.

Nonetheless, as Andrew Barry has shown in the context of the "measured impacts" of the disputed Eurasian Baku-Tbilisi-Ceyhan pipeline, transparency may ironically inscribe absences.[62] In the initial EIS that SAERI produced for Premier Oil, the degrees of confidence ratings did not have a major impact on the overall statement. To arrive at the degree of confidence for assessing the impact of a project activity, including accidental events such as an uncontrolled spill, SAERI considered how the nature of the effect corresponded to its magnitude and the sensitivity of the "environmental receptor" based on available data. The effect's magnitude and the sensitivity of the "receptor" informed the rating of impact significance: Low (no action required); Moderate (action required to reduce risk through mitigation); or High (also categorized as Major, meaning risk unacceptable and immediate action required). This in itself did not differ greatly from previous EISs, but SAERI added a confidence rating to the formula. Degrees of confidence were based on (1) how clearly the activity was defined, (2) how well understood the sensitivity of the receptor was based on availability of data, and (3) how well understood the nature of the impact was. Yet, when Premier held public consultations, it remained unclear in which areas the confidence rating affected the assessment, if at all. One audience member pointed out that there was no area that was so data deficient that it significantly changed the impact. Out of the possible categories of confidence—Certain, Probable, or Unlikely—all aspects of the EIS were rated as either Certain or Probable.[63] Where the word unlikely did appear, it was not a rating but an adjective, used to minimize the likelihood of an uncontrolled oil spill.[64]

The confidence ratings, penguin tracking sensors, and benthic data curation thus contributed new layers of data infrastructure that perform scientific rigor, yet these measures also produced a "bandwagon effect" for the FIG to continue issuing unilateral licenses for drilling.[65] In some cases, pointing to "many unknowns" signaled caution, but in other areas, a lack of confidence softened the severity and likelihood of a potential spill.[66] In turn, this lowered the impact significance from Major to Moderate.[67] Moreover, uncertainty in risk evaluations have been used as a political strategy to limit regulation and allow markets to manage ecological problems.[68] Here, uncertainty rationalized the need for new data gap analyses that would inform future environmental management plans and set acceptable performance standards and monitoring methods.

Scaled Out

Unanswered questions of geographical scale highlight gaps in the EIA process that the marine fauna tracking project and confidence ratings have not been able to address. First, and perhaps most alarming, is that the contested regional politics of the sovereignty dispute and Argentina's contradictory claims over maritime zones are not discussed. The penguin tracking data indicate that the highest overlap of species, such as the Rockhoppers represented in figure 2 of the scientific report published by Baylis and colleagues, migrate directly through the Sea Lion drilling area north of the Falklands, and then far into Argentine (and even Uruguayan) waters.[69] From the perspective of the Argentine government, this would merit enforcement of domestic laws against drilling, yet the FIG and its oil licensees viewed the baseline and overall proportion of penguins as properly monitored. This renders the seabird specimens "boundary objects" that maintain their identity as they navigate multiple public and private networks.[70] Because of this boundary object status, the Gap Project's repository of penguin data possesses interpretive flexibility for the respective sovereignty assertions of the Argentine government and the FIG to overlapping maritime claims in the disputed South Atlantic territory or the incompatible objectives of biodiversity conservation groups and oil firms. Crucially, it also affected the scale at which environmental governance work could be arranged and organized.[71]

Nonetheless, this highly controversial aspect of the disputed maritime territory was not factored into SAERI's assessment of drilling's "transboundary impact." The FIG has acknowledged that a spill passing into Argentina's zone would be a political disaster. But Argentina claims the entire continental shelf as its national maritime area (the Pampa Azul). Legislators have not sought transboundary consent because they view it as unachievable when the actual boundaries are not even agreed upon. Given that their drilling activities have been considered illegal according to Argentine domestic law, operators are not likely to consult the neighbors. They are active members in a global network called Oil Spill Response Limited, which has an office in Brazil, but it could take weeks before support vessels arrive to control an accidental spill and rescue wildlife. In published scientific reports, SAERI's researchers have argued implicitly for more integrated data across the region. They seek to show that the "distribution of marine predators on the Patagonian Shelf are not constrained by national jurisdictions," claiming that their insights "transcend national boundaries."[72] Nonetheless, by performing transparency and expressing confidence in unlikely harm to the natural environment of the region in the EIS, the data also consolidate exclusive claims to territory and resources for the FIG's British licensees.

Second, despite Falkland Islanders' demands that little or no development occur onshore, Premier Oil engineers decided that they would need to use an inshore transfer system to export the oil to market. Concerns over the harsh ocean, offshore safety, and restrictions from accessing a sheltered location near the continent due to the sovereignty dispute influenced Premier to plan for inshore ship-to-ship transfer located in Berkeley Sound, just outside the Falklands' Stanley Harbour. It was rumored that the inshore transfer option may be removed if Premier had the technology to manage an offshore transfer within health and safety parameters. The inshore transfer condition left a "gray box" in the EIS: coastal communities or intertidal zones.[73] Premier's consultants considered what kinds of "important bird areas," particularly penguin colonies, might be impacted. However, available data on these areas were very limited: again, despite our efforts, the poor spatial resolution of GLS trackers used does not offer information about penguin movements within 100 km, so there is a very large margin of error.[74] In these areas, indicator species such as kelp geese (*Chloephaga hybrida*), logger ducks (*Tachyeres brachypterus*), and passerine land birds serve as proxies for the ecological health of individual islands and coastal zones.[75] Yet, because SAERI's Gap Project has focused on more "charismatic megafauna," and farther offshore areas, much of the onerous work that went into penguin tagging did not factor directly into the EIA of inshore areas.[76] An updated EIS that is outside the scope of the present study incorporated more modeling, testing, and data results.[77] The assessment appears exceedingly thorough, but some inconsistencies remain, particularly regarding whether an inshore oil spill during transfer of crude would have a High or Moderate initial impact on coastal communities. In either instance, such a spill was considered to be Unlikely, with an Uncertain confidence rating.[78] In sum, data gaps and degrees of confidence operate on selective scales of exclusion that are designed to guarantee safety from harm, yet they may also enable the "slow violence" of toxic contamination if they are based on inherently partial mapping technologies.[79]

Reflecting on these unfortunate absences, one SAERI researcher told me that they hoped that the companies never develop oil in the Falklands. "For just 2,000 people, they can survive without oil." This statement came as a surprise, for this scientist had contributed extensive research in support of the government and its licensees in preparation for the oil development projects. It demonstrated how unnecessary offshore oil development had begun to seem—even to those involved in the forum—particularly in a remote island chain with a small population of settlers enjoying relatively high standards of living. After all, many islanders have already amassed considerable wealth from other, less toxic enterprises like agriculture, tourism, and fishing. Participants in the Gap Project and the broader forum have much to celebrate, for their extraordinary scientific rigor and attention to detail is uncommon in other oil frontiers. Yet some found the difficulty of prioritizing risks and collecting new data to fill gaps in a limited timeframe before first oil a frustrating structural problem. While it may have contributed new knowledge informing environmental monitoring and management, the new data infrastructure accelerated the drive toward oil exploitation, and thus may ultimately heighten the risk of environmental despoliation, despite individual researchers' earnest motivations.

This chapter analyzes the development of science in the South Atlantic since the establishment of new regimes of fishing and offshore oil exploration. Recent campaigns by the FIG Fisheries Department and the Argentine government have attempted to leverage their respective sovereignty claims by demonstrating environmental stewardship through scientific research. New talks have now occurred on sharing scientific data on the South Atlantic, representing a significant step for more sustainable comanagement and conservation of squid, particularly *Illex argentinus*.[80] Nonetheless, Argentina's government has maintained its sovereignty claim over the Malvinas, and in doing so, has reconceptualized the South Atlantic as an oceanic national frontier imaginary: Pampa Azul. This scientific program clashed with the establishment of SAERI, through which the FIG has sought to anchor knowledge production of the South Atlantic in the Falklands archipelago. From these contending vantage points, what is the role of scientists in the materialization of disputed geopolitical claims?

While Pampa Azul reproduces a particular South Atlantic universal through its mission of "science in the service of national sovereignty," SAERI seeks to appeal to the islanders' self-determination claim. These contradictory modern principles demonstrate how science comprises not one but rather diverse visions of objectivity.[81] On the one hand, scientists found that the government's Pampa Azul imaginary opened up a public tool for epistemological freedom by offering future generations of Argentine citizens new ways of understanding the sea. SAERI, on the other hand, is a scientist-led public/private initiative steeped in a British imperial tradition that has used islands as stepping stones for territorial expansion and knowledge production. Ultimately, these competing representations of science as either oceanic or archipelagic reinforce ideals of environmental stewardship and political stability that risk closing off collaborative ways of addressing ecological problems exacerbated by the sovereignty dispute, such as transboundary impacts from depleting fishing stock and potentially hazardous oil development. The knowledge frontier condition in the South Atlantic has produced contradictory rather than complementary socio-technical imaginaries. Both Pampa Azul and SAERI offer grand visions that depend on invented horizons of knowledge, as well as new enclosures for extractive industries.

This chapter's ethnographic observations suggest that scientific data collection may run counter to environmental governance when extractive capitalism is embedded in local resource management regimes.[82] In spite of considerable technological innovation with regard to data collection, the course of resource exploitation remains undeterred without social intervention rooted in environmental concern, political commitment, or geopolitical cooperation. The UK has now declared a "climate emergency," and there is an increasing probability that offshore oil projects will become stranded assets. But at the time of my research, the global price of oil and the efficient availability of extractable materials in British-controlled territories still outweighed risk avoidance. SAERI researchers have made a sincere effort to avoid the common problems of misassessment typical of third-party consultants. They viewed oil consultancy as a pathway toward enhancing the region's digital information technologies and governance standards. Nonetheless, the new tracking data and EIS confidence ratings serve the immediate interests of distant private investors at the possible expense of the local environment as a public good. This chapter shows how data infrastructure performs classification and compliance, yet it may do so in a way that boosts rather than challenges extractive development projects.[83] Moreover, it is crucial that we consider how the historical condition of imperialism and settler colonialism is prerequisite to ecological modeling in contested resource frontiers.[84] Like the Noble Frontier, data infrastructures are nested within uneven scales of power, conditioning whose voice counts for conservation. Their effectiveness for environmental governance is especially contingent on historical and localized circumstances, such as ethnonational citizenship in a disputed territory.[85]

In the case of the Gap Project, the customization of scientific research for extractive industries suggests that analyzing data gaps may not be enough to change political-ecological outcomes. Filling gaps with geospatial facts about penguin migration patterns and general ecological sensitivity not only fails to alter the trajectory of oil exploration but, in this case, also advances the path toward exploitation. As they discover new resources, oil firms and petro-states are continuously making new frontiers; as some gaps appear to close, new ones open. There remains a discrepancy between what the data gap analyses reveal and what is important for sustainable governance and environmental justice.[86] The transparency measures examined here authorize unnecessary harm and may fly in the face of violent realities, namely, transboundary impacts exacerbated by the sovereignty dispute and undervalued aspects of local ecology. This chapter thus offers a parable for human and nonhuman complicity in geopolitical conflict, local-level pollution, and, ultimately, the global climate and ecological crisis.[87]

In sum, this chapter showed how the geopolitical dispute in the South Atlantic takes on new contours in the realm of science. Like the offshore oil infrastructures themselves, both Pampa Azul and the Gap Project are frontier-making schemes that depend on imagined horizons of knowledge, as well as new enclosures for easy rents of extractive capital. Similar to the points required for permanent residence discussed in chapter 3, the confidence ratings in the EIS function as degrees of imperial sovereignty. Designed to prioritize the survival of biodiverse species threatened by the climate and ecological breakdown, they also help to ensure an afterlife of empire fueled by oil. Tracked for the purposes of science, statecraft, and industry, penguin data have become actors of environmental governance, yet in this case they serve corporate and national interests. Given the potential hazards of oil exploitation on the ecosystem, these feathered specimens have become active participants in their own potential extirpation.

COLONIZING WITH NATIVES

Toward a general theory of *settler indigeneity*, this final chapter ties together threads from the previous chapters to tell a synthetic story of environmental stewardship that selectively reinforces the imperial agroindustrial and techno-scientific norms and values described thus far. Situated at latitude 51°S, with London located nearly 8,000 miles away at 51°N, the Falklands may be considered a quintessential example of the temperate zones that Alfred Crosby termed "Neo-Europes."[1] Through ethnographic research with white settlers in the Antipodes and Southern Africa, social anthropologists have begun to examine how attachments to place in such frontier areas may constitute not just extensions of Europe but also assertions of indigeneity or autochthony.[2] While there are certainly parallels, unlike those enclaves, there is no historical evidence of an initial colonial encounter with a precolonial Indigenous population within the Falklands. The unusual nature of the islanders' self-determination claim unsettles our assumptions of how its principles usually apply. This not only incites geopolitical debates over sovereignty but also prompts new political-ecological questions about the relationship between settler colonialism and natural resource management: What constitutes an Indigenous or settler ecological landscape? How have the islanders not only salvaged but naturalized their colonial British heritage, as stewards of the land with authority over the environment? How are they negotiating asymmetries in national, ethnic, and racial identity as they assert their "inalienable" rights to territory as "a people"? And what makes a non-native population of a UN-designated "non-self-governing territory" either a settler colony or an alien invasion in the first place?

Through examination of quotidian practices to ensure biosecurity, this chapter offers an anthropological perspective on how settlers of the Falklands have reinvented themselves as natives through environmental management. First, I reconsider how the islanders' assertion of peoplehood is rooted, paradoxically, in the dehumanization and genocide of Indigenous peoples of Tierra del Fuego. Examining a series of human–nonhuman interactions, I then build on "multispecies anthropology" to consider how the course of eradication changed over time in the islands, from extermination of "native pests" in the Falklands and Patagonia (1833–1982) to defense against "alien invasion" (1982–present).[3] The chapter describes how the rising importance of biodiversity conservation and the intensification of biosecurity converged in an extraordinary experiment in "pure" environmentalism to eradicate feral reindeer from South Georgia Island, the site of a former whaling station administered from the Falklands. Toward a conclusion, I analyze how islanders have begun to deracinate their ecological imperialist past and colonize *with* native species, through removal of British-introduced invasive species, as well as habitat restoration methods designed to protect globally important birds and reintroduce endemic plants.

The logic of enclosure and improvement of the commons described earlier served to legitimize colonialism throughout the British Empire based on notions of wilderness, the desert and *terra nullius*.[4] Multispecies ethnography offers a generative methodology for analyzing what counts as wild, feral, or domesticated in contested terrains like the South Atlantic.[5] This approach encourages an ecological optimism that revels in how nonhumans not only survive but even flourish in increasingly toxic environments.[6] In light of seemingly insurmountable crises of climate change, global capitalism, and settler colonialism, this comes as a refreshing challenge to the current age of destruction that geologists call the Anthropocene.[7]

However, while scholars of animal studies and environmental humanities have begun to probe ethical questions of species commodification, endangerment, extinction, and eradication, relatively few have been attuned to the political motivations of such ecological transformations.[8] An instructive exception is the envelope-pushing work of anthropologist Stefan Helmreich.[9] This chapter draws on Helmreich's insights into how biologists and Native Hawaiians think about species they categorize as "alien" or "native" in the settler colonial context of Hawaii, where the word "native" also suggests Indigenous rights to sovereignty.[10] This approach requires refocusing our ethnographic attention to practices that have institutionalized some nonhumans into systems of racial and colonial classification while making others appear natural in particular periods and locations.

What follows connects the contemporary eradication of non-native nonhumans and restoration of native habitats in the Falklands to the attempted elimination of native life-forms in the broader region of Patagonia, with the aim of demonstrating that, even in a fiercely disputed territory, the boundaries that categorize races and natures are powerful yet permeable, and unstable over time.

Settler Self-Determination and Indigenous Erasure

Identifying as "British to the Core" in a locality positioned geographically and politically on the global periphery requires a particularly abstract geohistorical imagination. When I inquired as to how the Falkland Islanders see their claim of self-determination relative to that of other postcolonial populations, one MLA punctuated his response with repetitions of the phrase "I don't know how to say this without sounding racist." He explained that the islanders are "essentially very, very, very British . . . the essential core of the Falkland Islands is very British, and we don't have that divergence of views that maybe other countries have." Of course, "British" in this response not only signifies national identity, it is also, given the councilor's self-conscious qualifier, an obvious euphemism for whiteness.

Predictably, this paradox of self-determination for white British settlers has proven to be a difficult obstacle for the islanders' international diplomacy efforts.[11] To counter diplomatic resistance, islanders and their British administrators employ two different rhetorical strategies that circumvent the history and politics of race. The first is to argue for a deracinated universalism that considers no one Native. Former governor Nigel Haywood reasoned: "You can come at it from a number of directions. One is, it would be very, very difficult to find anywhere other than possibly the center of Africa who you could claim has an Indigenous population because the whole history of humankind is migration."[12] Moreover, the governor's benign image of migration contrasts with the worldview of the Falklands' so-called liberator: Thatcher, a staunch nativist, crafted authoritarian populism based on the xenophobic fear of Britain being "swamped by people with a different culture."[13] Haywood's reasoning is, nonetheless, consistent with the rhetoric of the FIG. Decentering the British from colonial history, they point to "the Mongols and the Chinese and Polynesians" who colonized peoples' lands in previous eras, or the Spanish who "were absolutely brutal" in the Americas.

The second strategy is to prioritize local residence over European descent. Governor Haywood's successor, Colin Roberts, insisted, "They *are* natives! In relation to this place, they are natives." Here, Roberts invokes a relational notion of "becoming indigenous" akin to James Clifford's concept of "diasporic natives," a "rooted experience of routes."[14] Whether or not they constitute a "people," it might be more precise to say that the islanders are part of an "imperial diaspora."[15] In this sense, they have "indigenized" British colonial rule not only through population but also through possession, commerce, administration, and military defense.[16]

While I have discussed strategies that islanders use to circumvent the history and politics of race in order to claim self-determination, not all Falkland Islanders are interested in passing as native or denying its possibility. Some embrace the idea that the Falklands is a kind of settler utopia. When Argentina's history of violent national expansion came up in a dinner party conversation, one former government director commented: "That's true, but we used to go over to the mainland and shoot Indians like pheasants." The director's spontaneous confession surprised this naïve white male American anthropologist, which was not a typical occurrence during my fieldwork. I had many fond interactions, and islanders in my experience were generally welcoming and gracious with a charming British sense of humor. Yet, despite my own upbringing in perhaps a more normalized North American settler nation with a shared pastime of nonchalant genocidal humor, this moment made me feel uncannily positioned in a living, breathing colonial situation.[17] I had presupposed incorrectly that islanders might attempt to erase the role of some of their ancestors in appalling colonial encounters to enhance claims to indigeneity. Instead, I found that, rather than identifying as native in a self-conscious political strategy, islanders approached the contradictory flux of settler indigeneity precisely through such banal affirmations of human and nonhuman decimation.

This person was not the only islander who revealed to me that their ancestors were paid mercenaries who "hunted Indians" in Patagonia. This bloodthirsty professional sport was obviously more common among those who settled permanently in the continent. During the migration of Falkland Islanders to Patagonia in the late nineteenth and early twentieth century, "animals were swapped and people were swapped," as one islander put it. In the process, landowners regarded both upland geese and Indigenous people as "native pests" to be "exterminated."[18] It was ordinary and unremarkable for sheep farmers to string together savings literally by collecting goose beaks and human ears alike for bounty. The FIC paid its employees ten shillings per hundred beaks.[19] In Patagonia, landowners such as José Menéndez and mining investor Julius Popper paid one pound sterling to clans of "Indian hunters" led by Alexander Mac Lennan, aka *Chancho Colorado* (Red Pig), for each pair of a deceased Selk'nam (Ona) person's ears.[20] With the complicit support of the Argentine Republic and Chile, settlers thus valued a Native human's life equal to the lives of two hundred native geese. What does this metric tell us about the hierarchical ordering of race and nature in the South Atlantic?[21]

A key figure bridging the social formations and ecological assemblages of the South Atlantic and Southern Cone in the region's early stages of European colonization is Charles Darwin. In the conclusion to *The Voyage of the Beagle*, Darwin ranked the Yagán people of Tierra del Fuego as "man in his lowest and most savage state."[22] Darwin considered the Yagán less intelligent or capable of semiotic communication than domesticated animals. Like the parodied anthropologist in Raoul Ruiz's absurd 1986 film *On Top of the Whale*, Darwin assumed erroneously that the Yagán possessed a lexicon of fewer than 100 words. Yet missionary Thomas Bridges would later list at least 32,000 words and inflections in his English dictionary of the Yagán language.[23] Moreover, this misconception was contradicted by the existence of individuals like O'rundel'lico, who learned English during the "civilizing" process to which he was subjected as a hostage of FitzRoy. Nonetheless, unlike the Iroquois, Inca, or Aztec, whom colonizers viewed as civilizations worthy of limited diplomacy, Darwin compared the Yagán with exotic nonhuman animals in ecological niches: "the lion in his desert, the tiger tearing his prey in the jungle, or the rhinoceros wandering over the wild plains of Africa."[24] From Darwin's imperial scientific perspective, the Yagán were not just at home in their ancestral territory but also relegated to a nonhuman, "wild" condition.[25]

Legal scholar William Ian Miller uses Darwin's encounter with a singular Yagán Indigenous person of Tierra del Fuego as a point of departure for theorizing the emotion of disgust.[26] Darwin reflects: "The native touched with his finger some cold preserved meat which I was eating at our bivouac and plainly showed utter disgust at its softness; whilst I felt utter disgust at my food being touched by a naked savage, though his hands did not appear dirty."[27] Here, Darwin acknowledges that there is something illogical about feeling a threat of pollution after his food was handled cleanly. Miller explains that "the native recoils at the idea of what manner of man could eat such stuff, whereas Darwin fears ingesting some essence of savagery."[28] While Miller does not make explicit the racist classificatory work that the emotion of disgust has done for genocidal political regimes, Sara Ahmed examines the "sticky" qualities that make objects designated as disgusting seem like they are invading one's bodily space.[29] Ahmed notes that disgust performs a slippage: the sickening and hateful textures of a foreign object can take the form of the bodies of others in order to construct them as nonhuman or lower than human.[30] The disgusting qualities of such bodies thus justify their expulsion.

The following sections of this chapter aim to track a reverse process, examining how the historical disgust that Darwin and European settlers felt toward

Indigenous humans of Tierra del Fuego became constitutive of contemporary desires to expel nonhuman bodies and objects in the Falklands.

A Sheep in Wolf's Clothing

We may recall that Darwin visited the islands in March 1833, just months after the British ship *Clio* arrived to reclaim and settle the Falklands permanently, and then again one year later, in March 1834. As examined in chapter 1, "El Gaucho" Rivero had led his famous violent uprising, which Argentine nationalists have mythologized as a struggle for liberation. As in other cattle frontiers during this period, the largely *mestizo* gaucho workforce embodied barbarous danger and criminality for Darwin.[31] He described the islands as "miserable," and its treeless, rocky landscape and windy climate, "wretched."[32]

One particular species mystified the otherwise unflappable Darwin: the Falkland Islands wolf (*Dusicyon australis* or in local vernacular, warrah; see figure 22).

Canis antarcticus.

FIGURE 22.   Warrah (*Dusicyon australis*). Reproduced with permission from Van Wyhe 2002.

Source: Waterhouse 1838.

Darwin found no evidence that the wolf species had been discovered anywhere else. In a moment reminiscent of Darwin's encounter with the Yagán person— both mediated by meat—he describes a gaucho technique of luring the wolf with a piece of meat in one hand and sticking it with a knife with the other. Thanks to this method, the warrah would become extinct by 1876.

During the week of the 2013 vote, the wolf threatened to steal media attention away from the referendum when a scientific article was published solving Darwin's mystery. Based on a comparison of the wolf's teeth with that of other South American species, scientists dated its origin in the Falklands back to a period when an ice bridge connected the archipelago to the rest of the continent. The researchers hypothesize that the wolf followed seals and seabirds as food sources, but that no other land mammal was able to make the journey at that time.[33]

Before its extinction, the warrah had become one of the archipelago's first native pests.[34] Considered an audacious threat during the lambing season, the warrah was remarkably tame.[35] Selk'nam Indigenous people in Tierra del Fuego used relatives of the same genus *Dusicyon* for hunting, pointing to the likelihood that humans had domesticated the warrah before its isolation and extinction in the Falklands. If the warrah did not arrive in the islands by crossing an ice bridge, it is possible that humans transported it on canoes.[36] Despite its decimation, the warrah still haunts the islands' bucolic landscape like Sherlock Holmes's hound of the Baskervilles, through stories of hunting accidents and place names.[37]

Introduced feral cattle and horses have now followed in the wolf's destiny. After Lafone's FIC imported gauchos from South America to tame and kill wild cattle in Darwin, the first main settlement outside of the port town of Stanley, FIC growers noticed that cattle had "degenerated" beyond the possibility of improvement.[38] Their destruction ushered in the monopolistic land tenure regime focused on sheep farming. As discussed earlier, parallel to this shift in livestock, British shepherds, primarily from the Scottish Highlands, displaced the South American gauchos, whom the British colonial government always considered "alien." In addition to the gauchos, British colonial administrators viewed *Mestizo* sheep imported from Río de la Plata as a racially inferior breed and therefore crossbred them with English rams to build an internal stock managed by the FIC.[39]

The rising socioeconomic valorization of sheep produced a new assemblage of native pests in the islands. Farmers began to view birds of prey and scavengers, such as the striated caracara (*Phalcoboenus australis* known locally as the Johnny rook), the crested caracara (*Caracara plancus* or carancho), the giant petrel (*Macronectes giganteus* or stinker), the gull-like skua (*Catharacta antarctica*), and the turkey vulture (*Cathartes aura*), as dangers to lambs.[40] Settlers also accused rare marine fauna, particularly the burrowing Magellanic (*Spheniscus magellanicus* or jackass) penguins and sea lions, of causing soil erosion and grass degradation.[41]

Colonists and transient sealers dating back to the French, British, and Spanish colonies of the late eighteenth century had slaughtered elephant seals and sea lions, melting their flesh for oil and preserving their skin.[42] They also collected penguin and albatross eggs on a mass scale, accelerating depletion.[43] Reclassifying these animals as native pests, Falkland Islanders rationalized the continued practice of boiling them down for oil: quite literally, rendering native pests.

Targeting native fauna as pests did not result in an immediate boom in wool production. The operation struggled to meet its goals until 1867, when, under the direction of Frederick E. Cobb and Wickham Bertrand, a visiting sheep farmer from New Zealand, the islanders initiated their first non-native species eradication: ticks (or ked) and the dermatitis caused by their feces (scab).[44] The scab disease that one governor called "the bane of sheep-farmers" prevented sheep from growing wool, so it was treated as a serious economic affliction.[45] To eliminate scab, Bertrand advanced what would become a local custom of dipping freshly sheared flocks in solutions of tobacco and sulfur, lime, or arsenic and smearing them with tar, oil, and ointment.[46] In an effort to kill two birds with one stone, one resourceful grower even requested permission from the governor to dress scabbed sheep with penguin oil in 1875.[47] When a scab ordinance went into force, the *Falkland Islands Magazine and Church Paper* published infringements to publicly shame offenders who mixed flocks or failed to properly dispose of carcasses.[48] Once the islanders removed scab and expanded enclosures, wool flourished as the staple commodity.

However, the wool industry introduced new antagonism due to the uneven land tenure system and sharp class division between growers and workers. While workers pined after the possibility of raising their own flocks, growers culled sheep.[49] Again, as the proletarianized settlers began to self-identify as native Falkland Islanders, colonial administrators started to treat them as subjects rather than citizens. Bearing remarkable likeness to Darwin's comparison between the Yagán and undomesticated animals, one British medical officer said of the undereducated children of the remote settlements on West Falkland Island: "In cunning and cruelty they resemble animals and birds of prey, and they are essentially wild creatures."[50] Ironically, in the eyes of British colonialists, the Scottish shepherds recruited to displace the "alien" gauchos had thus descended the social-natural ladder, becoming non-white "feral" settlers. One visiting inspector went so far as to say that, despite the abundance of land for grazing, the islands had been "overstocked" with Scottish shepherds.[51]

This resonates with Shackleton's contradictory taxonomy of the islanders as both people of "indigenous 'kelper' stock" and people of "British stock." Colonial administrators' use of this term originates in the meaning of a stock as a piece of wood, or trunk of a family tree. This metaphor for relationality of human genealogy carried another significance throughout the British Empire for breeding and capital accumulation in sheep raising.[52] In Shackleton's reports, "stock" refers primarily to different ways of conceiving the origins of kelpers' ancestry, which indicates the difficulties colonial administrators had in conceiving of these people in typical oppositional terms of settler and native. Having removed the gauchos and introduced cattle, as well as select native nonhumans regarded as pests, settlers were labeled non-white yet still "of British stock." In response, islanders crafted a white ethnic community construct that mixed British national identity with local origin as "indigenous kelpers," offering key advantages for the labor relations associated with agribusiness. Tangling themselves in an improvised multispecies ensemble of introduced livestock and native seaweeds thus conditioned the possibility for Falkland Islanders to seize control over factors of production, even though their "stock" as a major sheep station had fallen.

It may seem like a relic of the colonial past now that commercial fishing, oil exploration, and tourism have come to yield more local revenue than agriculture, but the Falklands' current coat of arms (figure 23) reveals the continued importance of sheep to islanders. Combining biology and history in national symbolism, the seal of the colony illustrates the Falklands' "layers of naturecultures."[53] The species of animal on the coat of arms has changed twice to reflect more accurately its centrality for the local economy. The colony's earliest badge

FIGURE 23.   Coat of arms of the Falkland Islands.

Source: Open Clip Art Library (available online: https://commons.wikimedia.org /wiki/File:Coat_of_arms_of_the_Falkland _Islands.svg#file).

depicted a bull, representing the abundance of wild cattle during colonization. Traces of cattle farming are evident on the next version in a leather gaucho horse-gear strap on which the motto is inscribed, but the animal is changed to a sea lion. This is indicative of how the islands were once a principal destination for animal oil. There are even local rumors that the "right" in the motto "Desire the Right" was actually a pun, referring to the species of whale of that name. Finally, once wool production became staple in the mid-twentieth century, the sea lion was replaced by a ram, floating precariously on a patch of tussac grass (*Poa flabellata*) atop the submerged English ship *Desire*, which islanders claim discovered the islands.[54] One of the ironies of the current seal of the colony is that, while settlers cut tussac for horses and cattle, the sheep are largely responsible for the degradation of the native grass.[55] Similar to other colonial situations in the Americas and the Antipodes, the irruption of "ovine colonizers" damaged the landscape.[56] Embodying both "biological control" and "biology out of control," as Sarah Franklin put it, sheep became a domesticated economic resource as well as a dominant socio-natural force in the Falklands.[57] Still, when I discussed the coat of arms with a local townsperson, he suggested, "They ought to update the image of the sheep and ship once more to a squid and an oil rig."

Defending against Alien Invasion

Settlers disregarded penguins as native pests historically, but a widespread local embrace of conservationism has reclassified the Falklands' seabird populations as globally important.[58] The adored seabirds also draw in thousands of tourists annually. Once boiled down for oil by early settlers, penguins now enjoy the distinction of being immortalized by the very same taxidermist who melted the Iron Lady into a bronze bust to mark Stanley's Thatcher Drive. As a testament to their increased local value, presently, farmers protect penguin colonies located within their properties *from* their sheep rather than the other way around. This reclamation of local seabirds as natural assets signals a wider shift in the islanders' environmental management: not only do they prize livestock or exotic native species, but they have also come to regard introduced species, including some of British origin, as alien invaders. In the postwar era, the naturalization of sheep intensified the eradication of particular alien plants after the Argentine invasion, and islanders' extermination of alien insects seems to have become a way to manage postinvasion trauma. Without suggesting direct causation, I propose that the condition of having been invaded animates settler assertions of indigeneity, and habitat restoration has offered a way for settlers to reclaim home within native surroundings.

The interest in sustainability in the Falklands is not solely internally driven. Wider global environmental movements and UK policy for British Overseas Territories externally inspired these changes. Besides SAERI's research on marine ecology and environmental science, Falklands Conservation, the nonprofit affiliate of Birdlife International that employed Mike Bingham, raises public awareness through a watch group and penguin rehabilitation center. During my fieldwork, I participated in collaborative projects with Falklands Conservation. These organizations primarily employ experts on temporary contracts from the UK, Europe, or the Commonwealth, but their particular Western environmental conservationist values have enhanced local assertions of indigeneity. The combined efforts of these external forces have influenced Falklands residents, especially those residing in Camp settlements that now function as tourist lodges, to support biosecurity and biodiversity. Many farmers were once at odds with environmentalists; they gave them the pejorative label "turkey lovers" for advocating against shooting native pests such as vultures. However, several Camp residents who might still be skeptical about the liberal values of conservation now take an active role in eradicating aliens and collecting data on native habitats for visiting researchers.

In addition to eradication campaigns against various introduced mammals, new waves of alien plant eradication have altered the islands' countryside since the war.[59] Calafate or Magellan barberry plant (*Berberis microphylla*), used in Patagonia for fruit and red dye, has been subject to elimination in the Falklands because sheep get caught in its thorns. A sensational headline from the British newspaper *The Independent* read: "Calafate Invasion: Falklands Natives under Threat from an Argentinian Force of Nature: Invasive—and Tenacious—South American Shrub Affecting Plants and Damaging Sheep." The connection between calafate plants and the 1982 Argentine invasion is rather tenuous, because they were introduced well before that as a garden ornamental. However, according to the FIG's Environmental Planning Department, Argentine soldiers did introduce another alien plant: mouse-eared hawkweed (*Hieracium pilosella*). Farmers bemoan mouse-eared hawkweed's unassailable force in a way that is reminiscent of Rodolfo Fogwill's fictional depiction of pichiciegos (*Chlamyphorus truncatus*): native pink Patagonian armadillos that came to represent Argentine combatants burrowing into the islands' soil during the war.[60] Mouse-eared hawkweed spreads over large areas and chokes out other vegetation more palatable for sheep. These South American species are considered invaders, even though 85 percent of flora are also native to Patagonia and Tierra del Fuego, and all endemic plants in the region share affinal ties.[61]

Even though it is not of South American origin, the Scottish thistle (*Onopordum Acanthium*) has also become a priority species for eradication in the

Falklands. Thistles' thorns taint sheep's wool and make native bird and animal habitats impenetrable. Thom, a Scottish construction manager stationed in the Falklands under a temporary contract, signed up reluctantly with the NGO Falklands Conservation to go "thistle bashing." Anyone who has viewed the prominent place of the thistle in Scotland's heritage sites, such as James V's palace at Stirling Castle, would understand how Thom felt insulted by islanders complaining about "the dreaded thistles." Pulling up his sleeve, Thom revealed that he even has a tattoo of a thistle emerging out of a "tribal" design on his left bicep. In repeated visits to Saunders Island, on the northwest edge of the archipelago, islanders and expat volunteers like Thom pull the plants out by hand to stop them from seeding. The neighboring Keppel Island, where British missionaries once tried to "civilize" captive Yagán Indigenous people, has become overrun with thistles. Civilians are not usually granted access to Keppel, so a group of officers from the Royal Air Force went by helicopter on a voluntary mission to rid the abandoned settlement of thistles. Meanwhile, the soldiers' own base at Mount Pleasant is also on thistle alert, as are Stanley residents' lawns. One FIG councilor warned that any private property could be a source of a thistle outbreak, so "everyone has to do their part." The islanders must remain vigilant because thistle seeds stay dormant for up to fifteen years and can still potentially germinate.

These alien plant eradication programs reveal how the political relations between human and nonhuman species in the Falklands shape perceptions of the landscape as an invaded territory (figure 24). The islands' British settlers, and even stationed military officers, are no longer cleansing their land of so-called native pests. Rather, they seek simultaneously to protect introduced sheep and native habitats by removing both South American alien flora and select vestiges of the British Empire. The islanders may continue to consent adamantly to British sovereignty through self-determination, but in the process of defining the nature of their polity, they have begun to decolonize the landscape.[62]

While the eradication of alien plants has become a priority, the most high profile of the Falklands' alien species are insects. Apparently introduced via South America, the European earwig (*Forficula auricularia*) is invading Stanley residents' vegetable gardens and homes. Woody, an especially cheerful organizer of Stanley's annual horticulture show, spent decades honing his gardening skills in Camp, first in a remote corner of West Falkland and later at the Goose Green settlement on East Falkland. Notwithstanding the occasional glut of cabbage, Woody is able to produce fresh vegetables year-round. However, he told me, "earwigs are a problem. And the earwig debate is probably about as big a debate as the referendum was. Probably bigger, because the referendum nobody talks about. We talked about it; we've got the results. But we're still talking about

FIGURE 24. Invasive species awareness poster. The vegetation indicates that the landscape depicted is not the Falklands.

the earwigs." Eradication methods range from pesticide spray and liquid dish-washing soap to biological agents like chickens and tachinid flies.[63]

Islanders negotiate daily irritation from the sovereignty dispute through talk of insect invasion. Conversation topics among Stanley residents switch seam-lessly between "the bloody earwigs" and "the bloody Argies!" Echoing the pre-vious headline about calafate, the FIG's biosecurity officer published a story on invasive ants in the local paper under the headline "Falkland Islands' Biosecu-rity Officers Thwart Argentine Ant Army Invasion."[64] There were no known na-tive ants in the islands, and UK experts identified the newcomers as *Linepithema humile*, a species found in Argentina. However, a New Zealand expert found them to be from the different *Ochetellus* genus. In the report, the biosecurity of-ficer acknowledged the confusion, but added, "I doubt you would be reading this article had I named it 'The Ochetellus Invasion' would you?"

Taking such discursive practices seriously, the earwigs or ants may be un-derstood as more than mere symbols for Argentine antagonism. A common perceived threat of invasion entangles insects with the South American neigh-bors, both having occupied homes and eaten food without permission. One is-lander I interviewed during the referendum had been locked in a hall in Goose Green for twenty-nine days during the 1982 invasion, given daily rations of three spoons of baked beans and crackers. He told me, "There are some good Argen-tines; they're the ones in Darwin." He was referring to the cemetery of Argentine combatants near the Darwin settlement. This cemetery has been at the center of the sovereignty controversy regarding vandalism and exhumation for identifica-tion of human remains, which Argentine supporters of the Malvinas cause have used to stake territorial claims.[65] In turn, traumatized islanders have petitioned for laws banning Argentine flags, and signs that say "No Argies" wreathe many doors. Most islanders take a moral high ground and frown upon xenophobia, but earwig or ant extermination provides an outlet for fortifying homes through what they see as acceptable forms of violence. By associating insects' bodies with the sickening qualities of foreignness assigned to human enemies, islanders thus harness disgust to authorize nonhuman invasive species eradication. The eradi-cation demonstrates how experiences of invasion inspire new forms of disgust and racial exclusion that enhance settler claims to belonging.[66]

Pure Environmentalism

Like the sheep of the Falklands, another introduced species—reindeer (*Rangi-fer tarandus platyrhynchus*)—stands atop the seal of the colony for South Geor-gia and the South Sandwich Islands, which had been a Dependency of the

Falklands throughout much of the twentieth century. But far from having naturalized residency in the South Atlantic like sheep, the reindeer have instead
been rounded up and slaughtered for meat. One does not expect to find "Ru-
dolph" quite so close to the South Pole, let alone anywhere near their dinner
plate, but it should come as no surprise that the government is going to great
lengths to govern the population. Scandinavians first introduced reindeer to
South Georgia between 1911 and 1925 as a source of protein for whaling crews.[67]
After their population irrupted to an unsustainable scale, the British government
of South Georgia, in conjunction with the Norwegian Nature Inspectorate, recruited a group of a dozen Indigenous Saami of Finnmark to fence, herd, and
butcher the feral reindeer for the first round of eradication in 2012. Initiated by
the same British expat who would later become the FIG public relations/media
manager responsible for wording the actual referendum question, the rationale
was that (1) culling the reindeer was too expensive in such a remote location,
(2) their grazing caused significant land degradation, and (3) the government
needed the reindeer and any carrion gone to efficiently eradicate rats, which prey
on the seabird population.[68]

To learn more about the reindeer eradication, I visited the South Georgia Government's Office in Government House, Stanley, where Anne, a British environment officer, met with me in a storage room with walls bare of everything
but a framed portrait of Margaret Thatcher. Anne explained to me that people
find rats *disgusting*, and British people in particular are bird lovers, so they de-
spise South Georgia's rats for predating on rare burrowing seabirds. Reindeer,
however, possess the Christmas "cuddle factor," and thus, like penguins, they have
more social capital.[69] Coupled with the fact that South Georgia has no human
natives, this made it so that selling the idea of reindeer eradication required a
purely environmentalist logic.[70]

This is a different kind of abstract geohistorical imagination from that of the
Falkland Islanders' self-determination claim, because the UK removed South
Georgia from the governance of the Falklands' native settlers by turning it into a
separate BOT in 1985.[71] According to Anne, biosecurity in the case of near-pristine
South Georgia is "a way to emphasize how unique and special the place is." She
explained, "South Georgia is great in a way because we don't have a human population, so there's that whole side of it that you can take out and you can just look at
what's going to be best for the environment." Surrounded by the world's largest
sustainable use marine protected area, the territory is owned entirely by the
Crown, enabling resource managers like Anne to experiment with extreme biosecurity measures for total eradication, going beyond population control.

While this pure environmentalism might help to endear the eradication project to Western wildlife tourists, the Saami reindeer herders did not find this

reasoning to be compatible with their own social values. No one had ever tried to eradicate reindeer anywhere else in the world, so there is no proven methodology to do so, and the South Georgia government ultimately chose the Saami through a preliminary process of elimination. Animal welfare experts, vets, scientists, and eradication experts from around the globe who have carried out comparable projects targeting goats, camels, pigs, and other mammals put their minds together and decided against aerial shooting for a variety of reasons, electing instead to find people who are used to herding the animals, understand how they behave, and know how to slaughter them.

Accustomed to preserving and underproducing the herd as pastoralists—not simply hunters or ranchers (let alone eradicators!)—the Saami wanted to proceed at a more moderate pace than the cost-conscious government.[72] After all, reindeer or caribou populations have become at risk of extirpation in their native range habitats of the boreal forest and tundra as a consequence of settler colonialism and extractive industries like logging and mining in the Northern hemisphere.[73] Methods the Saami were used to employing in the Arctic tundra also faltered in the sub-Antarctic. For example, the Saami tried using a ribbon to funnel the herd, but this method accidentally gathered fur seals along with the reindeer. Moreover, due to the strict regulations in the pristine territory, the Saami were not permitted to use quad bikes. According to Anne, "this was a shock" to the Saami, who had become accustomed to a developed setting where automobiles were the norm, as well as working on more level terrain and staying in more comfortable accommodations. Summoning the techniques that gauchos had employed to herd cattle in the Falklands during the nineteenth century, as well as the method SAERI researchers used to tag penguins, the Saami eventually put up temporary corrals for the reindeer. They fenced the herd in at the tip of a peninsula, where they killed the reindeer with a captive bolt and slaughtered them.

Unlike the British government officers, who viewed the reindeer as a "problem" that needed to be eradicated, Anne told me that the Saami saw reindeer as an asset, so they did not want to waste any of the animals. "In Finnmark, they're working with their own animal, and their animals are their *wealth*, their inheritance, and what they're going to pass on to their children." In contrast, "These were our reindeer, they were South Georgia government reindeer, so [the Saami] didn't have that same personal investment." The collaboration between British conservation officers and Saami reindeer herders thus presented an opportunity for bridging different cultural understandings of property and value. However, despite their extraordinary efforts, more than half of reindeer in this phase were shot rather than herded, to meet government deadlines.[74]

Reminiscent of the "factory ships" of the earlier era of whaling in South Georgia, the Saami butchered the reindeer they had herded within a makeshift, floating abattoir: a converted fishing vessel that the government chartered. To recover some of the government's costs, the Saami produced about four tons of meat, which was sold to cruise ships and Falklands residents; the Malvina House Hotel in Stanley offers reindeer carpaccio on its menu. Yet, while some Falkland Islanders indulged in the apparent luxury as guilt-free "conservation in action," others found the reindeer eradication tragically hard to stomach. The reindeer eradication underscored how Falkland Islanders harbor resentment toward the UK for "taking away" South Georgia from local management by the Falkland Islands Government in 1985. Bobby, an elderly Falkland Islander who worked on Christian Salvesen's Leith Harbour whaling station more than fifty years earlier, called the reindeer eradication "the biggest sin and the biggest disaster." Bobby invited me to his home to discuss his experiences whaling in South Georgia and his opinion on the reindeer eradication. Seated at his kitchen table, next to the stove, was a margarine jar full of empty, broken shells of eggs from gentoo penguins (*Pygoscelis papua*). Bobby supported the rat eradication, but he thought the reindeer should have been allowed to die out more naturally, by, say, castration rather than hiring Norwegian "marksmen."

South Georgia's reindeer had cultural value that nativist settlers of the Falklands associated with the legacy of whale hunting. Bobby and others were outraged at the reindeer elimination, because the antlered ungulates serve as a living link to the history of industrial whale destruction from the earlier twentieth century, when South Georgia was a Dependency of the Falklands and the local government regulated whaling stations through a licensing regime.[75] This contended with the government's environmental reasoning for eradication in a different way from the Saami's opposition. The British expat who initiated the reindeer eradication project reflected:

> Your people population is a global one, and on an island like that where there isn't a passport holder saying, "I am a South Georgian," it attracts massively strong associations for people, whether they're historical or "I went there and I just loved it," or whether it's aspiration—you know, this "frontier" at the end of the world type thing—so it's something that generates amazingly strong and passionate views from a huge diversity of people. So certainly, whilst the reindeer thing as an example, you didn't have to go out and canvas the locals on South Georgia and say, "How do you feel about this?" but you've got the heritage aspects to consider.[76]

In Trouillot's sense, the extreme eradication of reindeer "silences the past" of the near-total decimation of whales, in which whalers like Bobby took pride.[77] Bobby was comforted by the fact that, before the eradication had even been planned, local residents salvaged reindeer from South Georgia and, without a formal "business case," they introduced them to areas of outer Falkland Islands.[78] He preferred to procure his reindeer meat directly from one of these residents, refusing the more expensive meat that the South Georgia government offers Stanley residents.

South Georgia may have no native humans, but the former whalers and Saami contractors serve as counterpoints to the ostensibly pure environmentalism of the eradication project. In this case, Saami Indigenous knowledge of and settler attachments to reindeer defy the logic of environmentalism, offering a wider view of life on the frontier.[79] Examining how these different values converge in an unprecedented biosecurity experiment may be helpful for considering the unanticipated blowback of imperial salvage.

Native Regeneration

Islanders and their British trustees are making themselves at home not only through alien removal but also through native regeneration. This includes an initiative to reprioritize the ecological value of tussac grass with respect to the sheep that have damaged it. Tussac grows in tall clumps, typically reaching two meters high. The plant hugs coastlines, providing cover for breeding habitats of seabirds and marine mammals for the last 5,000 years.[80] Unfortunately, the tussac grasslands have been overgrazed and eroded. Sheep damaged the roots of tussac, particularly on the perimeters of the larger, settled islands. However, Falklands Conservation and independent landowners have begun fencing off and/or replanting tussac, which may help to restore the linkage between terrestrial and marine ecologies that has been vital for seabird populations.[81] Alongside this effort is a widening move away from set stocking—long-term grazing in a single location, which causes desertification—toward more carefully planned procedures of rotational grazing methods.[82]

However, local conservationists argue that the effectiveness of land restoration or "holistic" farming depends on the proportion of native to invasive plant species already on the land. Therefore, complementary to replanting tussac and redesigning farms, Redd, a habitat restoration manager with Falklands Conservation, began to develop a business plan for a native seed hub. The Falklands' ubiquitous exposed clay and sand patches are "hard to recolonize" with native vegetation, Redd explained, so he is creating a seed bank to reclaim the landscape and, as he put it, "colonize" with native species.[83] With assistance from the Royal Botanic Gardens at Kew and the Natural History Museum in London, Redd conducted a series of restoration trials on seventeen of the most eroded patches of land on East Falkland Island, as well as collecting more native specimens of fodder grasses for the seed stock. Tussac is a massive grass, making it easy to replant by hand if one removes the tillers (aboveground stems) and allows them to take root. However, this method is ineffective for growing smaller plant species, so Redd's alternative was to have a multiplicity of native seeds, as well as tussac, available to sow. In addition to spoil heaps and nutrient-rich penguin colonies, one of the areas in which this experiment is taking place is newly cleared minefields. Fenced off since 1982, the land has had time to replenish. Redd was therefore hopeful that replacing land mines with native seeds will support indigenous microorganisms in the soil.[84] By making their postwar environment more native, islanders and their British trustees seek to cultivate an image of themselves as responsible stewards.

Finally, in addition to practices of colonization with native nonhumans, in partnership with SAERI, visiting scientists have now found discernible evidence of prior Indigenous human presence in the Falklands' soil.[85] Previous palaeoecological analysis of carbon-dated charcoal raised the possibility of precolonial Indigenous human activity in the Falklands.[86] In addition to containing signs of possible burning by earlier non-Europeans, peat cores include fossils of warrah and evidence of interaction among seabirds, soil, vegetation, pollen, and metals that mark previous human settlements.[87] Moreover, isotope analysis of peat records formed by tussac grasses offers new insights into global climate change: seasonal temperature and humidity, how the last Ice Age formed, what kinds of life used the islands as refugia, and other paleoclimate phenomena.[88] As mentioned in chapter 1, archaeologists have also found middens, shell waste piles that prove Indigenous people inhabited the islands before Europeans.[89] Contrary to the patriotic motto, "British to the Core," promoted during the referendum, these findings reveal that, at their material core, the Falklands may not be so British after all.

This chapter shows how patterns of species eradication and habitat management offer a cross-temporal view of a settler society's ethnogenesis. As I discussed what constitutes an Indigenous or settler ecological landscape with Falkland Islanders, they regularly joked that "the ultimate invasive species here is humans."[90] Nonetheless, by organizing the landscape in hierarchical categories of race and nature, Falkland Islanders asserted *settler indigeneity*: a popular authoritarian conjuncture of British colonial heritage and constructed "native" status. In the early period of permanent British occupation, settlers attempted to annihilate what they considered to be native pests—humans and nonhumans of Patagonian and South Atlantic origin—that posed perceived threats to efficient exploitation of wool from introduced sheep. As sheep naturalized and islanders claimed a bolder sense of belonging, the targets of eradication shifted from na-tives to non-native "aliens." The Argentine military invasion and occupation in 1982 intensified local hostility toward plants and insects introduced from South America. Yet conservation and ecotourism inspired islanders to assign greater value to nonhumans formerly known as native pests. Indigenizing colonial rule ultimately entailed displacing nonhuman plants originating in Europe with re-stored native habitats.

Shifts in the valuation of the native serve as a proxy for the ways islanders have negotiated asymmetries in national, ethnic, and racial identity as "a people." Early colonists harnessed their disgust of "savagery" to rationalize violence against Indigenous humans and native nonhumans. Yet, as British settlers and sheep displaced South American gauchos and cattle, they began to assimilate into colonial administrators' perceptions of nature as non-white "wild creatures." In response, nativist settlers constructed a white ethnic community that en-hanced job security and made them doubly legible to government officials as "indigenous 'kelper' stock" and "British stock." Global shifts in relation to the value of being native had a greater impact than islanders' history of colonial sub-jection in raising local concern for biosecurity and biodiversity conservation. Nonetheless, the broad transformation in environmental stewardship, from the elimination of the native to the eradication of the non-native, demonstrates how islanders have come to hold their hybrid mixture of local peoplehood and Brit-ish citizenship in tension.

What do these ecological practices tell us about the nature of a "non-native" population of a UN-designated "non-self-governing territory?" Are the Falk-lands better understood as a legitimate occupation or an alien invasion, and what anthropological concepts offer insights on this distinction? Without ad-vocating for claims of settler indigeneity, taking it seriously requires that we abandon absolutist racial notions yet throw critical light on political uses of self-determination. To UK administrators, islanders have been viewed as both rela-tionally indigenous and resolutely British.[91] Nearly all consent to remaining British, but their "emergent autochthony" has become a source of capital for green extractivism and ecotourism.[92] If we accept elements of the islanders' au-tochthony as "local people," their criteria for Indigenous status remains dubi-ous because they are neither confirmed prior inhabitants nor confined tribal sovereigns.[93]

This chapter reveals another layer to the sedimentation of racial, political, and ecological wreckage in the perpetual frontier of the South Atlantic. The histori-cal and ethnographic evidence suggests that the Falklands' indigenized settler landscape constitutes more than a making of the colonial world in Europe's ecological image.[94] While nativism did play an important role in the formation of Falkland Islanders as a white ethnic "people," the scientific practices that categorize the fit of particular humans and nonhumans in the islands are based on both colonial organizing principles and native logics of inclusion and exclusion.[95] By colonizing with native species, settlers have sought to assert indigeneity and salvage empire.

UNSETTLED CLAIMS

This book describes how settlers of the Falkland Islands have constructed themselves as "natives" in attempts to salvage resources from the wreckage of the British Empire in the South Atlantic. Even though there is no historical evidence of an initial colonial encounter between Europeans and Indigenous people in the islands, our point of departure was dispossession. Contending claims over the emergent cattle frontier led Buenos Aires to lose its young colony before the Argentine Republic had expanded to conquer Patagonia. However, when we take the coercive "civilizing" practices of British missionaries and the participation of islanders in "Indian hunting" into consideration, the South Atlantic archipelago may be understood as a colonial staging ground for the dispossession of Indigenous peoples in the continent.

Despite the lack of a prior Indigenous population within the islands, the South Atlantic was no safe haven for settlers. Even after permanent British colonial occupation in 1833, subjugated South American gauchos led by Antonio "El Gaucho" Rivero and, later, Yagán Indigenous people, including O'rundel'lico of Tierra del Fuego, revolted against their respective systems of debt peonage and settler colonialism. In contemporary Argentina, proponents of the sovereignty cause have drawn on parts of this history of dispossession to press the case for territorial integrity. Within the islands, the only remaining vestiges of the cattle venture are corral ruins, which proud British islanders have surprisingly appropriated as local cultural heritage products. Classical liberal property ideologies took a forceful hold under British colonial administration. Despite new practices of enclosure, a monopolistic land tenure system focused the economy increasingly on sheep farming through uncontrolled grazing. The settlers embraced frontier ideals of the self-sufficient, rugged individual, as well as collaborative mutual improvement programs. These localized norms and values manifest in lasting forms of horticulture and barter of goods and services. Nonetheless, the hegemony of the Falkland Islands Company (FIC) went virtually unchecked until a sweeping subdivision of land occurred subsequent to the 1982 war.

As settlers formed a distinct social class from British colonial administrators, struggles for the right to work and hold office resulted in an alliance of nativist "kelpers." Islanders consented to British sovereignty and military defense through popular claims of inherited whiteness. Here, we began to observe the formation of a settler colonial protectorate that would become central to the islanders' postwar extractive economic boom. Concentrations of transient British contractors and military personnel have created an enclave effect, separating elite expats from rural local residents yet excluding migrants considered "undesirable" from political rights and other social programs. Birthright secures access to the settlers' peoplehood, as well as British citizenship via Falkland Islands Status. In addition to salient racial categories, long-term migration is contingent on degrees of imperial sovereignty, represented as points for permanent residence, which govern who may or may not "desire the right" to self-determination.[1] Nevertheless, migrants and seasonal workers have become increasingly integrated into the settler society. It is significant that all land mines have now been cleared from the Falklands' landscape, thanks to the courageous work of Zimbabwean de-miners, several of whom have found a new home in the South Atlantic.[2] In the wake of the global Movement for Black Lives, racist discrimination practices received heightened scrutiny in the Falklands, and in 2020, the FIG introduced legislation to hold employers, including the government itself, accountable.[3]

With British sovereignty cemented after the 1982 war with Argentina, islanders lobbied for rights to sell fishing licenses. This generated tremendous wealth for the FIG and local business elites. In the 1990s, the islanders also began licensing offshore oil exploration. The turn toward commercial fishing and oil drilling follows a long history of imperial salvage in the islands that can be traced back to previous modes of environmental wreckage and marine resource extraction, including ship repair, sealing, and whaling. Agreements on fishing and joint oil development with Argentina expired, but new forms of anticipatory politics materialized since the commercial discovery of oil at the Sea Lion well in 2010.

In spite of a new fiscal regime, supply chain management, the near-unanimous referendum, and oil diplomacy efforts, the islanders have entered a state of collective angst due to heightened dependence on tax revenue from overexploitation of land, territory, and resources. One especially cynical Falkland Islander reflected:

> When you look at the history of the islands: they raped the whaling—massive amount of tax from that—that was the main income for government; the farming basically sort of raping the land—they went up to over 700,000 sheep, they can't get anywhere near that now, the place is drying out; and now then you've got the fisheries, well they've raped that to a level . . . and now we're on the next rape, which is the oil.

Here, nature is viewed dispassionately as a passive victim of sexual assault, a political ecological affliction resulting from successive regimes of salvage accumulation.[4] In addition to the violent imagery, this observation points to a temporal crisis in the environment: whales have neared extinction, land is degraded, fish stock is precarious, and oil is a nonrenewable resource with a limited shelf life due to climate change.[5]

Through ethnographic description of the politics of infrastructure, we analyzed how this temporal crisis has materialized in the built environment during oil preparations. Temporary and makeshift forms of fixed capital suspended the islands in a perpetual frontier situation, and the construction of a temporary dock, the Noble Frontier, eclipsed islanders' desires for a permanent deepwater port.[6] Temporary infrastructure appears to act as a quick fix for managing the problem of sovereignty and the promise of oil. Yet the engineering of the temporary dock on a modular frontier model—based on the Gulf of Mexico—shows us that what might be suitable in one offshore frontier may not weather the conditions of another. The global infrastructure of oil in all its enormity is shot through with geopolitical, economic, and ecological differences, making adaptable technologies like the Noble Frontier impossible to customize according to local desires. An anthropological approach to infrastructure allows us to lay bare these precarious links in the hydrocarbon commodity chain, enabling a greater understanding of the perpetual frontier as an enduring yet unstable limit of power.

As a gateway to Antarctica, the South Atlantic may also be understood as a frontier of knowledge. The FIG and the Argentine government have hinged their respective claims to sovereignty and marine resources on scientific research in the South Atlantic. Argentina's government promoted an image of the ocean as Pampa Azul, an allusion to the historical agrarian frontier of the nation acquired through the attempted annihilation of Indigenous peoples. On the ground in the islands, the FIG established SAERI: a government-sponsored institute with close ties to the fishing and oil industries. As part of a consultancy project for the oil licensees, marine ecologists tagged penguins with tracking devices, yet the effectiveness of these data for environmental governance was limited. Ambiguities and contradictions in assessments of transboundary impact and intertidal zones suggest that the data gaps and transparency measures reinforce notions of environmental stewardship and political stability that fall short of addressing urgent matters related to the territorial dispute, key indicators of local ecological conditions, and, ultimately, global climate change.

Finally, to analyze how Falkland Islanders have naturalized settler colonial belonging through environmental management, we examined changing forms of authority over agriculture and biodiversity conservation, particularly species eradication and survival through habitat restoration (figure 25).

Settler Indigeneity

By consenting to British sovereignty through self-determination, Falkland Islanders have extracted not just resources but also indigeneity throughout their history in the South Atlantic. This process has taken the form of nativist assertions

FIGURE 25.   Restored tussac grass on Bleaker Island.

Photo by James J. A. Blair.

of settler belonging through territorialization: Lockean enclosures, restricted access to "the Falklands way of life," and exclusive control over marine resources. However, it has also involved the dispossession of gauchos and Indigenous peoples of South America, as well as the extraction of labor from globally marginalized workers. We observed this in the South American Missionary Society, the recruitment of seasonal labor from St. Helena and Chile, the contracting of de-miners from Zimbabwe, the sea servitude of squid fishing crews, and the repurposing of Saami herders' Indigenous knowledge for reindeer eradication on South Georgia.

Yet, while they were certainly not dispossessed in a colonial encounter and still have the protection of a UK military base, the islanders have endured their own periods of colonial subjection and foreign occupation. From the imperial perspective of UK administrators, the islanders have been considered kith and kin as well as "indigenous" kelpers. Most share British ancestry, and almost all consent to remaining a British Overseas Territory, but their emergent assertion of autochthony has become a way to claim ownership over resources.[7] In fact, if we follow Métis feminist anthropologist Zoe Todd in thinking of oil as "fossil kin," it is remarkable to consider how this industry is ironically built around excavating and burning the carbon of decomposed kelp forests—the very ancestors that the islanders have already identified with so profoundly as "kelpers."[8] From this critical Indigenous perspective, if oil is produced in the South Atlantic, then self-determination may entail self-destruction.

One way to parse the claims at hand would be to acknowledge elements of the islanders' belonging as a "local people" without Indigenous status, because they were not the original inhabitants and are not a sovereign tribal nation.[9] However, the archaeological discovery of the presence of Indigenous people in the islands before European colonization may unsettle such claims to local belonging. In an article published in the journal *Science Advances*, paleoecologist Kit Hamley and colleagues suggest that this Indigenous presence was most likely that of Yagán people, and they suggest that future research on cultural resource sites in the Falklands "should be done in collaboration with regional Indigenous communities, whose oral histories and traditional ecological knowledge represent millennia of expertise in the prehistory of Tierra del Fuego and the South Atlantic."[10] Presumably this would involve recruiting Yagán participants from Tierra del Fuego for archaeological research, but what might a sincere intercultural exchange also mean for sovereignty claims and natural resources in the Falklands?

The Yagán Indigenous struggle did not end with O'rundel'lico. In fact, the vice president of Chile's 2021–2022 constitutional convention was Yagán leader Lidia González Calderón, the daughter of Cristina Calderón, who was the last known fluent speaker of the Yagán language before her death in 2022.[11] In May 2019, Cristina Calderón led a group of Yagán protestors in a confrontation with the king and queen of Norway over the controversial development of salmon farms in the Beagle Channel.[12] What if a similar territorial claim were applied to self-determination, commercial fishing, or oil drilling in the Falklands? Taking obligations for ensuring Indigenous rights seriously may require a more radical reckoning with settler colonialism.

Scholars have conceptualized settler colonialism as a structure, not an event.[13] In other words, it is a complex social formation governing the seizure of land by means of exclusion or elimination, which lasts through time and is restructured by Indigenous refusal.[14] Most frontier zones like the South Atlantic, where authority is unstable, have been conquered or narrowed. Settler peoplehood in a British Overseas Territory without a historical encounter between colonizers and a precolonial Indigenous population in the islands may be thought of as a different sort of dynamic relation. The perpetual frontier is not necessarily a past occurrence, rigid structure, or linear process, but rather a suspended state of colonial nostalgia that makes imperial power present in a new constellation.[15] The Falkland Islanders have conjured past colonial utopias through heritage practices, speech habits, and naming practices, which we have explored through stories of the *Allen Gardiner*, gaucho corrals, settler colonial diglossia, and Port Smyley. Yet preparations for oil have repositioned the islanders in a new global network of corporations, consultants, and experts.[16] This reconfiguration of the frontier brings complementary aspirations but also incompatible aims to the islanders' long-term wishes and interests.

In this context, if settler colonialism is a structure, not an event, then settler indigeneity may be understood as an event, rather than a structure. The current layering of an oil frontier on a disputed imperial territory not only builds foundations for new forms of corporate personhood, it also has a double effect of keeping the seemingly closed structure of the settler colony open. Temporary infrastructure investments, changing engineering designs, and check-box environmental impact assessments may be common for companies staking out acreage on resource frontiers.[17] Yet the prospects of new economic booms in the South Atlantic heighten geopolitical division, reminding us that the settler colonial protectorate's task of salvaging empire is never complete.

Salvaged or Stranded?

Speculating on potential resolutions to the sovereignty controversy can be as unpredictable as investing in wildcat drilling. The dispute over land, maritime territory, and natural resources in the South Atlantic takes on new contours of contestation on a regular basis. The 2015 election of Mauricio Macri as president of Argentina offered an unexpected window of opportunity for mitigating and potentially resolving the sovereignty dispute, or collaborating on shared governance over commercial fishing and oil. Unlike his predecessors, the neoliberal Macri promised less contentious relations with the UK over the Malvinas, and the FIG sent him congratulations in the hope that they may be able to build peaceful dialogue.[18] The dispute made little movement toward resolution, however, because while the Argentine Foreign Ministry has insisted on bilateral negotiations with Britain, the UK has refused to discuss the issue without the presence of Falkland Islands representatives. In 2016, the UN Commission on Limits of the Continental Shelf ruled in favor of extending the Argentine continental shelf, which Macri's administration interpreted as an affirmation of its claim over the subsoil and mineral resources of the South Atlantic. With a permanent seat in the UN Security Council and powerful military forces installed in the islands, the UK dismissed this ruling as nonbinding. Nonetheless, the return to power of left-Peronist Cristina Fernández de Kirchner as vice president with Macri's successor President Alberto Fernández breathed new life into the Malvinas Secretariat's mission to restore territorial integrity in the South Atlantic.

British citizens voted to leave the European Union in 2016, yet "Brexit" has also made colonial holdouts like the Falklands more vulnerable.[19] Leading up to Brexit, licenses for commercial fishing contributed to a 43 percent rise in GDP from 2007–2016, comprising two-thirds of the FIG's corporation tax receipts and one-third of government revenues in 2016.[20] This has been possible because up to 89 percent of exports entered Europe through Vigo, Spain, where Falklands-based companies enjoyed tariff- and quota-free access.[21] Thus, because the UK's deal with the EU excludes exports from the Falklands, it makes the overseas territory subject to World Trade Organization tariffs of 6–18 percent, which put the lucrative calamari export industry at risk of losing up to 16 percent of fishing industry revenues.[22] Brexit also has significant repercussions for the Falklands' agricultural exports of meat and wool, as well as funding for biodiversity conservation science and environmental governance in the South Atlantic. To ensure profitability and sustainability, the islanders might be inclined to build better neighborly relations in the region, but the Malvinas Secretariat has viewed the exclusion of the islands from the Brexit deal as a signal of increased European support for Argentina's sovereignty claim.[23]

With fishing's future in question due to Brexit, islanders have rested hopes for salvaging empire on the prospect of offshore oil in the South Atlantic. However, after all the exploration drilling, financial de-risking, infrastructure planning, and environmental impact assessment, the Sea Lion project may become a stranded asset. After the construction of the Noble Frontier temporary dock facility in Stanley Harbour, Noble Energy abandoned its interests in the South Atlantic. Noble was eventually bought by the supermajor Chevron for $5 billion in response to the plummet in demand for oil after the outbreak of the COVID-19 pandemic.[24] Similarly, after Israel's Navitas Petroleum had agreed to farm into 30 percent interest in Sea Lion in 2019, the operator Premier Oil decided to suspend the first phase of production amid the global glut of oil supply in 2020.[25] Under the control of private equity firm EIG Global Energy Partners, Premier became part of Harbour Energy and merged with Chrysaor, making it the top oil and gas operator based in the UK.[26] However, Harbour backed out of the Sea Lion project when the Argentine government announced sanctions in 2021.[27] This has left the small exploration firm Rockhopper with by far the largest share of oil in the Falklands.[28] These delayed decisions, mergers, and acquisitions signal the possibility that oil may never leave the seafloor of the South Atlantic surrounding the islands. Oil prospects seem especially dubious in the context of the global energy transition needed to mitigate climate change through national and international commitments to decarbonization. Both the British and Argentine governments have declared a climate emergency, and on December 11, 2020, Brexiteer prime minister Boris Johnson announced that the UK would no longer subsidize fossil fuel production overseas.[29] This may impact UK-based supply chains like the one designed to bring oil from the Sea Lion well to market.

Argentina's Malvinas Secretariat has repeatedly referred to the British Overseas Territory as an "anachronistic colonial situation."[30] Falkland Islanders have rejected this characterization, asserting that they have a "modern" relationship with the UK, which respects their wishes for self-determination.[31] Nonetheless, the emerging challenges presented by Brexit and climate change test the endurance of settler indigeneity as a long-term strategy of salvaging empire through development of an oil frontier. The authoritarian populism that Thatcher originally crafted to defend the Falkland Islanders in the 1982 war with Argentina has now encouraged the UK to leave the EU. In an ironic twist of fate, this latest instance of British imperialist nostalgia may have also left the Falkland Islanders and their assets stranded in the South Atlantic.

Notes

INTRODUCTION

1. This book attempts to remain true to the ethnographic or historical context for top-onymic conventions. While I do use both names for the islands in ambiguous contexts, slashes interrupt the flow of reading. Labels for the islands and individual settlements should not be misconstrued as support for a particular cause in the sovereignty dispute.

2. The FIG is a group of locally elected representatives and British contractors legislating on all matters other than defense and foreign affairs.

3. On assertions of belonging, indigeneity, creolization, and autochthony, see inter alia Dominy 2000; Diaz 2006; Geschiere 2009; Sturm 2011; Zenker 2011; Clifford 2013; TallBear 2013; Gressier 2015.

4. For analysis of how this image was used in a government pamphlet for the purpose of diplomacy, as well as its significance for performances of memory, see Benwell and Pinkerton 2022.

5. South Georgia is a former Falkland Islands Dependency, now a separate British Overseas Territory (BOT).

6. Weszkalnys 2008, 2016.

7. "Sea Lion," *Rockhopper Exploration PLC* (blog), accessed June 21, 2022, https://rockhopperexploration.co.uk/operations/falkland-islands-2/north-falkland-basin/sea-lion/. See also Livingstone 2022.

8. Blair 2017. The "totem pole" has a central role in how Falkland Islanders tell stories about their heritage (StoryFutures Creative Cluster 2020). To be consistent with international legal norms, I capitalize Indigenous or Indigeneity when referring to unambiguously Native peoples but not disputed settler assertions of belonging.

9. Clancy 2021.

10. Smith 1973. See also "The Shipwreck of *Lady Elizabeth*," *Atlas Obscura*, http://www.atlasobscura.com/places/the-shipwreck-of-lady-elizabeth.

11. The UK government preferred to call the violent events of 1982 a "conflict" to avoid losing allies in a formal declaration of war, but I use the terms interchangeably. See Moore 2015, 679.

12. *Mercopress* 2014c.

13. United Nations 2022.

14. Besides former colonies that gained independence, the continuity of British support for the Falklands during the era of decolonization stands in contrast to abandoned Crown dependencies or territories, particularly Diego Garcia, where the Chagos Islanders were displaced for a US military base. See Madeley 1982; Jeffery 2011; Vine 2011; Gott 2022.

15. Wolfe 2006.

16. Connell and Aldrich 2020.

17. As ethnographic studies of democracy have shown, the temporal frame of a vote throws conceptions of civic duties and rights into sharp relief. See Paley 2001; Coles 2007; Banerjee 2008.

18. The Referendum International Observation Mission (RIOM)—Misión Internacional de Observación del Referendo (MIOR)—was funded and organized by a Canadian

organization called CANADEM, but the head and deputy head were both US Americans with ties to the Republican party. One of the outside observers, Emeritus Professor of Global Politics Peter Willetts, gave a controversial talk before the voting day at Stanley's Chamber of Commerce in which he argued that the referendum was a moral assertion rather than a legal right of self-determination. Giving advice to voters in a public forum made Willetts ineligible to serve the role of observer. Even though he agreed to being disqualified, the Argentine government construed his removal as an act of censorship.

19. *La Nación* 2012. For her own coverage of the referendum, see Sarlo 2013a, 2013b, 2013c, 2013d, 2013e.

20. Dodds 2013.

21. Hall 1980, 1985, 2016. See also Hart 2019; Jessop 2019; Davey and Koch 2021.

22. See also Koram 2022; Silverwood and Berry 2022.

23. Swyngedouw 2019; Edelman 2020.

24. Blair 2016.

25. Hall 2016.

26. Rosaldo 1989; Mitchell 2021. In disagreement with Thatcher, Jeane Kirkpatrick, US ambassador to the UN under Ronald Reagan, considered the Argentine dictatorship to be a tolerable authoritarian state to maintain alliances against communism in Latin America (Moore 2015, 684).

27. Guber 2001; Palermo 2014; Lorenz and Vezub 2022.

28. Guber 2001; Verbitsky 2002; Moro 2005; Rozitchner 2005; Lorenz 2009; Palermo 2014.

29. Scholarly literature on the Falklands/Malvinas focuses primarily on the 1982 armed conflict or the ongoing sovereignty controversy. These are important concerns for global politics and international law. However, relatively little has been written about the settler society of the Falkland Islanders. Journalists, naturalists, insider hobbyists and independent scholars have produced a number of historical sketches and travelogues. While these accounts provide useful insights on cultural and ecological aspects of the lifeways of Falkland Islanders, by excluding the sovereignty controversy from their work, they tend to reproduce imperialist nostalgia by essentializing the Islanders as "a people." See Boyson 1924; Cawkell, Maling, and Cawkell 1960; Gleyzer 1966; Taylor 1971; Strange 1972; Cawkell 1983, 2001; Foulkes 1987; Miller 1988; Wigglesworth 1992; Bound 2012; Smith 2013; Lorenz 2014; Niebieskikwiat 2014; MacFarquhar 2020. Two previous US social scientists attempted to study cultural dynamics of the disputed islands through field research; unfortunately, their doctoral dissertations remain unpublished. See Melchionne 1985, which is in press in Spanish, and Bernhardson 1989, whose examination of historical and environmental changes in the islands served as a valuable reference.

30. The Falklands had also been the scene of a violent battle during World War I (Pascoe 2014).

31. Bound 2006, 143.

32. The exact number, particularly of Argentine deaths, is under dispute.

33. In spite of its brevity, the war has inspired countless battle histories, diaries, memoirs, song tributes, video games, and other cultural products. In addition to official versions of the war, Falkland Islanders have given captivating insider accounts of poise, mischief, and sabotage under occupation. Authoritative accounts, though disputed, include Freedman and Gamba-Stonehouse 1990; Freedman 2004; Hastings and Jenkins 2012. For Falkland Islander–authored memoirs, see J. Smith 1984; Bound 2006; Betts 2012; Fowler 2012. Works of fiction, creative nonfiction, art, music, and film have also offered imaginative reflections. See Kamin 1984; Bauer 2005; Herrscher 2010; Daversa 2012; Fogwill 2012; Lorenz 2012; Bonomi 2014; Gamerro 2014; Carassai 2021;

Guber 2022. Scholars have also begun to examine rituals of memory and everyday nationalism in homes and classrooms (Benwell and Dodds 2011; Panizo 2015; Benwell 2016b; Benwell, Gasel, and Núñez 2019). For further analysis of declassified documents regarding the British government's relations with dictatorships of Argentina and Chile see Livingstone 2018. For reflections on the fortieth anniversary of the war, see the special issue of the *Journal of War & Culture* edited by Woodward et al. 2022.

34. Borges, quoted in *Time* magazine, February 14, 1983.

35. Gilroy 1987; Dodds 2002, 2–7, 118–41; Ho 2004; Mohanram 2007; Olusoga 2023. This book contributes to a growing body of scholarship that has begun to rethink the Falklands, and particularly the 1982 war, through imperial history (Mercau 2019).

36. Thompson 1982; Hobsbawm 1983a; Gilroy 1987, 51–55; Hall 1988; Jackson and Saunders 2012; Milne 2014.

37. Argentina was informally subject to British imperial rule through financing of its national infrastructure in the late nineteenth and early twentieth centuries. See Ferns 1960; Rock 1975; Wright 1975.

38. Kon 1982; Lorenz 2006; Guber 2009; Niebieskikwiat 2012; Chao and Guber 2021; Guber et al. 2022; Natale 2022.

39. On connections between Peronism and populism, see Laclau 1977; Finchelstein 2017. On cultural and political particularities of Peronism not reducible to populism, see James 1988; Torre 1990; Auyero 2000; Levitsky 2003; Karush and Chamosa 2010; Elena 2011; Healey 2011; Halperín Donghi 2012; Milanesio 2013; Lazar 2017; Semán 2017. While some see Peronism as the original ideology of populism, others trace populism (and antipopulism) back earlier to the late-nineteenth-century US (Frank 2020). On the economic crisis in Argentina and responses from social movements, see Colectivo Situaciones 2011; Sitrin 2012; Muir 2021.

40. See also Benwell 2017.

41. Falkland Islands Government n.d.

42. Samuel Johnson (1771) was one of the early theorists of sovereignty in the South Atlantic.

43. In international law, self-determination serves both as a criterion for expressing "juridical statehood" (Jackson 1993) for constituencies with "primordial attachments" to places (Geertz 1963b) and a tool for recognizing "lived sovereignty" (Agamben 1998; Hansen and Stepputat 2005). Given the concept's flexibility, broad tendencies in scholarly literature and postcolonial thinkers swing between championing anticolonial nations' uses of its Wilsonian notion of "nonintervention" to support new state independence (Manela 2009) and critiquing its imperial uses for maintaining dependency through integration (Anghie 2004; Getachew 2019). Social scientific research on Indigenous politics describes a symbolic use of self-determination with a "relational" meaning of dignity, which has inspired new transnational movements. See Kingsbury 2000; Tully 2000; Muehlebach 2003; Niezen 2003; Barker 2006; Allison III 2015. On the tension between self-determination and territorial integrity, see Blay 1986; Gustafson 1988; Szasz 1999; Elden 2006.

44. Even though this resolution was not binding—the UK argues that it became moot when the Argentine military junta seized the islands by force in 1982—all subsequent resolutions in favor of the Argentine position can be traced back to this one. For a critique of Argentina's argument supporting this resolution, see Pascoe and Pepper 2012. See Gustafson 1984, 2015 for analysis of the different interpretations of the "self" in the applicability of self-determination since this resolution. See Fernández 2014 for a collection of resolutions passed in favor of Argentina.

45. Freedman and Gamba-Stonehouse 1990, 8, 122, 206, 278, 291, 305, 413. For the Falklands lobby, see Ellerby 1990; Dodds 2002, 118–40.

46. Thatcher (1983) 2015, 24.

47. Moore 2015, 695. Nonetheless, when British forces overtook Argentina's military occupation, the term "unconditional" was deleted from the statement of surrender, and the subsequent UN resolution used the word "interests" in lieu of "wishes" (Freedman and Gamba-Stonehouse 1990, 410, 413).

48. Braslavsky 2013; Bustamante 2013; Pastorino 2013.

49. During his first few months in office, President Mauricio Macri all but eliminated this branch of the Foreign Ministry (Política Argentina 2016).

50. Members of the Malvinas Secretariat told me they would not visit the islands because they felt that getting their passports stamped by the FIG would amount to de facto recognition of British jurisdiction. The Foreign Ministry has, however, recruited defected Islanders as spokespersons for the Argentine cause. The divergent paths of two defected Islanders in Argentina, whose stories have been well publicized, are instructive. See Betts 1987; Peck 2013, 2015. On the Foreign Ministry's position, see Pastorino 2013.

51. Ivanov 2003.

52. While no regional trade embargo truly limited the free flow of goods with Chile, Uruguay, or Brazil during my fieldwork, Falklands-flagged vessels are banned from stretches of the Argentine coastline. The possibility that container vessels may be impounded affects many transactions in the islands.

53. Interview with Falkland Islander, March 2, 2013.

54. Methodological nationalism is defined as "the assumption that the nation/state/society is the natural social and political form of the modern world" (Wimmer and Glick Schiller 2002, 302).

55. On resource nationalism, see Koch and Perreault 2019.

56. This framing builds on a broader field in history of the present, stemming from poststructuralism as well as world systems theory and colonial studies (Braudel 1982; Sewell 2005; Fassin 2011; Garland 2014; Stoler 2016). In this case, it draws on emic descriptions of the Falklands' arguably anachronistic geopolitical situation. While Argentina's government calls the islands an "anachronistic colonial situation," the MLAs suggest that the UN C24 is, itself, an "anachronism" for privileging the interests of member states over the people of Overseas Territories. See Blair 2013b.

57. Hancock 1997; Bailyn 2005; Elliot 2007; Greene and Morgan 2008; Armitage and Braddick 2009; Belich 2009; Benton 2009; Bowen and Reid 2012. While Mentz (2015) stretches globalization earlier in oceanic history, also by centering global ecology on shipwrecks (1550–1719), this book takes a complementary approach to extend it further (1770–present).

58. Smith 2004.

59. Tomich 2020; Scanlan 2022. NATO also continues to govern based on a twentieth-century Atlantic imaginary (Trouillot 2002).

60. Williams (1944) 1994; Holt 1991; Gilroy 1993; Maurer 2000.

61. Stoler 2013, 2016; Gordillo 2014.

62. On the role of plunder in imperialism, see Mattei and Nader 2008; Colwell 2019; Docherty 2021; Hicks 2021; Philips 2022.

63. Smith 1984; Cronon 1991.

64. Blumenberg 1996.

65. Smith 1973; McKee 1985; Miller 1995; Bound 1998a, 1998b; Pope 2007.

66. Tsing 2015a, 2015b, 63. On alternate visions of "salvage commoning," see Linebaugh and Rediker 2000; Oka 2017; Allinson et al. 2021; Anderson and Huron 2021.

67. For critiques of Tsing's broader conceptualization of the "Plantationocene," see Davis et al. 2019; Jegathesan 2021.

68. Chapman 2010, 443. There are different naming conventions for Indigenous peoples in Patagonia and Tierra del Fuego used in Argentina and Chile and among community members. I have tried to use names preferred by respective Indigenous groups, when known. See Chapman 2010, xxi; Ogden 2021, 2.

69. Linebaugh and Rediker 2000, 111.

70. Stoler 2013, 2016; Yarrow 2017; Ureta 2020; Du Plessis et al. 2022.

71. See also Giminiani 2018 on powerful, embodied connections between the politics of Indigeneity and sentient aspects of land and property.

72. Benjamin 1968, 257–58; Glissant 1997, 33, 174; Löwy 2016.

73. Stoler 2006.

74. Stoler 2016, 177.

75. Anghie 2004.

76. Getachew 2019. These aspects derived from the Anglophone Atlantic World are distinctive from that of the French former colonies reconstituted as federal nation-states. See Cooper 1996, 2014; Wilder 2015.

77. For general theories of imperialism, see Hobson 1902; Lenin (1917) 1939; Schumpeter (1919) 1951; Du Bois (1946) 1965; Arendt (1951) 1968; Padmore (1936) 1969; Bukharin (1929) 1973; Marx (1867) 1976; Arrighi 1978; Rodney (1972) 1981; Nkrumah (1965) 1987; Hobsbawm (1968) 1990; McClintock 1995; Hardt and Negri 2001; Harvey 2005; Fanon (1963) 2007; Luxemburg (1913) 2015. For the history of British Imperialism, see Cain and Hopkins 2002; Newsinger 2013; Gopal 2020; Sathnam 2021; Elkins 2022; Gott 2022; Koram 2022. For histories of empire from below, see Linebaugh and Rediker 2000; Grandin 2006; Davis 2017. See Lutz 2008; McGranahan and Collins 2018 on ethnography of empire.

78. Following concerns from Native and Black studies expressed by King 2019, who draws on Indigenous scholar Joanne Barker's blog (website no longer accessible), this book seeks to reconnect the focus on white settlers—not just with their relationship to land but also with their "parasitic and genocidal relationship to Indigenous and Black peoples" (King 2019, 68). See also Kelley 2017; Cordis 2019.

79. Povinelli 2002; Wolfe 2006; Cattelino 2008. See Byrd 2011; Coulthard 2014.

80. Kauanui and Wolfe 2012; A. Simpson 2014.

81. Simpson 2016, 2017. See also Nichols 2019.

82. Mamdani 2001, 2020.

83. Pineda 2006; Bushnell 2008, 198.

84. Boas 1989; Baker 1998; Blackhawk and Wilner 2018; Anderson 2019; King 2019.

85. Boas (1911) 1938, 132. Boas's use of the possessive for the Indigenous individual associated with Darwin is symptomatic of a larger pattern of settler colonial grammar that tragically reproduces Indigenous dispossession through salvage ethnography (Simpson 2018).

86. See Chapman 2010, 141–42, for a thorough analysis of this event, as well as many other colonial encounters between Europeans and Yagán (Yámana) people in Tierra del Fuego. See also Hazlewood 2001.

87. Simpson 2020; Cooper 2022.

88. Social scientists interested in relations between societies and their environment have long analyzed "adaptive structures" for maintaining balance in fragile settings. Anthropologists took up "cultural ecology" to study alternative irrigational and agricultural institutions based on local knowledge. See Steward 1955; Geertz 1963a; Rappaport 1968. This book draws on interdisciplinary work related more specifically to "political ecology," which in its early years took ideas from political economy and used them to understand land degradation or famine as historically produced in underdeveloped areas. Wolf 1972; Watts 1983; See Blaikie 1985; Blaikie and Brookfield 1987. On the

connection between political ecology and STS, see Forsyth 2002; Goldman, Nadasdy, and Turner 2011.

89. On the convergence of authoritarian populism and environmental governance, see McCarthy 2019; Tilzey 2019. And on environmental politics, knowledge production, and extractivism in South America, see Bunker 1990; Lins Ribeiro 1994; Coronil 1997; Swyngedouw 2004; Tinker Salas 2009; Sawyer and Gomez 2012; Svampa 2013; Tironi and Barandiarán 2014; Bustos-Gallardo, Prieto, and Barton 2015; Henne 2015; Li 2015; Acosta 2017; Gómez-Barris 2017; Riofrancos 2017; Barandiarán 2018; Giminiani 2018; Gudynas 2018; Mendoza 2018; Renfrew 2018; Vindal Ødegaard and Rivera Andía 2018; Babidge et al. 2019; Enns, Bersaglio, and Sneyd 2019; Folch 2019; Gustafson 2020; Jerez, Garcés, and Torres 2021; Ogden 2021; Blair and Balcázar 2022.

90. See Auty 1993; Karl 1997; Humphreys, Sachs, and Stiglitz 2007; Ross 2012.

91. Coronil 1997; Onley 2004; Watts 2004; Ferguson 2005; Bridge 2008; Reed 2009; Zalik 2009; Mitchell 2011; Appel 2012b, 2019; Hecht 2012; Bond 2013; Huber 2013; Muzio 2015; Rogers 2015a; Leonard 2016; Malm 2016; Wylie 2018.

92. Mitchell 2011. See also Nader 2010; Nuttall 2010; Smith and Frehner 2010; Boyer 2011; Rogers 2012; Strauss, Rupp, and Love 2013; Weszkalnys and Richardson 2014.

93. Rogers 2014, 2015b; Appel 2019.

94. Onley 2004; Watts 2006; Hecht 2012.

95. Peet and Watts 1996; Watts 1987.

96. Scientific facts circulate as powerful cultural norms among relational networks of human and nonhuman actors, objects, and institutions. See Nader 1981; Callon 1986; Latour and Woolgar 1986; Latour 1987; Fujimura 1992; Knorr-Cetina 1999; Law and Mol 2002; Jasanoff 2006.

97. See Haraway 1988; Mitchell 2002; Latour 2004.

98. Barry 2006, 2013.

99. Liboiron 2021, 132. See also O'Brien 1993.

100. On green extractivism and colonialism, see Dunlap and Jakobsen 2019; Riofrancos 2019; Jerez, Garcés, and Torres 2021; Voskoboynik and Andreucci 2021; Blair and Balcázar 2022. On critical perspectives on conservation, see West 2006; Brockington 2009; Büscher, Dressler, and Fletcher 2014; Martínez-Reyes 2016.

101. Crosby 1986; Anker 2002.

102. Malinowski 1922; Radcliffe-Brown 1922.

103. Asad 1973; Wolf 1982; Trouillot 1991; Smith 1992; Swyngedouw 1997; Tsing 2005.

104. Nader 1972; Harding 1991; Hughes 2017; Carey 2019.

105. Blair 2013a, 2013d.

106. Blair 2013b, 2013c.

107. Blair 2019b.

108. On the ethics of social scientific field research on natural resource industries, see Johnson et al. 2020.

109. Rediker 2010.

1. SETTLER SAFE ZONE OR COLONIAL STAGING GROUND?

1. Paleoecological analysis of carbon-dated charcoal raised the possibility of precolonial Indigenous human activity in the Falklands (Buckland and Edwards 1998). Archaeologists have now also found middens—piles of discarded sea lion bones—and fossil evidence that shows more definitively that Indigenous people inhabited the islands before Europeans (Hamley 2016). Hamley et al. (2021) argue that the contemporaneous connection between middens and burning demonstrates that humans were present on the Falklands' New Island at 550 B.P. (before present).

2. Anon. 1832, "A Visit to the Falkland Islands," *The United Service Journal and Naval and Military Magazine.*

3. On theories of dispossession, as well as the related Marxian concept of primitive accumulation, see Marx 1976; Perelman 2000; Federici 2004; Harvey 2005; Coulthard 2014; Luxemburg (1913) 2015.

4. Legal scholar Julius Goebel's (1982) study is a foundational analysis of the discovery claims and eighteenth-century settlements of the islands. Laurio Destefani's (1982) work is also highly regarded, but it is perhaps more partial to the narrative that supports the Argentine sovereignty cause; Gough (1992) provides a complementary analysis from British Commonwealth sources. Falkland Islanders and their British trustees defer to V. F. Boyson's (1924) study, which is faithful to local knowledge and difficult to find outside of Stanley. The work of Mary Cawkell (Cawkell, Maling, and Cawkell, 1960; Cawkell 1983, 2001) derives largely from Boyson's work (see also UK FCO-sponsored literature, such as Levick 1982). The Argentine Foreign Ministry (Fernández et al. 2014) and its supporters have produced their own narrative, deriving from the government-funded research of Paul Groussac (1934). See Muñoz Aspiri 1966. Meanwhile, in debates with international legal scholars supporting Argentina's claim (Kohen 2013, 2014), independent researchers Graham Pascoe and Peter Pepper (2012) have sought to develop an exhaustive account sympathetic to the UK.

5. "Right" in "Desire the Right" also carried the connotation of whale oil in the early twentieth century, as in the species southern right whale (*Eubalaena australis*).

6. A series of Portuguese and/or Spanish sailors may have "discovered" the islands between the expeditions of Magellan and Davis, particularly the bishop of Plasencia on the vessel referred to as the *Incognita* in 1540. See Destefani 1982, 37–46; Goebel 1982, 16–34. It is unlikely that Amerigo Vespucci sailed to the islands on his controversial voyage of 1501–1502.

7. Martinic 2002; Chapman 2010, 33. On the differences between and among Fuegian Indigenous peoples, see Chapman 2010, 116.

8. The French may have arrived earlier. A detailed map demonstrates that Amedée François Frézier landed there in 1711.

9. On the French outpost, see Bougainville 1772; Pernety (1771) 2013.

10. Bernhardson 1989, 87–99; Pernety (1771) 2013.

11. Goebel 1982, 231–40.

12. Penrose 1775; Clayton 1981.

13. Defenders of British sovereignty argue that even if occupation was not continuous, the claim never expired because, upon departing, Commanding Officer S. W. Clayton left a plaque asserting that the islands are property of King George III.

14. Bernazani, "Relación que manifiesta al Estado de las tripulaciones y guarniciones de las fragatas . . . Liebre y Esmeralda . . . , February 24, 1767," AGN IX, 16-9-1, cited in Bernhardson 1989, 113–14. Bernhardson (1989, 124) also notes that the Spanish had proposed sending a "multitude of free married and single blacks" from Buenos Aires to populate the islands. For further details on the Spanish governors of the Malvinas, see Destefani 1982, 53–72.

15. "Nota del Sr. Pedro de Mesa y Castro al Ecmo. Sr. Marqués de Loreto referente al recuento y entrega de ganado en las Islas Malvinas, 24 marzo 1787," AGN VII, 127, doc. 5, 2-3-3. "Inventario de las materías y edificios en Malvinas, 1 marzo 1792," AGN VII, 127, doc. 11, 2-3-3.

16. Ruiz Puente, "Untitled Memorandum, May 1, 1768," AGN IX, 16-9-2; Bernhardson 1989, 115. On the modern history of the establishment of an early-twentieth-century prison, integrated with development of a local forestry industry in Ushuaia, Tierra del Fuego, see Edwards 2021.

17. Domingo de Recaunte, "Relación que manifiesta los individuos que quedan exis-tentes en el día de la fecha en este establecimiento, February 27, 1782," AGN IX, 16-9-5, cited in Bernhardson 1989, 129.

18. Figueira 1985. Vernet, "Doc. 19, 1820, Lista de algunos oficiales y tropa a bordo de la nave 'Heroína' cuando tomó posesión de las Malvinas," AGN VII, 127, 2-3-3.

19. Vernet, "Doc. 22 25 agosto 1823. Concesión del usufructo de las Islas Malvinas a favor de Jorge Pacheco," AGN, Buenos Aires, Sala VII, legajo 127 (2-3-3; 1768–1908).

20. Vernet, "Doc. 23 18 diciembre 1823. Concesión de terrenos al Sr. Jorge Pacheco," AGN, Buenos Aires, Sala VII, legajo 127 (2-3-3; 1768–1908).

21. "Doc. 33 [20 Mayo 1845]. Copias de las colecciones de documentos referentes al coste de la colonia fundada en el Puerto de San Luis." See also "Doc. 56 31 diciembre 1825. Copia de la concesión de Jorge Pacheco para la explotación de ganado en la Isla Soledad." A11 note from 23 de diciembre de 1823 from Pablo Areguati, AGN, Buenos Aires, Sala VII, legajo 127 (2-3-3; 1768–1908).

22. Pablo Areguati, of Guaraní descent, assumed this role during the initial coloni-zation campaign in 1823–1824.

23. "Doc. 55 noviembre 1825. Contrata celebrada con el personal que fue a Malvinas en el bergantín inglés 'Alert' para las faenas de hacienda, includes Contrata General a peones y Capatanes ante la Policia 1825," AGN, Buenos Aires, Sala VII, legajo 129 (2-3-4; 1670–1831).

24. Vernet, "Doc. 33 Copias de las colecciones de documentos referentes al coste de la colonia fundada en el Puerto de San Luis," AGN, Buenos Aires, Sala VII, legajo 127 (2-3-3; 1768–1908), files C1 and C3. For a different perspective, see Bernhardson 1989. Here, the author draws on clauses in contracts, which give the appearance that the slaves entered into work in the Malvinas freely as indentured servants (AGN series 2-3-6). This is inconsistent with most contemporary accounts. In Vernet's own documents, these in-dividuals are referred to in racialized terms as "*negros y negras*" (Vernet, "Doc. 33," AGN 2-3-3, files C1, C3, C12). In letters written to Vernet in English, they are referred to as "your negroes," implying ownership (Henry Metcalf, "Doc. 33," AGN 2-3-3, file F1). One of Vernet's documents even gives a price: "*un negro . . . 60 pesos*" (Vernet, "Doc. 33," AGN 2-3-3, file C11). Vernet's accounts also list them specifically as "*negros escla-vos*" (Vernet, "Doc. 33," AGN 2-3-3, file C4). See also Anon. 1832.

25. Delaney 1996; De la Fuente 2000.

26. See, for example, extracts from the diary of settler Thomas Helsby (Academia Na-cional de la Historia 1967, 55). See also Gallo 1983; Freidenberg 2009.

27. A passenger in the visiting Thomas Laurie ship described the gauchos of Vernet's colony as gambling tricksters with short tempers who do little or nothing. The passen-ger wrote, "No greater proof of the laziness of the men generally, need to be adduced, than the following—very good potatoes are grown by Don Vernet, and sooner than raise them themselves (though offered them by him for seed gratis) they pay him ten pence per pound for them" ("A Visit to the Falkland Islands," 1832, 309). For a more chari-table description, see the diary of Vernet's wife, Maria Saez de Vernet (1829). For other cattle frontiers, see also Baretta and Markoff 1978; Slatta 1992; Rausch 1993; Chasteen 1995; Guy and Sheridan 1998; Sluyter 2012.

28. "Docs. 70 and 71, Vernet's letters to Woodbine Parish of 19 and 21 June 1829," AGN legajo 129, 2-3-4.

29. Vernet, "Doc. 32, 23 marzo 1831, Plan propuesto al gobierno por el comandante político y militar de las Islas Malvinas," AGN, 2-3-3.

30. A sample cargo for Vernet's colony lists, in addition to passage settlers and peons from Montevideo: crockery, spades, hinges, locks, ponchos, clothing, paper, medicines, seeds, trunk of ready-made clothes, groceries, hides, a watering machine, biscuits, beans,

window frames, nails, twine, coats, casks, shirts, stockings, "Monkey jackets," saddlecloth, pair of pantaloons, drawers, sugar, coffee, cask whitening, stamp paper, tent, shot, book, pipes, barrel whitening, horseslings, chain dog, timber dogs, boathooks, curtails, bottom drawers, guitar, earthen pots, sweet wine, vinegar, yerba, sweet oil, small articles, beef barrels, buckets, water funnel copper pump, pringborer, riverts, keg, halfpipes for lime, pipes for bran, chalk, water casks, flags, ham, ringbatt, Carpinters work on the Elbe, horses, hay, lumber for platform for horses, bran, biscuit, rope, green tar, arrabic [sic] coffee, arabic sugar, gin, fowls, ducks, small pigs, stamp for permit, empty bags, large iron rings, smaller ones, brushes for whitening, horse bits, "Indian corn" for horses, hide rope for halters for horses, pipe brandy, lime, coals, barrel flour, rice, rum, fresh beef, empty cask, peons vegetables, clothes, preserved fruit, tobacco for gauchos, large iron pot, small articles, large trunk, tin pots and spoons, horses, and many other services (Vernet, "Doc. 33 Copias de las colecciones de documents," D6, AGN, 2-3-3).

31. Saez de Vernet 1829; Anon. 1832.

32. Anon. 1832.

33. The *Lexington* and other American sealers who sacked the colony in the wake of its destruction not only flattened Vernet's home but also senselessly killed several horses. Vernet calculated the total balance of damaged owed at 207,728 ¼ Spanish dollars (Vernet, "Doc. 112 balance of cost or value of the colony remains after its destruction by the Corvette *Lexington* on 31 Dec. 1831," AGN, legajo 129, 2-3-4. On the role of Andrew Jackson in the violent frontier expansion of nativist settlers in the Americas, see Grandin 2019.

34. De la Fuente 2000.

35. Martiniano Leguizamón Pondal (1956) published a widely distributed book on Spanish place names in the islands, which embellished the role of Rivero as an Argentine patriot, building on the resurgence of interest in the cause for Malvinas among intellectuals José Hernández and Alfredo Lorenzo Palacios (Guber 1999a). The Condor Group of left-Peronist militants famously took up the myth when they hijacked and landed a plane in the islands in 1966. They renamed Stanley "Puerto Rivero," which the dictatorship interestingly changed to Puerto Argentino during the 1982 invasion. In spite of their nationalist cause, the military junta did not want to support the values of those connected ideologically to others persecuted in the Dirty War (Guber 2000; Lorenz 2014, 129–32). In 2012, the Kirchner administrations invoked the folk hero by passing a 2012 "Gaucho Rivero" bill that banned British-flagged ships or those calling in the Falklands from docking in the province of Buenos Aires and areas of Patagonia. In 2015, Cristina Fernández de Kirchner tried to immortalize the hero by putting a new fifty-peso note into circulation depicting Gaucho Rivero waving the Argentine flag on horseback.

36. Moderate to conservative Argentine historians suggest that the violent event was a crime stemming from a dispute over payment. See Fitte 1966, as well as work by Ricardo R. Caillet-Bois and Humberto F. Burzio, who published a translated archive of British documents relating to the Gaucho Rivero incident (Academia Nacional de la Historia 1967). Amateur historian Peter Pepper and colleague Graham Pascoe (Pascoe and Pepper 2012), whose work is unabashedly biased in favor of the British claim, draw on these sources to argue that the uprising had nothing to do with Argentine sovereignty. Left-nationalist Mario Tesler (1971, 2013) argues that these authors rely heavily on sources that detract from the sovereignty cause.

37. For the cause of the assassinations being attributed to the money form of payment, see the direct testimony of José María Luna, one of the accused gauchos, PRO, Admiralty 1/42 (Academia Nacional de la Historia 1967, 84). See also the statement of Antonio Rivero, himself, AGN, doc. 242, 2-3-6 (Academia Nacional de la Historia 1967, 97). This

theory has also been accused of being overly materialist (Tesler 1971, 322–23). However, it reflects trends of similarly disenfranchised peoples staging hidden revolutions (Linebaugh and Rediker 2000; Grandin 2014).

38. See, e.g., the map drawn by Luis Vernet's brother Emilio, "Doc. 5 24 marzo 1787. Nota del Sr. Pedro de Mesa y Castro al Ecmo. Sr. Marqués de Loreto," legajo 127, 2-3-3. See also "Doc. 33 Copias de las colecciones de documentos referentes al coste de la colonia fundada en el Puerto de San Luis" and "Doc. 112 balance of cost or value of the colony remains after its destruction by the Corvette *Lexington* on 31 Dec. 1831." See also Spruce and Smith 2018, 102.

39. Chapman 2010, 74–88.

40. Chapman 2010, 219.

41. The inscription follows both the British liberal and imperial opposition of "civilization" versus "barbarism" (Mehta 1999). Ironically, it also echoes Argentine statesman Domingo Faustino Sarmiento's racialization of gauchos specifically as barbarous, even though he sought to recover the Islands for the Argentine Republic (Sarmiento 2010).

42. Chapman 2010, 218.

43. Darwin 1997, 180.

44. Darwin, Kohn, and Montgomery 1985, 380, quoted in Chapman 2010, 224.

45. Darwin, FitzRoy, and King 2015, 328–32, an earlier edition of which is referenced in Chapman 2010, 225.

46. When the British came to detain the rebels, Rivero tried to strike a bargain to "capture [and hand over] the Indians" in exchange for a pardon. This bargain was unsuccessful, but Rivero's "Chilean Indian" (likely Mapuche) comrade Luna did provide testimony to the king, which Tesler (1971, 80) argues constituted a hidden act of betrayal. See "Extract from the Diary of Lieutenant Commander Henry Smith" (Academia Nacional de la Historia 1967, 69, 72). See also AGN Sala VII, 127, doc. 33; "Doc 46 14 noviembre 1836. Carta en inglés del sr. Graham e. Hamond al sr. Luis Vernet."

47. Vernet was quite candid about his claims being about personal commercial interests lost from his settlement, and not a question of sovereignty over the islands. He confessed to accepting the role of political and military commander from Buenos Aires with "aversion." See "Letter from Luis Vernet to Woodbine Parish, July 23, 1834, PRO, F.O., Argentina (6/501) Falkland Islands, Document pertaining to the archive of Diego Luis Molinari." For Spanish, see Tesler 1971, 381.

48. Despite the debt peonage system officially in place in the islands under Vernet, gauchos generally practiced a status-based common resource regime comparable to usufructuary rights over chattel, including cattle in other societies (Lattimore 1940; Gluckman 1965; Baretta and Markoff 1978; Edelman 1992).

49. Research Aid: People, 1832–1879, JCNA.

50. Tatham 2008, 471.

51. Tatham 2008, 471.

52. Edward Villiers and T. F. Elliot, "Letter to James Stephen, August 22, 1840, Correspondence Respecting the Colonization of the Falkland Islands," JCNA.

53. Captain Frankland, "Letter to King, September 17, 1841," PRO, CO, 78/4, cited in Bernhardson 1989, 210.

54. See Bernhardson 1989, 272.

55. Moody, "Despatch to Stanley, July 6, 1843," JCNA, B1. See also Rennie, "Despatches to Grey, May 4 and May 26, 1849," JCNA, B6 and Rennie, "Despatch to Grey, November 30, 1850," JCNA, B7.

56. For example, the Blue Book for 1848 lists "whites" (101 males, 44 females) and "coloured" (5 males, 4 females); "Aliens and Resident strangers" not included in preceding Columns: 83 (not categorized by gender).

57. See Bernhardson 1989, 291.

58. Lane, "Despatch No. 4, November 11, 1858," JCNA.

59. This was consistent with a broader trend in the region of racist eugenics implemented through immigration, among other social policies (Stepan 1991).

60. Lane, "Despatch No. 10, February 3, 1859," JCNA.

61. Mackenzie, "Despatch to Cardwell, November 16, 1864," B13, JCNA, quoted in Bernhardson 1989, 351. See also Spruce and Smith 2018, 92–93.

62. Trouillot 1997. See also Scott 2020.

63. Trouillot 1997, 49.

64. Bridges (1948) 2007.

65. See Gusinde 1961, 143; Taussig 1993, 87; Bridges 2007, 314–15; Chapman 2010; Marchante 2014. On "Indian hunting" elsewhere in South America, see Bessire 2014; Bjork-James 2015.

66. See Philpott 2009, 126–27. This detailed survey draws not only on extensive archaeological field research but also on historical analysis of periodicals produced by the Society, especially the *Voice of Pity* (1854–1862), later renamed *A Voice for South America* (1863–1866), and the *South American Missionary Magazine* (1867–).

67. Ogden 2021, 151.

68. Chapman 2010, 142.

69. Chapman 2010, 127.

70. Chapman 2010, 193.

71. Chapman 2010, 233.

72. Chapman 2010, 265–66.

73. Chapman 2010, 270.

74. Hyades and Deniker 1891, quoted in Chapman 2010, 287.

75. Chapman 2010, 296.

76. Chapman 2010, 308.

77. Marsh and Stirling 1867, 79.

78. *Voice of Pity*, Vol. 9, 1862, cited in Philpott 2009, 6–7.

79. Chapman 2010, 312.

80. Philpott 2009, 6–12. Subsequent colonial governors, such as Governor D'Arcy, gave more positive reports about the settlement, despite knowledge that a number of the displaced Natives had died from consumption (tuberculosis). (The death registry at JCNA also lists "visitations by God" that may not have been natural.) D'arcy wrote, "This little Arcadian settlement will probably be the means of eventually civilizing the remaining few barbarous inhabitants of Tierra del Fuego.... It cannot be questioned that if these good missionaries are to succeed in making a man a Christian out of a mere savage and wild man, they must first make him a rational being, and it must contribute a great deal to forward his conversion if he can by degrees be brought into a settled mode of life, and to this most desirable end the mission is practically working by the mode so wisely adopted" ("Colonial Report for 1871, Despatch from Governor D'Arcy, February 27, 1872").

81. Snow 1857, 363, cited in Philpott 2009, 9.

82. Chapman 2010, 330.

83. Snow 1857, Vol. 2, 35–36, cited in Chapman 2010, 328, 332.

84. *Voice of Pity*, Vol. 9, 1862, quoted in Philpott 2009, 26. For further speculation about whether to exterminate or proletarianize Indigenous peoples of Tierra del Fuego, see "The Aborigines of South America from a Missionary and Commercial Standpoint," *Falkland Islands Magazine*, September 1911–January 1912.

85. Chapman 2010, 337.

86. Chapman 2010, 358.

87. Chapman 2010, 340. On theories of refusal, see Simpson 2014, 2017; McGranahan 2016.

88. Chapman 2010, 345–46.

89. Chapman 2010, 345–46.

90. Despard, "Letters of the Rev. G. P. Despard," *Voice of Pity*, 6, 1859, cited in Philpott 2009, 19.

91. Chapman 2010, 352.

92. Grandin 2014; Horne 2018.

93. Chapman 2010, 364–65.

94. Chapman 2010, 369–73.

95. Chapman 2010, 377.

96. Chapman 2010, 377.

97. Chapman 2010, 380.

98. Chapman 2010, 383.

99. Chapman 2010, 383.

100. Chapman 2010, 309.

101. Smyley returned to recover the *Allen Gardiner* on a subsequent voyage. Governor Moore, "Despatch to Duke of Newcastle, May 8, 1860," JCNA, B/11.

102. "Enclosure 1: Evidence of Alfred Cole, Cook of the *Allen Gardiner,* March 10, 1860," JCNA, B/11. See also *Voice of Pity,* Vol. 7, 1860, quoted in Philpott 2009, 24.

103. "Enclosure 2: Evidence of Jemmy Button, Tierra del Fuegian, March 12, 1860," JCNA, B/11. See Chapman 2010, 389 for more evidence for this hypothesis.

104. Governor Moore, "Despatch to Duke of Newcastle, March 15, 1860," JCNA, B/11. See also Philpott 2009, 24.

105. Charles Bull, "Letter to Robinson, October 1, 1866," JCNA, H/24.

106. Chapman 2010, 403.

107. Chapman 2010, 349.

108. *Voice of South America*, Vol. 12, 1865, cited in Philpott 2009, 39–41. On the "racist anti-racism" of this approach to anthropology, see Simpson 2018; Baker 2021.

109. Chapman 2010, 421.

110. Philpott 2009, 44; Chapman 2010, 422. Later descendants, including Pinoense, a resident at Keppel, were also baptized with the name Allen Gardiner (Chapman 2010, 526).

111. Chapman 2010, 407.

112. Chapman 2010, 426.

113. *South American Missionary Magazine*, Vol. 6, 1872, quoted in Philpott 2009, 46. See also "1880 Colonial Surgeon Report" quoted in Spruce and Smith 2018, 224.

114. Marks 1991; Hazlewood 2001; Canclini 2009; Chapman 2010. The Keppel settlement also bears remarkable resemblance to a science fictional scenario in Ursula K. Le Guin's (1972) *The Word for World Is Forest*, in which the *Shackleton* transported a group of sixty-two colonized Creechies—whom the imperialist Terrans had dehumanized, enslaved, raped, infected with disease, and attempted to "exterminate" like rats for causing trouble to sheep and colonists—from their homeland of New Tahiti to the "New Falkland Isles."

115. Ogden 2021, 37.

116. For a well-researched, but less critical, analysis of the labor conditions at Keppel, see Bernhardson 1989, 354–56.

117. Linebaugh and Rediker 2000; Chapman 2010, 443.

118. O'rundel'lico passed away between 1863 and 1864 from one of several epidemics that devastated Indigenous people in Tierra del Fuego (Chapman 2010, 407).

119. In addition to the incident at Wulaia, there have been other instances of Indigenous resistance that defy the common notion that the Indigenous people of Tierra del Fuego were passive victims of disease and genocide. See Bridges 2007, 204, 314–15, 496.

120. Milun 2011.

121. *Falkland Islands Magazine*, September 1901.

122. Bernhardson 1989, 441. See also Mainwaring 1983; Foulkes 1987; Niebieskikwiat 2014. *Falkland Islands Magazine* featured a number of fascinating articles and reports about the pioneers of Santa Cruz, Tierra del Fuego, and Punta Arenas. See *Falkland Islands Magazine* December 1890, January 1899, December 1900, April 1901, September 1901, October 1901, October 1902, November 1902, May 1903, November 1904, August 1905, February 1910.

123. Cristina Fernández de Kirchner also created a fifty-peso banknote with Gaucho Rivero's image, but her successor Mauricio Macri, seeking friendlier relations with the UK, erased references to Malvinas from the currency.

124. See Spruce and Smith 2018, 47, 151.

125. Hamley et al. 2021. On the disavowal of indigeneity in settler memory, see Bruyneel 2021.

126. For continued correspondence between the Falkland Islanders and the South American Missionary Society, even after the sale of Keppel Island, see the extensive extracts on catechism and attempted extermination of Natives of Tierra del Fuego from *Falkland Islands Magazine* May 1892, July 1892, October 1892, August 1893, September 1898, September 1904, May 1905, November 1907, September–December 1911, January 1912, June 1914, May 1920.

127. See Kerr, "Despatch to Kimberley, June 27, 1881," JCNA, B/20, discussed in Bernhardson 1989, 437. Some settlements also hired missionaries, so they became part of broader settler colony (Spruce and Smith 2018, 181).

2. COMPANY ISLANDS

1. This catalog is now published in book form, which is why I have used Joan's name (Spruce and Smith 2018).

2. Spruce and Smith 2018, 40.

3. One cable and wireless provider has had a monopoly in the islands and charges exorbitant fees, which may be due in part to the reliance on satellite to make up for the lack of underwater cables connecting to South America.

4. Spruce and Smith 2018.

5. Collins 2011.

6. On pilgrimages of veterans back to the islands, see Jenkings and Beales 2022.

7. Spruce and Smith 2018, 31.

8. Spruce and Smith 2018, 31.

9. Written in a time of great political and social upheaval in late seventeenth-century England, Locke's (1964) *Two Treatises of Government* argues that labor generates value for establishment of property. Enclosure and improvement of land, as well as consumption of its natural goods, were based, for Locke, on what is properly one's own, according to ownership of the laboring person and duties to the moral authority of God. This understanding of private property does not hinge exclusively on ownership or notions of "possessive individualism" (Macpherson 1962). See also Pocock 1985. Locke naturalized property without deriving it from consent by locating the abundance of his imagined state of nature geographically in "America." Political philosopher James Tully (1993) has shown convincingly that this move exposed a hidden agenda in Locke to rationalize English colonialism. Given Locke's introduction of appropriation without consent, his theory on property served as a justification for the dispossession of Indigenous peoples and their land by slave traders and planters. Locke's vision of America in a state of nature both dismantled Indigenous political organization as non-self-governing and deterritorialized the land on which Indigenous peoples dwell by deeming it uncultivated

212 NOTES TO PAGES 51–54

terra nullius. This ethnocentric norm gave land perceived as being "un-improved" the false moniker "waste," making it available for appropriation through British imperial conquest and occupation (Tully 1993, 163, 156). See Mehta 1999; Gidwani and Reddy 2011; Milun 2011. Moreover, through domestic enforcement of the Enclosure Acts and Game Laws, as well as global strategies of accumulation by dispossession, British estate owners used this seemingly benign laissez-faire ideology to expropriate rural peasantries, control women's bodies, and extract common resources. See Thompson 1975; Goldman 1998; Perelman 2000; Federici 2004; Harvey 2005.

10. Verdery 2003. Rather than speculating in a Lockean mode of analysis about rights to cultivation and extraction, an anthropological approach draws meaning from historical and ethnographic moments of objectification and obligation, through a broader understanding of consubstantiality between chained and unchained things. See Malinowski 1935; Mauss 1950; Gluckman 1965; Strathern 1984; Weiner 1985.

11. Li 2007.

12. Grandia 2012.

13. These questions, and the anthropological framework on property used here, are adapted from and indebted to the work of Katherine Verdery (2003).

14. I have tried to provide a localized analysis that should not be interpreted as either a moral success story about British "gentlemen" who spread the gospel of improvement to grow commercial markets (Hancock 1997) or a strict historical materialist origin story about how capitalist property relations in the South Atlantic should be traced back to English agrarian reforms (Wood 2002).

15. Bernhardson 1989, 177–79.

16. Luis Vernet, "Map of Isla Soledad or East Falkland," AGN VII, 2-3-9. See also Bernhardson 1989, 175, 212.

17. "Declaration of José Báez," AGN VII, doc. 18, 2-4-1 (Academia Nacional de la Historia 1967, 115).

18. "Declaración que da el peón Antonio Rivero sobre los motivos que han ocurrido en esta isla y atraso del trabajo, a pedimento del Capataz Dn. Juan Simón," AGN VII, doc. 242, 2-3-6 (Academia Nacional de la Historia 1967, 94).

19. Eckert 1980; Ferguson 1985; Gal 1987; Woolard and Schieffelin 1994.

20. Rodríguez, Elizaincín, and González 2022.

21. Makihara 2004.

22. See Makihara's (2004) discussion of Ferguson 1959; Fishman 1967; Eckert 1980.

23. Rodríguez, González, and Elizaincín forthcoming.

24. British scholars have produced interesting interview-based socio-linguistic research on the Falkland Islanders' English dialect. See Wells 1982; Trudgill 1986; Sudbury 2001, 2004; Britain and Sudbury 2008. Such work neglects to take seriously their significant "low" vernacular uses of Spanish. This research operates under a teleology that positions islanders' speech at a halfway point toward an ostensible destiny of full Southern Hemispheric English dialect koinéization. Instead of diagnosing the underdevelopment of Falkland Islander English, here I wish to take up the fascinating syncretisms directly.

25. Rodriguez (2022) debunks the misconception suggested by Woodman (2006) that Spanish place names are not used in the islands. Instead, Rodriguez distinguishes the gaucho-derived place names from the ones used in Argentina. See Pondal 1956.

26. See Solari Yrigoyen 1998; Canclini 2000; Pastorino 2013. Argentine government functionaries I interviewed were not aware that islanders call one another "che." If we are to analyze the speech practices of islanders, it behooves us to take into consideration extended linguistic-cultural circumstances, which are tied closely to peoplehood and nationality. Take, for instance, Argentina's own settler colonialism and its effect on lan-

guage that is sharply split between the older *"voseo"* form of Spanish in the capital of Buenos Aires and Guaraní-Spanish bilingualism in the northern provinces (Hirsch 2003), not to mention the "ethnic revitalization paradoxes" to which Spanish monolingualism gives way (Rindstedt and Aronsson 2002). Furthermore, research on English language use in infrastructure, trade, and investment spheres of the Argentine economy, as well as newspapers, theater, and sports, indicates that there may be a "high" English and "low" Spanish diglossia—based on British and American dispossession—that informs nationalism, perhaps ironically, in Argentina (Nielsen 2003).

27. Massey 1993; Harvey 1995, 2009; Escobar 2001; Brown 2005; Tsing 2005. As linguistic anthropologist Keith Basso (1996) found from research among the Apache in the Southwestern US, "placenaming" enables people to appropriate physical environments with local wisdom. Unlike the "linguistic alienation" that Garret (2007) observes in a comparable diglossic context in St. Lucia—drawing on Jaffe 1999—the particular place of the Falklands/Malvinas may be best understood not as "pure," or even as "periphery," but rather porous.

28. On the significance of such family romance for hybrid identity formation, see Vergès 1999.

29. Louis Vernet, "Letter to Secretary of State for the Colonies, September 6, 1852," AGN VII, 2-3-7. See also Bernhardson 1989, 269.

30. Governor Richard Moody, "Colonial Report of 1846, March 9, 1847," JCNA.

31. "Some Account of the Falkland Islands. To which is added a Preliminary Sketch for the Formation of a Company, to be called The Royal Falkland Land, Cattle, Seal and Whale Fishery Company, London," JCNA, Governor's Library, (1850) 1949. See also Governor George Rennie, "Blue Book for 1852, January 8, 1853," JCNA.

32. Rennie, "Letter to Grey, February 6, 1850," JCNA, G/4 (Bernhardson 1989, 292–292A); Rennie, "Letter to Pakington, September 7, 1852," JCNA B/8. For details on small capitalist competition for grabbing land, see Moore, "Despatch to Newcastle, July 26, 1860," JCNA, B/11; Moore, "Despatch to Newcastle, March 8, 1862," JCNA, B/12. See Bernhardson 1989, 320, 323, 337 for a thoughtful discussion of the dynamics of the local "elite" class during this period, who, other than FIC managers, did not hold very much land.

33. Dean and Packe would have more influence locally, while George Thomas Whitington was more invested in making commercial interests in the Falklands dovetail with colonization efforts elsewhere. See G. T. Whitington, "1834 Prospectus of the Falkland Islands Association," AGN VII, doc. 129, 2-3-4.

34. In a report to London, FIC manager James Lane wrote in 1862, "our Sheep extend now over the whole Choiseul Sound District viz from our leased District No 8 to the Government reserve at Swan Inlet—& that there are Shepherds Houses & corrals erected at Mackinnon Creek, Teal Creek & at Swan Inlet River, North of Mackinnon Creek—all these on unleased Districts, formerly in 'temporary occupation'" (Lane, "Despatch 113, November 24th, 1862," JCNA).

35. Lane, "Despatch 25, July 26, 1859," JCNA. Having focused mainly on the East Falkland Island, colonists did not settle the other large West Falkland Island or other western "outer islands" until 1866, when James Waldron claimed 87,000 acres at Port Howard. See *The West Falklands, History of Farms and Farmers*, Dean/Chorley Collection, JCNA.

36. Relatively small enclosures had provided initial "proofs of progression" in agriculture in the early stages of the colony, but it would take decades before this practice became common throughout the islands (Governor George Rennie, "Blue Book of 1851, February 25, 1852," JCNA). According to FIC manager James Lane, as of 1858, the FIC did not have a single grass enclosure, and "The words 'cultivation' & 'enclosures' were a

signal for laughter" (Lane, "Despatch 4, November 11, 1858," and "Despatch 15, March 14, 1859," JCNA).

37. Dale, "Despatches to Directors of Falkland Islands Company, 1852–1853" (Bernhardson 1989, 313–15). See also Lane, "Despatch 30, September 28, 1859," JCNA.

38. See, e.g., Lane, "Despatch 28, August 11, 1859," and "Despatch 113, November 24, 1862," JCNA. For a description of the weekly rodeos to keep herds tame, see "Blue Book of 1860, February 12, 1861," JCNA.

39. Guillebaud 1967, 4, 15.

40. On the role of wire cables in gaucho culture in Uruguay and their broader significance for resourceful inventiveness, see Renfrew 2018, 19.

41. For a breakdown of fencing costs, see Lane, "Despatch 24, June 27, 1859," and Lane, "Despatch 38, November 29, 1859," JCNA. See also Bernhardson 1989, 414–15, 461.

42. *Falkland Islands Magazine*, January 1913, JCNA.

43. See also Rasmussen 2021.

44. For these terms, see Blake, Cameron, and Spruce's (2011) dictionary of Falkland Islander vocabulary, as well as Munro's (1998) booklet on place names and Spruce and Smith's (2018) book on rural heritage. For a sophisticated analysis of place names and gaucho heritage loanwords, which may include relevant enclosure-related terms, see Rodriguez 2022 and Rodriguez, Elizaincín, and González 2022.

45. Moody, "Despatch to Stanley, April 14, 1842," cited in Bernhardson 1989, 242–43, 250.

46. Drayton 2000.

47. Munro 1924; Guillebaud 1967, 18–24.

48. Munro 1924; Bernhardson 1989, 508.

49. Governor T.S.L. Moore, "Blue Book of 1857," JCNA. See also Lane, "Despatches No. 3, November 10th, 1858," "No. 21, June 14th, 1859," and "No. 38, November 29th, 1859," JCNA. Colonial administrators and FIC managers referred to sheep imported from South America as "Mastiza" or "Monte Videan" and contrasted them with the superior attributes of the English "Cheviot" or "Southdown" breeds. Late nineteenth- and twentieth-century sheep farmers also introduced Romney Marsh, Lincoln, Merino, and Corriedale (a cross between the latter two), among other breeds (Guillebaud 1967, 8–9).

50. "Blue Book of 1859, January 28th, 1860," and "Blue Book of 1860, February 12th, 1861," JCNA.

51. Governor D'Arcy, "Colonial Report of 1872, March 22nd, 1873," JCNA.

52. Governor D'Arcy, "Colonial Report of 1871," JCNA.

53. Governor William Robinson, "Blue Book of 1866, July 3rd, 1867," JCNA.

54. Governor D'Arcy, "Colonial Report of 1870, April 21st, 1871," JCNA.

55. Governor George Rennie, "Blue Book of 1852, January 8th, 1853," JCNA.

56. "Blue Book of 1866, July 3rd, 1867," JCNA.

57. Crosby 1986.

58. "Petition of Stanley Residents, Reviewed January 1859," JCNA, H/15. During my fieldwork, there was a comparable campaign called alternately "Friends of Cape Pembroke" or the "common sense group." The collective seeks to monitor the commons in an area just outside of Stanley that is currently used as an unofficial garbage dump. Here, rather than pushing for more pasturage, the group is motivated by environmentalist values of conservation.

59. *Falkland Islands Magazine*, October 1894, JCNA. See also Bernhardson 1989, 388.

60. "Camp Life," *Falkland Islands Magazine*, January 1896, JCNA.

61. *Falkland Islands Magazine*, June 1889, July 1889, November 1911, August 1912, JCNA.

62. *Falkland Islands Magazine*, October 1890, and "A Visit to Stanley," August 1894, JCNA.

63. *Falkland Islands Magazine*, April 1891, September 1924, JCNA.

64. See Malinowski 1922; Mauss 1950; Weiner 1985; Strathern 1988; Graeber 2012; Gregory 2015. On the distancing of meat and animal and the relation between sheep and imperialism, see Woods 2015.

65. Thompson 1971, 1991; Scott 1977; Edelman 2005. The difference in this case is that the islanders are now farmers who own their own property, as well as the means of production.

66. Guillebaud 1967, 12–13; Shackleton 1976a, 1:333–34.

67. Wright 2006, 20.

68. Wright 2006, 21.

69. After my fieldwork, he was elected as a member of the FIG Legislative Assembly.

70. Shackleton 1976a, 1976b, 1982.

71. Shackleton 1976a, 1:10, 12, 73; 1976b, 2:1, 4; 1982, 3:41, 45.

72. Shackleton 1982, 3:60.

73. Shackleton 1976b, 2:46–47; 1982, 3:9, 96.

74. Wannop 1961; Davies 1971.

75. Shackleton 1976a, 1:76, 80–81; 1982, 3:41.

76. Shackleton 1982, 3:16.

77. Falkland Islands Governors Goldsworthy (1891–1897), Allardyce (1904–1915), Middleton (1920–1927), Henniker-Heaton (1923–1924; 1935–1941) and particularly Cardinall (1941–1946) had each crafted plans to diversify the islands' economy and subdivide land decades earlier. However, all failed to persuade either London or the FIC. Governor Allan Wolsey Cardinall developed an especially fascinating "revolutionary" utopian plan to transform the islands into a rural society modeled on Danish cooperatives. Governor Alan Cardinall, "Despatch to Viscount Cranbourne, February 2, 1942," JCNA, CSO 8/42, discussed in Bernhardson 1989, 579–87; Tatham 2008, 140–41.

78. Shackleton 1982, 3:30. In 1981, government bought Roy Cove, a farm on West Falkland Island, for subdivision. Subsequent to the 1982 conflict, the following farms subdivided: Fox Bay East, Dunnose Head, and Port Howard in 1983; San Carlos in 1985; Fox Bay West in 1986; Port Stephens and Teal Inlet in 1988; and Port San Carlos in 1989. Finally, in 1992, the FIG purchased Goose Green, North Arm, Fitzroy, and Walker Creek from the FIC (Pompert Robertson 2014, 207).

79. Pompert Robertson 2014. See *The West Falklands, History of Farms and Farmers*, Dean/Chorley Collection, JCNA.

80. Gurr 1996, 24.

81. Bernhardson 1989, 685. See Bernhardson (1989, 466) for a map of smallholders, and Bernhardson (1989, 652) for a map of the subdivision.

82. Edelman 1992, 96.

83. Pompert Robertson 2014, 208.

84. Pompert Robertson 2014, 210.

85. Shackleton 1976a, 1:45.

86. Graeber 2012.

3. IMPERIAL DIASPORA

1. "Undesirable" is an emic term used locally in the Falklands. For broader context on discrimination against "undesirable immigrants," see Agier 2010; Bryce 2019; Goodfellow 2020; Rosenberg 2022. It is outside the scope of this book, but Falkland Islanders have also experienced a sex panic crisis regarding multigenerational sex and pedophilia.

2. Even well-written popular representations of the islanders evoke a charming mixture of liberal multiculturalism and imperialist nostalgia that flies in the face of scandalous racism and xenophobia within the islands. See MacFarquhar 2020.

3. Gilroy 1987, 51–55; Dodds 2002. For an original, eccentric philosophical treatise on "Falklandness" and British notions of "Motherland," see Fynn 2014.

4. Dodds 2002, 2–7, 118–41.

5. Unlike the Falklands, the British military did not defend former colonial subjects against invasions of East Timor, Cyprus, or Angola (Gustafson 1984, 86). Perhaps the most obvious contradiction is the rejection of the Chagos Islanders' right of return to Diego Garcia. The Chagossians, whom the UK FCO racialized as Black, were forcibly removed for the construction of a US military base. See Madeley 1982; Jeffery 2011; Vine 2011; Sands 2022.

6. Mercau 2019.

7. Du Bois 1998, 700; Getachew 2019, 22.

8. Hall 1980, 1985, 2016; Mullings 2005.

9. On diaspora and imperialism, see Ho 2004; Mohanram 2007; Olusoga 2023; on diaspora more generally, see Gilroy 1987; Hall (1990) 2021; Patterson and Kelley 2000; Edwards 2003; Brown 2005; Brubaker 2005; McKittrick and Woods 2007; Cohen 2008; Thomas 2009; Rana 2011; Clifford 2013; Figueroa-Vásquez 2020; Tamarkin 2020; Ogden 2021.

10. Wallerstein 1987, 379.

11. Mamdani 1996.

12. As Andreas Wimmer and Nina Glick Schiller (2002) have shown, migration offers a key field of inquiry for interrogating common instances of ignorance, naturalization, and territorial limitation in methodological nationalism. See also Rouse 1992; De Genova 2005.

13. Hall 2002.

14. Stoler 2006.

15. Burkett 2013; Godreau and Bonilla 2021.

16. Extract republished in *Falkland Islands Magazine*, May 1915, JCNA.

17. While the British media racialized the islanders as "a rough, uncultivated people" and regarded the islands as a far-flung "little blob" on the map, the prevalence of racist cultural products, such as blackface minstrel shows or "red skin village" wild west shows within the islands celebrated a white settler colonial sensibility. See *Falkland Islands Magazine*, February 1902, September 1915, December 1915, October 1918, June 1922, July 1922, April 1923, and July 1923, JCNA.

18. Once they had spent enough time in the rural Camp, the English and Scottish laborers, whom colonial managers recruited to replace the gauchos, were considered to be similarly barbarous and degraded. Outside British observers described the settlers as an unhappy "race of beings" that "ought to be happy," an "Anglo-Saxon race" or a "race of white men" with working-class values. *Falkland Islands Magazine*, May 1891, March 1895, May 1913, JCNA.

19. In spite of the Club's efforts, the appointed legislators would consist entirely of managers and elites for many years to come. Other key grievances were the lack of access to credit, perpetuating debt for store goods, and land redistribution. *Falkland Islands Magazine*, January 1921, September 1922, and December 1922, JCNA. See also James Wilson, "Letter to H.B.L. Jameson, Acting Colonial Secretary, September 15, 1887," JCNA, H42, quoted in Bernhardson 1989, 391. For further discussion of the Reform League, see Bernhardson 1989, 393–98.

20. Bernhardson 1989, 545–60. The Reform League became a more formal labor organization, the Falkland Islands Labour Federation, which ultimately took the form of a General Employees' Union that still exists, albeit in a much less active state.

21. The FIC London Chairman described it as Bolshevik-type, although the Governor downplayed any serious rebellion. J. S. Barnes and McAtasney, "Circular to 'Dear League Member,' August 20, 1937," JCNA, CSO C2/38, quoted in Bernhardson 1989, 555.

22. J. S. Barnes, McAtasney, F. Allan, and Brechin, "Memorial to Secretary of State for the Colonies, October 3, 1940," JCNA, CSO 160/40, quoted in Bernhardson 1989, 583.

23. For percent "Native" in population trends from 1881–1946 and 1946–1986, see Bernhardson 1989, 539, 604.

24. Social scientists have traced the formation of the "white ethnic community construct," extending the scope of analysis to the fields of gender studies and cultural theory. See Frankenberg 1993; Harris 1993; Leonardo 1994; Brodkin 1998; Jacobson 1999; Lipsitz 2006; Moreton-Robinson 2015.

25. Harrison, "Letter to A. I. Fleuret, October 26, 1933," JCNA, CSO 195/33, cited in Bernhardson 1989, 560. See also Miller 1988, 59. See *Falkland Islands Magazine*, May 1901, December 1906, January 1907, February 1907, and April 1907, JCNA.

26. Colonial Ordinance No. 7, 1935; "Interview with Governor Henniker-Heaton, July 11, 1936," JCNA, CSO 135/34. See Bernhardson 1989, 561. Interestingly, Bernhardson (1989, 564, 597, 614) notes that after World War II, there were proposals to (1) settle 2,000 refugees in Port Stephens, (2) settle Iraqi Syrians, and (3) recruit a team of German laborers, part of a religious group that ended up in South America who did in fact arrive and build Stanley's main Ross Road.

27. The petition was also directed against Clifford's wife, who was viewed as abusing colonial power. See Dodds 2002, 118–40 and Tatham 2008, 155–58 for an overview of Clifford's actions as governor.

28. "Letter to the Secretary of State for the Colonies from Falkland Islands Residents, June 4, 1948," PRO CO 78/241/1, quoted in Dodds 2002, 118.

29. Dodds 2002, 118–41. Islanders developed a custom of classifying South Americans, including Argentines, as non-native "coloured" people or "aliens." In Argentina, white settlers of Río de la Plata displaced the urban African-descendant population, colonizing and annihilating the Native people of the Chaco and Patagonia regions, through an extended national project of ethnic cleansing. See Rutledge 1977; Andrews 1980; Stepan 1991; Shumway 1993; Isla 1998; Gordillo and Hirsch 2003; Adelman 2006; Chamosa 2008; Salvatore 2008; Borucki 2011; Johnson 2011; Anderson 2014; Bryce 2019; Edwards 2020. In the twentieth century, the rise of Peronism influenced a form of Argentine multiculturalism that has reproduced divisions of class and status by racial classification. See Ratier 1971; James 1988; Guber 1999b; Joseph 2000; Gordillo 2004, 2016; Briones 2005; Segato 2007; Telles and Flores 2013.

30. Du Bois 1998; Roediger 2007.

31. See Saxton 1990; Allen 1994, 2005; Ignatiev 1995; Buck 2001.

32. H. Tempany, "Memorandum, July 10, 1941," PRO, CO 78/213/13; J. J. B., "Minute, July 8, 1952," PRO, CO 1024/80, cited in Bernhardson 1989, 577, 618.

33. On the Falklands lobby and debates over its importance in the era of decolonization, see Ellerby 1990; Dodds 2002, 118–40; González 2013; Donaghy 2014; Livingstone 2018; Mercau 2019.

34. Dodds 2002, 130.

35. Reprinted in Smith 2013, 205.

36. Getachew 2019.

37. See Shackleton 1976a, 1:10, 12, 73; 1976b, 2:1. The usage of the word "stock" parallels the way in which the colonial government had long reported data on sheep raising. See, e.g., "Blue Book for 1859, Government House, Stanley, 28th January 1860," JCNA. See also Franklin 2007, 46–72.

38. Shackleton 1976b, 2:1, 4.

39. Shackleton 1976a, 1:266.

40. Tatham 2008, 439–41; Smith 2013, 210.

41. For similar trends of "neglect" by the British Empire, see Taylor 2018.

42. While the local government expressed a preference for "Europeans in Uruguay," a small but consistent portion of the labor force derived from Chile. "Minutes of the 86th Meeting of the Boards of Directors of Bertrand and Felton Ltd, November 7, 1970," and "Minutes of the 103rd Meeting of the Board of Directors of Bertrand and Felton Ltd, August 2, 1976," quoted in Bernhardson 1989, 658–59.

43. Benwell and Pinkerton 2020. The relatively few Argentines that remain are primarily spouses of islanders born in the Falklands before 1982. While they are subject to a fair amount of scrutiny, some of these Argentine spouses have owned and managed significant businesses.

44. See Shickell 2009.

45. See Koutonin 2015; Kunz 2019.

46. One former MLA was even born in Argentina but had several generations of "Falkland Islander" descent on one side of their family and Anglo-Argentine descent on the other.

47. The wine bar's managing owner is a Falkland Islander and does have considerable rapport with working class residents of Stanley. They are also a leader in the local LGBTQ community, which receives wide support generally.

48. Gilroy 1987. On nations and nationalism, see also Gellner 1983; Hobsbawm 1983b; Anderson 1991; Chatterjee 1993; and especially Nairn (1977) 2015 for an influential perspective on nationalism in Britain, in which the latter argued influentially that the UK is a Janus-faced antiquity fated for being split.

49. For experiences of islanders during the war, see Smith 1984; Bound 2006; Fowler 2012; Watson 2012.

50. Brown 2005, 72, 230. See also Olusoga 2023.

51. On envy and the structures of feeling in British working-class conservativism, see Steedman 1986. In her ethnography of Black Liverpool, Brown (2005, 6) shows how "diaspora attends to the production of affinities and the negotiation of antagonisms among differently racialized Black subjects."

52. See *Mercopress* 2014a for one iteration of this sexist proposal by an otherwise angel of the progressive left in Latin America, former president of Uruguay José "El Pepe" Mujica.

53. On August 8, 2020, the Falkland Islands Executive Council announced that it would double the yearly quota of permanent residence permits (PRP) (*Penguin News* via *Mercopress* 2020).

54. For analysis of the construction of race and race-mixing in St. Helena, see Yon 2007.

55. Gordillo 2016.

56. The local *Penguin News* investigated instances of racial discrimination in the Falklands in the wake of the global Black Lives Matter movement in 2020, which influenced the government to legislate on racism (*Penguin News* 2020a).

57. Mendoza (2018) thoughtfully examines a similarly critical role of seasonal service workers in the development of ecotourism in Patagonia.

58. FIG Ordinance Title 52 Nationality and Immigration.

59. On the flaws of points-based immigration systems, see *The Economist* 2016.

60. FIG Ordinance Title 52 Nationality and Immigration, 2. See also Falkland Islands Constitution Order 2008, section 22(5).

61. For discussion of the racial politics of a different "locally born" classification in the context of Black Liverpool, where "Liverpool-born Blacks" try to separate themselves from Afro-Caribbeans or ex-colonial West Africans, see Brown 2005.

62. The FIG has crafted a "vulnerable persons strategy" that may include those at the bottom of the wage scale for pensions and other additional social services.

63. Rosaldo 1989; Mitchell 2021.

64. Falkland Islanders claim to have witnessed Argentine helicopters randomly scattering mines, but the Argentine former colonel responsible for mining contends that they were mapped out according to grids, and that some may have been put there by the British. See *Mercopress* 2014b.

65. Falkland Islanders told me that since they are so well marked, the minefields have actually had a surprisingly positive impact. Minefields have effectively closed off areas of the countryside that had been overgrazed, and they have helped to conserve rare wildlife because birds are too light to set off the explosives, and humans or livestock are restricted from entering and disturbing their habitats. Islanders have even extended "minefield" fencing beyond areas of known mining in order to assist in conserving neighboring habitats or to keep people from poking around their rubbish dumps. There has not been an accident in the Falklands other than from people trying to remove them. And islanders point out there were several places around the world with genuine humanitarian considerations that should have been prioritized for mine removal, but it was not their decision.

66. Shingai explained that there are at least three ways in which they were trained to do de-mining: manual, mechanical, or animal-assisted (usually involving dogs). In the Falklands, they did both manual and mechanical mine removal. The machines they use have chains that are powered to hit and destroy the mines to a depth of 20 centimeters. Because of the shifting terrain of the Falklands, the majority of the mines are deeper than this, so to ensure that 100 percent of the mines are cleared, the work has been primarily manual. This involves calmly and cautiously digging 1.2-meter trenches, at least 20 centimeters deep, in 4–6 rows a meter apart, according to a pattern laid out in the original maps provided by Argentina's military.

67. Hughes 2010; Scoones 2010; Moyo and Chambati 2013.

68. Stoler 2006.

4. DOES THE SEA LION ROAR?

1. Gillis 2004, 124.

2. Carse 2014, 5, 14.

3. Blumenberg 1996.

4. Bushnell 2008, 198.

5. Governor George Rennie, "Blue Book for 1852, February 25, 1852," JCNA. See also Governor George Rennie, "Blue Book for 1854, January 3, 1855," JCNA; "Extortion at Port Stanley," *Falkland Islands Magazine*, December 1896.

6. Even today, wrecks carry value as cultural heritage products, and an FIG customs officer holds the title "Receiver of Wrecks." The position consists mainly of informing local wood lathing hobbyists when pieces of decomposing wrecks come ashore. During my research, wrecks also factored into an Environmental Impact Assessment for an oil exploration campaign by Noble Energy in 2015. Planned drilling areas were believed to be in the vicinity of "war graves" from World War I, although seismic surveys were not effective in locating them. One wreck from the 1982 war, the British ship *Atlantic Conveyor*, rests near one of the wells. In 2015, abandoned Argentine yacht *La Sanmartiniana* made its way to the islands and became the center of controversy. Its journey caused a minor air incident when the British military sent an RAF Typhoon from the Mount Pleasant Complex (MPC) to warn away Argentine search aircraft. No blood was spilled, but much to the delight of vitriolic British trolls in Falklands-themed Facebook groups, the FIG's "rules of salvage" did apply. See *MercoPress* 2015a, 2015c; "Falklands' salvage

rules apply to Argentine yacht La Sanmartiniana; contacts for its return," *Penguin News*, October 19, 2015.

7. Mentz 2015.

8. See also Baucom 2005; Rediker 2007; Hartman 2008; Levinson 2008; Tinsley 2008; Brown 2010; Khalili 2021; Leivestad et al. 2021; Markkula 2022.

9. King 2019; see also Jackson 2012.

10. Benjamin 1968, 257–58. For further analysis of Benjamin's "On the Concept of History," see Löwy 2016.

11. See Glissant 1997, 33, 174; Stoler 2016, 346. Similarly, in the context of the Chaco region of Argentina, Gordillo (2014) examines such "rubble" as haunting absences that become felt as a presence with different meanings (e.g., wealth for settler *criollos* or legacies of violence for Indigenous Wichí people).

12. For Tsing (2015a, 2015b, 63) nonhumans like matsutake mushrooms are not produced under conditions of capitalist control, but capitalists may take advantage of their scalable value by extracting them from noncapitalist value regimes and circulating them in the market.

13. Smith 1984.

14. Spruce and Smith 2018, 151, 176, 210.

15. Governor T.S.L. Moore, "Blue Book for 1855, January 16, 1856," JCNA. For a specific story about ship jumpers, see Spruce and Smith 2018, 121–22.

16. Such legal or licit forms of capturing wealth were perhaps more akin to plunder than salvage (Mattei and Nader 2008).

17. Here I do not follow the social contract model of personhood, but rather personhood's significance for modern progress deriving from ethnographic and historical research. See Strathern 1988; Li 1998; Mosko 2010; Ferguson 2013.

18. This may be referred to elsewhere as the Dutch disease or devil's excrement. See Auty 1993; Coronil 1997, 321; Karl 1997; Watts 2004; Humphreys, Sachs, and Stiglitz 2007; Ross 2012; Rolston 2013; Appel, Mason, and Watts 2015, 3.

19. Watts 1992; Coronil 1997.

20. Coronil 1997. While it is important to draw attention to the relationship among global inequality, state power, and energy regimes, as Gisa Weszkalnys (2011) has shown, the resource curse often acts as a performance of economic theory, a black box that explains the unexplained. Much of the literature lacks attention to local knowledge and history, so it remains unclear whether oil is the root cause of problems or, instead, resource dependence exacerbates already existing afflictions. Moreover, with so much emphasis placed on the power of the state, proponents of the resource curse hypothesis remove blame from corporations.

21. Americans also built a hut and active settlement for illicit whaling, sealing, and egging on the western New Island.

22. For detailed anecdotes of sealing, see Edmund Fanning's (1924) *Voyages & Discoveries in the South Seas*. See also Bernhardson 1989, 162, 216, 695–96.

23. "The Sealing Season," *Falkland Islands Magazine*, June 1903.

24. For Vernet's plan to regulate sealing and whaling, see Vernet, "Introductory Plan for Sealing and Whaling on and about the Coasts of Patagonia and Its Dependent Islands," AGN VII, 2-3-6, cited in Bernhardson 1989, 199.

25. "Statement of William Edward Murphy, January 9, 1854," JCNA, B/8, and Rennie, "Despatch to Earl of Newcastle, March 16, 1854," JCNA, B/8, discussed in Bernhardson 1989, 698.

26. For a French naturalist's account of sealing in the early settlement at Port Louis, see Pernety (1771) 2013.

27. See Lane, "Despatch 95, January 29, 1862," JCNA. The British colonial government passed a Seal Skin Ordinance in 1903, charging a duty not to exceed ten shillings for each skin. *Falkland Islands Magazine*, October 1903.

28. In 1846, settlers traded whale oil at 45s. or £562 per barrel ("Colonial Report for 1846," JCNA). For an account of sealing and whaling in the South Atlantic from this period, see Clarke 1854.

29. See Lane, "Despatches 22, June 15, 1859" and "Despatches 101–110," JCNA.

30. Lane, "Despatch 36, November 17, 1859," JCNA.

31. "Joseph Jenkins," *Falkland Islands Magazine*, September 1900, JCNA. For an incisive analysis of the contradictions of Christian optimism and "barbarism" in American sealing, see Grandin 2014. See also Rediker 1987. This also parallels how Western environmentalists have discriminated against Indigenous Arctic seal hunters through more overt racism, captured in the film *Angry Inuk* (Arnaquq-Baril 2016).

32. *Falkland Islands Magazine*, August 1908, JCNA.

33. Spruce and Smith 2018, 39.

34. Spruce and Smith 2018, 13. Raffles (2020, 123) discusses a similar phenomenon of whale remains embedded in rock called "blubberstone."

35. FIC Manager Lane wrote, "Mr Coleman mentions Penguin Skins as used for Ladies wear. Is there a market for 20,000 or 30,000 of them? & what price wd they fetch each or per lb in Cargo numbers?" Lane, "Despatch 103, July 22–30, 1862," JCNA.

36. Lane, "Despatch 106, August 11, 1862," JCNA. See also Lane, "Despatch 107, September 1, 1862."

37. Governor D'Arcy, "Colonial Report 1871, February 27, 1872," JCNA. For discussion of the whaling stations on South Georgia, see Robertson 1954; Basberg 2004; Gordon 2004.

38. "A Dinner of Whale," *Falkland Islands Magazine*, March 1911; "The Danger of the Whale Becoming Extinct," *Falkland Islands Magazine*, July 1912; "The Song of the Whalers," *Falkland Islands Magazine*, May 1920. See also shareholder reports in *Falkland Islands Magazine*, May 1922, August 1922, November 1922, and December 1922.

39. In 1912–1913, whalers at New Island killed eighty-seven whales, producing 2,128 barrels of oil, while in 1917–1918, those at South Georgia killed more than 3,000 whales, yielding 200,000 barrels of oil (Bernhardson 1989, 712).

40. *Falkland Islands Magazine*, 1922, JCNA. See also Bernhardson 1989, 708.

41. Allan 2010.

42. "The Falkland Islands: Memorandum on Potential Minor Industries," 1939, JCNA.

43. See Dodds and Benwell 2010 for maps and analysis of the overlapping zones. In 2016, the UN Commission on Limits of the Continental Shelf ruled in favor of extending the Argentine continental shelf, which Argentine media celebrated as a victory in the sovereignty dispute, but the UK dismissed the ruling as unbinding, and the commission had already decided not to consider areas subject to dispute, including the Falklands/Malvinas. See UN Division of Ocean Affairs and Law of the Sea 2016.

44. The 1986 Falklands Interim Conservation and Management Zone (FICZ) covers 150 nautical miles from the estimated center of the archipelago, and the 1990 Falkland Islands Outer Conservation Zone (FOCZ) extended the northern, eastern, and southern (but not western) edges of the previous zone to 200 nautical miles. The latter coincided with an agreement with Argentina on fisheries management.

45. Since the establishment of the fishing zones, the FIG has issued licenses to at least fourteen different firms and the FIG accumulated a total revenue of £10 to £29 million each year.

46. Falkland Islands Government 2012.

47. Gereffi and Korzeniewicz 1994; Gereffi, Humphrey, and Sturgeon 2005.

48. MacFarquhar 2020.

49. Watson 2014; Martinez 2015.

50. Urbina 2015.

51. An imaginary letter in a bottle from the future, published in the Islanders' local newspaper, tells of mineral riches greater than the vast Welsh coal deposits, a thriving peat briquette export industry, and tremendous petroleum reserves (*Falkland Islands Magazine*, October 1915, JCNA).

52. The Jane Cameron National Archives in Stanley lists the following reports:

–1921-07 Report by Dr. Herbert Arthur Baker, government geologist born in London, on possibility of occurrence of liquid petroleum in Falkland Islands
–1923-01-10 through 1923-04-19 Model oil prospecting licence and mining lease
–1947-05-24–1959-08-12 Oil and coal deposits
–1964-08-25 Mining legislation
–1968 New Zealand Petroleum Exploration Co Ltd, Dallas, Texas
–1968–1969 ACX Oil Co, Calgary, Alberta
–1968 Place Oil and Gas Co Toronto, Canada
–1968 Tenneco Oil Co, Houston, Texas
–1968 Sun Oil Co, Philadelphia, USA
–1968-06-26–1969-07-17 Atlantic and Oceanic Resources (Grynberg) Ltd
–Annexure to file
–1969-08-27–1971-10-27 Atlantic & Oceanic Resources (Grynberg) Ltd and Oceanic Exploration Co
–1969 Canadian Homestead Oil Co Ltd, Calgary, Canada
–1969 Global Marine, Los Angeles, California
–1969–1970 Offshore Exploration Oil Co, Philadelphia
–1970 Oil Resources Incorporated (Grand Valley Oil Ltd)
–1970 Western Oilfields Inc, Dallas, Texas
–1970 Peyto Oils Ltd, Calgary, Alberta
–1970 International Nuclear Corporation
–1970 Lewis G. Weeks Associates Limited, Connecticut
–1970 Sunningdale Oils Limited, Canada
–1970 Peter Bawden Drilling Services Ltd, Calgary, Alberta
–1970–1971 Chevron Overseas Petroleum Inc, San Francisco
–1971 Geocon Inc, Houston, Texas
–1976 Rangeroil (UK) Limited, London SW1E 5AG

53. Bernal 2011, 28–31, 94–96.

54. Phipps 1977.

55. Based on a November 1975 visit to the islands and Argentina with fellow MP John Gilmour, Phipps described the Falklands society as British but "almost feudal." He observed a significant problem of increasing emigration to the UK. Seeking to understand the motivations of the Argentine government's position, Phipps perpetuated the myth of Argentina as a white, "essentially European" nation and tried to reconcile the awkward alliance of anti-imperialist Peronists and the sizeable population of Anglo-Argentines.

56. Interestingly, Phipps had advised Thatcher on the potential of developing oil in the South Atlantic. While Argentine revisionists accuse Phipps of influencing Thatcher to declare war (Bernal 2011, 31), confidential sources informed me—based on direct correspondence with the former—that Phipps told the prime minister that he would *not* go to war over unproven oil reserves. The extent to which speculative oil prospects influ-

enced Thatcher's decision making is unclear, but it may be overly reductive to assert that the political economic, moral, and military logics for the war were reducible to unknown resource wealth. That said, as I outline, it was Phipps who would eventually carry the banner for oil exploration in the South Atlantic after the British victory. See also Silenzi 1983 for a perspective from Argentina.

57. Livingstone 2022.

58. "Oil and the Falkland Islands," Briefing paper for members of the Overseas & Defence Policy committee, attached to Hurd, D., 16 September 1991, FCO7/8312, cited in Livingstone 2022, 96.

59. *The Falkland Islands and Oil*, November 1993, FIG Secretariat, Stanley, Falkland Islands, JCNA. "Oil and the Falklands: A Progress Report by Falkland Islands Government," *Penguin News*, October 22, 1994. *Offshore Minerals Ordinance*, 1994, *The Falkland Islands Gazette Supplement*, Vol. 5 (28). See also Richards 1995. It is intriguing to note that the FIG had also licensed an onshore exploration license to Firstland Oil and Gas PLC in July 1984, but no drilling occurred, and no onshore oil has been found (Livingstone 2022, 95).

60. Tella 1982.

61. Ruzza 2011; Livingstone 2022.

62. Obituary of Colin Phipps (*The Telegraph* 2009).

63. Falkland Islanders and their British trustees have long held that the English captain John Davis first sighted the archipelago on his ship *Desire* in 1592, even though Iberian maps dating from up to fifty years earlier feature a comparable archipelago.

64. Fidler 1997.

65. Falkland Islander shareholders involved in the direction of Desire have been penalized for insider trading (Blackwell 2009).

66. Mason 2010.

67. Richard Visick, one of the founders of Rockhopper, gave the well the name Sea Lion after having owned Sea Lion Lodge, a popular tourist destination on Sea Lion Island in the Falklands.

68. Test drilling by the consortium of licensees in the south and southeast blocks of the conservation zones has revealed gas or oil "shows," but not commercial discoveries of hydrocarbons in exploitable reservoirs.

69. Blair 2014.

70. Approval procedures for mineral resources thus differ from most other largely devolved FIG legislation.

71. Swint 2013.

72. Rockhopper subsequently "bought" production through partnerships in the Mediterranean.

73. Terry found that at $13, the firms were potentially losing $3 per barrel. In case this did not convince the Treasury Department, Terry mapped the entire hydrocarbon commodity chain for the North Sea, including both offshore drilling and onshore support. Armed with the details of specific businesses sprinkled throughout Scotland and England, as well as the corresponding number of employees, Terry pressured MPs throughout the country with the number of jobs that their constituencies would lose if Parliament imposed further taxes on the industry.

74. To complement Falklink, the FIDC created a registry of Stanley suppliers, contractors, and oil and gas operators available to support the oil and gas industry, as well as an oil and gas certification program.

75. Anadarko Petroleum, for example, whose board of directors included former US military commander Kevin Chilton, reportedly flew a private jet to Stanley to bid for a partnership to develop Sea Lion, but when insiders decoded the plane's license and tipped

off tabloids, the ensuing embarrassment caused the company to issue a statement that they were backing out (Hawkes 2012). Petroleum geologists employed by super majors such as Shell have similarly received FIG mineral resources managers with enthusiasm, but because Shell has assets not only in Argentina but also in neighboring ally countries Uruguay and Brazil, the company did not pursue it further.

76. In turn, the FIG also has a blanket cover restriction on asset allocation parameters and benchmarks in Argentina.

77. Hugo Swire, UK FCO minister, press conference, February 12, 2014.

78. From the opposite end of the political spectrum, oil and gas have leveraged creative forms of diplomacy from Latin America as well. See Ellner 2007; Gustafson 2020.

79. Interview with FIC director, April 24, 2014.

80. The two polities also apparently share similar contradictory ideologies regarding climate change and fossil fuels (Hughes 2013).

81. "Noble lady is 'Inspirational' oil executive," *Penguin News*, March 21, 2014.

82. Nuttall 2010.

83. Shever 2012, 169.

84. Barry 2013.

85. Marx 1993; D. Harvey 2011, 47.

86. Wagner, Jones, and Dowse 2014.

87. Regeneris Consulting Ltd. 2013, 17.

88. FIG Policy Unit 2013.

89. FIG Legislative Assembly 2015.

90. Mitchell 2011.

91. Tsing 2015b.

92. Coronil (1997, 221–22) viewed this position critically as a subtype of "rentier bonapartism."

93. Blair 2016.

5. GROUNDING OFFSHORE OIL

1. Stoler, McGranahan, and Perdue (2007). The role of infrastructure in this imperial formation invites comparison with other disputed territories in which postcolonial sovereignty is experienced as graduated or exceptional (Bonilla 2013; Stoler 2016; Ong 2000; Hansen 2021).

2. On generational differences of politics and everyday memory among Falkland Islanders, see Benwell 2016a; 2017.

3. Wallerstein 1987.

4. Similar to Bonilla's (2015) protagonists, who struggle for labor rights without national liberation in Guadeloupe, my interlocutors claim self-determination in order to remain under the flag of their former colonial power. Yet, in contrast to decolonial workers defying linear temporalities of modern progress in the Caribbean (Scott 2014; Wilder 2015; Thomas 2016), the intensified commercial shift toward marine resources in the South Atlantic has formed a wealthy assemblage of white settlers, pursuing a different island dream of the "not yet" (Gillis 2004, 65). Claiming what Trouillot calls the "North Atlantic Universal" of self-determination, the islanders seek to insert their subjectivity into a particular regime of historicity (Trouillot 2003, 38–39, 44). As this chapter explores, this imaginary becomes localized in a constellation of infrastructural events that reinforce a frontier temporality (Benjamin 1968, 263). See also Folch 2013; Eilenberg 2014; Gordillo 2014; Stepputat 2015.

5. An ethnographic focus on the temporal politics of infrastructure has broader implications for the anthropology of settler colonialism. According to Patrick Wolfe (2006), settler colonialism is a structure—not an event—that reproduces imperial ideologies by eliminat-

ing or displacing Indigenous peoples from land. As J. Kēhaulani Kauanui (2016, 3) points out, "Understanding settler colonialism as a structure exposes the fact that colonialism cannot be relegated to the past, even though the past-present should be historicized." While there is no historical evidence of a colonial encounter with a precolonial Indigenous human population in the Falklands, I explored in part 1 how islanders have constructed "settler indigeneity" in assertions of peoplehood. Focusing here on commerciality, through the eyes of the islanders and their oil partners, I examine how settler colonialism presents itself not only as a permanent structure but also as an "accretion" of temporary infrastructures (Anand 2017, 13). In short, while scholars of settler colonialism and imperialism center on structural stability and durability (Wolfe 2006; Stoler 2016), here I uncover how the installation of temporary infrastructure designed to develop the Falklands into an offshore oil frontier reinforces a perpetual sense of instability for its onshore settler colonial frontier.

6. Appadurai 1981.

7. My use of the concept *perpetual frontier* differs significantly from that which A. Endre Nyerges (1992, 874) employs, deriving from precolonial African contexts that do not obtain in the Falklands.

8. While Coronil (1997) argued persuasively that the state consolidates power through magical illusions that oil is natural wealth rather than material substance, other studies highlight connections between carbon energy and the limits of democracy, as well as how communities hosting offshore oil experience hope and expectation in complex ways (Weszkalnys 2008, 2016; Mitchell 2011; Appel 2012a, 2019).

9. Larkin 2008.

10. Watts 2014, 194. See also Watts 2012, 445; 2015, 220.

11. See Sawyer 2004; Orta-Martinez and Finer 2010; Peel 2010; Cepek 2012.

12. Kemp 2011.

13. See Austin, McGuire, and Higgins 2007; Bond 2013; Breglia 2013; Watts 2015.

14. Nuttall 2010; Mason 2016.

15. Dodds and Benwell 2010; Benwell 2013.

16. Watts 2014, 193. Jusionyte (2015, 96) makes a similar observation about how the state uses mass media narratives of potential threats in border areas: "The frontier, though it connotes the process of state-building and the advancement of governmental control over 'no man's land,' is never completely closed."

17. Turner 1920.

18. Cronon 1991, 32, 150; Rifkin 2014; Jusionyte 2015; Grandin 2019.

19. See Baretta and Markoff 1978; Slatta 1992; Serje 2011; Bjork-James 2015.

20. Tsing 2005, 32.

21. Recent anthropological and social scientific research on the politics of oil and infrastructure in extractive frontiers has focused on the relations between material practices and socio-technical values (Bowker 1994; J. Ferguson 2005; Barry 2013; Harvey 2014). There is a powerful socio-technical tension between the visibility and invisibility of different infrastructures (Larkin 2013; Amin 2014). Across oil frontiers, energy companies attempt to make permanent infrastructure invisible, e.g., by burying pipelines underground (Barry 2013; Leonard 2016; Spice 2018). Yet other oil infrastructures, from "man camps" to rigs and barges like the Noble Frontier, make impermanence visible and contamination invisible (Limbert 2010; Appel 2012a, 2012b; Weszkalnys 2015; Baker 2016). It is telling that in contrast to the excitement surrounding the Noble Frontier's arrival, the Falklands' fishing industry is entirely absent from residents' field of vision, besides when squid jiggers enter Stanley Harbour to collect annual licenses.

22. See Guber 2009 for analysis of the Falklands/Malvinas as liminal and Gillis 2004, 76 on islands as liminal thresholds more generally. On infrastructure's suspension and colonial nostalgia, see Bissell 2005; Gupta 2015; Anand, Gupta, and Appel 2018; Ruiz 2021.

23. On the way oil may feel short-lived yet lasting, see Limbert 2010.

24. Argentines have called it either Puerto Rivero or Puerto Argentino (Guber 2000).

25. As Paul Carter (2010) analyzed astutely in the context of Botany Bay, Australia, this imaginative vision of a fixed settlement serves as a sturdy stage for imperial legitimation.

26. This parallels how Panamanians aspire toward the always unfinished interoceanic highway (Carse 2014, 179).

27. Spruce and Smith 2018, 37, 56–58, 65, 194.

28. On past utopias summoned as inspiration for social formations spanning centuries, see Galindo 1993.

29. To give me context on Noble's history of building oil frontiers, William lent me a book called *The Secret of Sherwood Forest: Oil Production in England During World War II* by Guy H. Woodward and Grace Steele Woodward (1973). This amateur history tells the story of a group of masculine "roughnecks"—the first generation of Noble Energy workers—sent to England in order to explore for oil and thereby rescue their vulnerable British "gentlemen" allies in need of fuel during World War II. An intriguing story, the book nevertheless suffers from the authors' absolute devotion to the company and its founder, Lloyd Noble. In this sense, the book is similar to the writing that Wallace Stegner was hired to do for ARAMCO in Saudi Arabia (Vitalis 2009).

30. Appel 2012a, 2019.

31. Levinson 2008.

32. This uniform national composition of the workforce contrasts with the intercultural rig described by Appel (2012a, 2019), which comprised 115 workers from twenty different nations, including underpaid Equatoguineans on rotating shifts.

33. Appel 2012a, 705.

34. Tsing 2005, 28.

35. Gandy 2003; Taussig 2004; Harvey and Knox 2015.

36. As Cymene Howe et al. (2016, 553) argue, "paradoxical infrastructures" require retrofit and rehabilitation.

37. Mason 2007. As Jane Guyer (2007) has shown, "prophecies" of long-run growth tend to thin out fluctuations in the near- and mid-term future.

38. Anon. 2016.

39. Krauss 2020.

40. *Penguin News* 2020b.

41. Ferry and Limbert 2008, 4.

42. Ross 2012.

43. Weszkalnys 2011.

44. Appadurai 1981, 2013, Koselleck 2004.

45. Howe et al. 2016.

46. By examining mundane experiences with the materiality of infrastructure, anthropologists and social scientists have captured how corporate elites, resource managers, or marginalized publics either enhance or challenge hegemonic regimes of nature, property, citizenship, and capital accumulation (Star 1999; Rogers 2012; Carse 2014; Weszkalnys and Richardson 2014; Harvey and Knox 2015; Swyngedouw 2015; Anand 2017; Jobson 2018). While the sustained focus on infrastructure's materiality is helpful for considering how oil shapes geopolitical imaginaries as a substance, I have been especially concerned here with its temporal work (Bowden 1985; Limbert 2010; Guyer 2015; Rogers 2015a).

47. Mukerji 2010.

48. Wolfe 2006; Simpson 2014.

49. Simpson 2014; Todd 2017b.
50. Cronon 1987.
51. Coronil 1997.
52. Tsing 2005; Bjork-James 2015; Ballvé 2020; Kröger and Nygren 2020.

6. THE GEOPOLITICS OF MARINE ECOLOGY

1. Bingham 2005. I use Bingham's real name here because he is a public figure who published this book and other articles.
2. Hall 1980, 1985, 2016.
3. This spectacle caught media attention. See Hobsbawm 1983a; Gilroy 1987, 51–55; Milne 2014; Hall 2016.
4. Bingham excelled at self-publicizing by appealing to popular affection for penguins. Yet *The Falklands Regime* has a conspiratorial tone, and Bingham's numerous critics point to inaccuracies, simplifications, and exaggerations.
5. Bingham 1998, 2002. FIG Fisheries has also adopted mitigation measures to reduce bird mortality, in accordance with the Agreement on the Conservation of Albatrosses and Petrels (ACAP). Bingham's critics suggest that, while there may be overlap between penguin foraging and commercial fishing, global changes in ocean temperature were the principal cause of penguin population declines. See Croxall, McInness, and Prince 1984; Clausen and Pütz 2002; Pütz et al. 2003; Dehnhard et al. 2013.
6. Bingham left his Falklands research behind after his triumphant court case. He had lived there for eleven years, from 1993–2004. Currently, he focuses his monitoring activities primarily in the Chilean Island of Magdalena, as well as Cabo Vírgenes, Argentina; both are no-fishing zones (personal communication with Mike Bingham, 2014).
7. On sustainability science, environmental politics and green capitalism elsewhere in the region of South America, see Henne 2015; Barandiarán 2018; Mendoza 2018; Mendoza, Greenleaf, and Thomas 2021.
8. The emergent literature on "geopolitical ecology" has primarily focused on the environmental impacts of military operations (Bigger and Neimark 2017; Belcher et al. 2019). Here, I focus primarily on the geopolitical implications of environmental science and governance.
9. UN Division of Ocean Affairs and Law of the Sea 2016.
10. Sábato 1975.
11. Babini 1954; Varvavsky 1969; Dagnino, Thomas, and Davyt 1996.
12. Alvarez León 2018; Blair 2019a; Goldstein and Nost 2022.
13. While research on military history, global politics, and international law is relatively saturated in the South Atlantic, overlooked aspects of science also become involved in territorial disputes as authorizing forces of technocratic governance (Sharp 2002; Callison 2014; Doel, Wråkberg, and Zeller 2014; O'Reilly 2017; Giraldo and Rosset 2018).
14. Barry 2006.
15. David et al. 2002.
16. Larson 2012. It is interesting to note that Mensun Bound, the accomplished maritime archaeologist who successfully found the wreck of the *Endurance*, was born in the Falklands (Bound 2022).
17. Stratford et al. 2011.
18. Dodds 2002.
19. The ecological niche of *Illex* squid crosses the Falklands exclusive zones, as well as adjacent Argentine, Brazilian, and Uruguayan waters. Stock assessment of *Illex* thus became a boundary object for the FIG and Argentine government to assert sovereign rights to common property resources as responsible managers of the exclusive marine environment (McCay and Acheson 1987; Ostrom 1990; Mansfield 2004). The FIG argued

that the Argentine government's refusal to share knowledge during the Kirchner administrations not only led to a depletion of *Illex* but also encouraged predatory overfishing to damage future stocks (Arkhipkin et al. 2013, 17). The interpretive flexibility of this particular squid species, which inhabits the shared space of a disputed territory, thus conditioned the possibility for different sets of public and private structural arrangements (Star 2010). See also Star and Griesemer 1989; Fujimura 1992; Forsyth 2002, 140.

20. Jasanoff and Kim 2009, 120. See also Jasanoff and Kim 2015; Barandiarán 2019.

21. Trouillot 2003.

22. Blair 2019a.

23. The UK's version of a South Atlantic universal resembles what Bonilla (2015), building on Trouillot, calls a "non-sovereign future," because Falkland Islanders seek self-determination in order to remain attached to former colonial authority.

24. Trouillot 2003, 38–39, 44.

25. Chari 2019.

26. Gillis 2004, 65. This chapter's examination of clashing South Atlantic universals also has broader implications for rethinking the contemporary realignment of center-periphery relations in the Southern Cone and the Atlantic World through STS research (Vessuri 1983; Kreimer 2007; Greiff 2012; Rodriguez-Medina 2013).

27. Crosby 1986; Drayton 2000; Briggs 2002; McCook 2002; Nieto-Olarte 2006; Delbourgo and Dew 2007; Harding 2008.

28. On the construction of the nation-state of Argentina based on attempted genocide, see Gordillo and Hirsch 2003; Briones 2005. Historians have often analyzed the Pampa as a source of easy rents due to highly fertile soil, though the rents for marine resources in the South Atlantic may be less accessible (Gallo 1983; Sabato 1987).

29. Website for Pampa Azul and the Ministry of Science, Technology and Productive Innovation, http://www.mincyt.gob.ar/accion/pampa-azul-9926.

30. Mendoza 2018. On emerging blue economies in other contexts, see Campero et al. 2022; Germond-Duret 2022; Kaur Hundle 2022.

31. Mendoza 2018.

32. Daston and Galison 2010.

33. Pueblos por Malvinas leader, "Malvinas: Ultima frontera planetaria," speaking event held at Instituto de Estudios Estratégicos y de Relaciones Internacionales (IEERI), June 5, 2014.

34. Guber 2001.

35. Pueblos por Malvinas leader, "Malvinas: Ultima frontera planetaria," speaking event held at Instituto de Estudios Estratégicos y de Relaciones Internacionales (IEERI), June 5, 2014.

36. This methodology has been informed by a broader field of public or engaged anthropology that encourages direct participation in activism or advocacy, as well as research that seeks to "study up" among experts and elites (Hymes 1972; Nader 1972; Scheper-Hughes 1995; Edelman 1999; Goldstein 2014; Hughes 2017; Kirsch 2018).

37. FIOHEF does not have any specific powers, but it holds biannual meetings for stakeholders, including oil companies, NGOs, and government officials. Minutes of the forum's meetings are not for public consumption and are not publicized, but a synopsis is produced. The meetings are not confidential, but these privacy protections were designed to encourage different social actors to speak candidly. Because I consented to this protocol when I was given permission to attend both FIOHEF meetings in 2014, I will not describe the events in detail, but the experience informed my discussion of the Gap Project and the other sections.

38. In addition to these internal projects, SAERI served as a central hub for visiting scientists researching or making policy on biodiversity, seabirds (penguins and alba-

trosses), marine flora (kelp and algae), cetaceans (whales and dolphins), and pinnipeds (seals and sea lions), as well as terrestrial ecologists, geologists, paleoarchaeologists, shallow marine survey divers, and nature documentary filmmakers.

39. See Smith 2013, 81–82.

40. See Harding 2014.

41. McAfee 1999; Fortun 2004; Smith 2006; Castree 2008; Dempsey 2016.

42. For a comparable island imaginary in the Galápagos, see Bocci 2019.

43. Weszkalnys 2015.

44. For a fascinating analysis of environmental managers' enrollment of higher (human) predators—anglers—for measuring fish stock, see Eden 2012.

45. Baylis et al. 2018.

46. Bedi 2013; Westman 2013; Dougherty 2019.

47. This adversarial relation has led to what Keith called "green-mailing," a form of blackmail with environmental science in which researchers demand funding in exchange for *not* tarnishing oil companies' environmental record. The threat lies in the ability of scientists to do the research independently and withhold data until they are published. In finance, greenmailing has a separate connotation as an investment strategy in which monopoly capitalists threaten to take over corporations but stop in exchange for smaller payments. Here, science has the power of money: not to directly acquire a firm's assets but rather to damage its public image.

48. Bedi 2013.

49. Mitman 1996; Benson 2010.

50. Latour 2007, 54.

51. This narrative derives from a presentation on February 25, 2014, by visiting researcher Jonathan Handley in Stanley's Chamber of Commerce. STS analyses of animal tracking, particularly with birds and marine animals, have shown considerable variation in the effectiveness of remote sensing for addressing the social and spatial consequences of industrialization and globalization (Mitman 1996; Benson 2010; Whitney 2014; Stokland 2015; Gabrys 2016). A significant divide exists between the values of animal tracking for wildlife biology on the one hand and environmental conservation on the other. Environmental "sensing practices" may serve diverse political goals and objectives for humans and more-than-human subjects (Gabrys 2019; Howe 2019), but their ethical and social origins lie in imperial hunting, domestication, and policing, rather than principles of reciprocity common to Indigenous hunting and trapping (MacKenzie 1997; Blaser, Feit, and McRae 2004; Berkes and Turner 2006; Whyte 2013). While some citizen scientists and advocates seek to use animal sensing data to situate nonhumans in broader relational dynamics or spread awareness (Haraway 2007, 249–63; Pritchard 2013; Ray 2014; Gramaglia and Mélard 2019), others may be motivated by state or corporate interests in surveillance programs designed to capture control and not necessarily to reduce harm.

52. Garnett 2016.

53. Li 2007; Li 2015.

54. Li 2015, 186.

55. Tironi and Barandiarán 2014; Horowitz 2015.

56. Bedi 2013; Westman 2013; Hoogeveen 2016; Spiegel 2017; Dougherty 2019.

57. O'Faircheallaigh 2017.

58. Arkhipkin, Brickle, and Laptikhovsky 2010.

59. Fairhead, Leach, and Scoones 2012; Rocheleau 2015; Goldstein 2016.

60. Taber 2016. Of course, the oil industry in the Ecuadorian Amazon has also been subject to a notorious legal controversy over toxic contamination in Indigenous territories (Sawyer 2004, 2022; Cepek 2018).

61. Premier Oil manager, presentation at the Narrows Bar, Stanley, October 14, 2014.

62. Barry 2013.

63. Premier Oil Exploration & Production Limited, Falkland Islands Business Unit 2014, 21.

64. In Premier's (2018) updated EIS, which is outside the scope of this research, an "Uncertain" confidence rating replaced "Unlikely," and several activities or events were rated as "Uncertain." The updated EIS also includes separate project-specific guidelines for assessment, with levels of "Likelihood" complementing the degree of confidence. Still, all "Accidental" activities assessed with an "Uncertain" degree of confidence are considered to be "Unlikely" or "Very Unlikely," even if the initial impact is rated "High," such as contamination or oil ingestion by seabirds during a subsea well blowout (Premier 2018, 1320).

65. Fujimura 1988; Bowker 1994, 2000.

66. Premier Oil Exploration & Production Limited, Falkland Islands Business Unit 2014, 302.

67. Premier Oil Exploration & Production Limited, Falkland Islands Business Unit 2014, 36.

68. Viatori 2019.

69. Baylis et al. 2019.

70. Star and Griesemer 1989; Forsyth 2002, 140.

71. Star 2010.

72. Baylis et al. 2019, 4, 8.

73. Fujimura 1992.

74. After my fieldwork, SAERI researchers were able to use more GPS data to perform generalized additive modeling to predict the distribution of species for shorter distances and unsampled colonies. See Baylis et al. 2019.

75. Tabak, Poncet, Passfield, and Martinez del Rio 2014.

76. Bowker 2000, 655.

77. Premier Oil 2018.

78. The initial impact of an inshore oil spill on coastal communities is listed as "Moderate" on p. 1276 but "High" in the overall summary on p. 1326 in Premier Oil (2018).

79. Robbins 2003; Nixon 2011; Liboiron, Tironi, and Calvillo 2018.

80. Niebieskikwiat 2018.

81. Harding 2015.

82. The increased affordability and availability of wireless sensors with enhanced storage capacity may have made data "key mediators," capable of making a difference in environmental management (Benson 2010; Ray 2014; Gabrys 2016; Ascui, Haward, and Lovell 2018).

83. See also Latour 1999; Mansfield 2003; Waterton 2003; Hinchliffe 2008; Eden 2012; Nost 2015.

84. Freidberg 2007; Hunt 2014; Todd 2016.

85. Sundberg 2003.

86. Edelman 2013; Zoomers, Gekker, and Schäfer 2016.

87. Fortun 2004; Thatcher, O'Sullivan, and Mahmoudi 2016; Hughes 2017.

7. COLONIZING WITH NATIVES

1. Crosby 1986. Limitations of Crosby's approach are that, despite being a powerful framing, it papers over historical ambiguities and gives little attention to Indigenous Peoples' environmental transformations or selective incorporation of introduced species. See Cronon 1983; Trigger 2008.

2. In New Zealand's high country, Michéle Dominy (1995, 2000) argues, European descendant "Pakeha" construct a sense of home through a discourse of "authenticity" in connections to land. From a more critical perspective, David Hughes (2010) finds that,

by affiliating with land and building dams, white conservationist large-scale farmers are also asserting indigeneity in the "failed neo-Europe" of Zimbabwe (Hughes 2006, 269). In the less egregious postcolonial context of Botswana, Catie Gressier (2015) has described how white "minorities" resolve political insecurity through claims to "experiential autochtony," that is, emplacement in the environment. Finally, Janet McIntosh (2016) analyses what she calls "structural oblivion" among white settler descendants in postindependence Kenya, who deny having a privileged social status in order to defend their belonging. Despite originality and rigor, Dominy drew ire from anthropologists who accused her of representing the settlers' claims as comparable to that of the historically disadvantaged Ngai Tahu Maori (NZASA et al. 1990). Gressier's work raises similar problems, namely, her interpretation of the environment as a sanctuary from alienation follows what Neil Smith (1984, 25–28) called a "romanticized" ideology of nature: a dualism of geographical externality on the one hand (the ostensibly hostile Okavango Safari) and internalized universalism on the other (the spiritual morality of experiential autochthony). Gressier dismisses Hughes's critical political position but fails to appreciate his treatment of dam-making as a production of nature that undergirds white settlers' constrained "disregard" for racialized others (Hughes 2010, 12). For related analyses in South Africa, see also Crapanzano 1985, 21, 39 and Tamarkin 2020. On debates over "aboriginal Europeans" and indigeneity in the Mediterranean, see Herzfeld 1985, 1987; Heatherington 2010.

3. Kirksey and Helmreich 2010; Blair 2017.

4. Tully 1993; Mehta 1999; Elliot 2007; Milun 2011.

5. Subramaniam 2001; Lowe 2006; Kirksey and Helmreich 2010; Neale 2017; Swanson, Lien, and Ween 2018; Tsing et al., 2020. Again, Anna Tsing has been especially influential in pushing for a collaborative analysis of the more-than-human from the vantage point of feral life-forms with outsized impacts on landscapes, particularly the matsutake mushroom, a gourmet fungus that thrives in the deforested ruins of industrial capitalism (Tsing 2015b; Tsing et al., 2020). Proposing a different version of "anthropology beyond humanity," Tim Ingold argues that there is "nothing new" about multispecies ethnography. Ingold rejects the fashionable multispecies turn, and even the concept of species itself, for being imprecisely plural (Ingold 2013, 19). In response, Eben Kirksey (2015) draws on ethnographic work among taxonomists to argue that framing studies around a multiplicity of species is useful for making sense of, and caring for, neglected creatures. In this debate, I defer to Donna Haraway's (2007) judgment that, while the word species carries dubious connotations of race and sex, it may still be a useful tool for critically analyzing structures and discourses of colonialism. As she puts it, "The point is not to celebrate complexity but to become worldly and to respond" (Haraway 2007, 18, 41). See also Hartigan 2017.

6. Kirksey 2015.

7. Chakrabarty 2008; Davis and Todd 2017; Dicenta and Correa 2021.

8. It bears reminding that like humans, nonhumans are not reducible to conscious, empowered "agent-subjects" (Asad 2003, 67–99). Even as they show how extraordinary nonhumans defy anthropogenic damage, multispecies ethnographies tend to downplay hegemonic structures of power, including Western environmentalism. For critical perspectives on environmentalism, see Barton 2002; Heatherington 2010.

9. Helmreich 2009, 145–249; 2015, 62–72.

10. Helmreich 2005, 2009, 145–70. Helmreich notes that the species category "invasive" is a metataxonomic classification that challenges what counts as natural or biological: its content is not purely or monolithically scientific but rather woven into social and economic "contexts" and "parameters" that biologists help produce (Helmreich 2005, 109–10, 125). Moreover, in Hawaii, experts face a dilemma of whether to classify species associated with Native Hawaiian culture and livelihood—but transported by Polynesians

before the arrival of Captain Cook—as "native" or "introduced" (Helmreich 2005, 113). Helmreich shows convincingly how the category "native" becomes either a difference in "kind" or "degree" (Helmreich 2005, 116). In addition to Helmreich 2009, 2015, see Shukin 2009; Lowe 2010; Van Dooren 2014; Clark 2015; Blanchette 2020.

11. One councilor complained to me that when they discussed the islanders' right to self-determination with minority leaders of St. Vincent's government, they were rejected, saying: "Self-determination was put in place for Black people, and as I was white, it didn't apply."

12. Ironically, this line of thought echoes Kuper's (2003) controversial anti-essentialist doubts about political criteria for Indigenous Peoples. In fact, autochthony has been a more salient category than indigeneity in Africa. African groups struggled to become involved in the global Indigenous movement after collectivities of the Antipodes or Americas had already gained rights (Niezen 2003; Hodgson 2009). See Geschiere and Nyamnjoh 2000; Geschiere 2009; Hilgers 2011; Zenker 2011.

13. Thatcher 1978. Not all Falklands residents are Thatcherites. It bears reminding that Nicholas Ridley, a member of Thatcher's administration, pushed for the islanders to consent to a leaseback deal with Argentina.

14. De la Cadena and Starn 2007, 213; Clifford 2013, 84.

15. Ho 2004, 214.

16. Cohn 1983. See also Taylor 1983.

17. See Ned Blackhawk's (2006) discussion of Mark Twain's racist humor about the Goshute Shoshone of the North American Great Basin, whom he relabels "Goshoot Indians" in his autobiographical novel *Roughing It*.

18. Dean Brandon, "The Goose Problem," *Falkland Islands Magazine*, September 1904, JCNA; J. J. Latorre, "The Ona Indians: It Is Resolved to Draft Them out of Tierra del Fuego," *Falkland Islands Magazine [with clipping from The Magallanes]*, September 1898, JCNA.

19. Governor William Grey-Wilson, "Despatch to Alfred Lyttelton, March 24, 1904," JCNA, FIGA, B/26. Allardyce, "Despatch to Lyttelton, January 18, 1905," JCNA, FIGA, B/26. Also cited in Bernhardson 1989, 493.

20. According to Gusinde (1961, 143), "Indian hunters" were offered one pound sterling, which was also the price of a puma. See also Taussig 1993, 87. Bridges (2007, 266) offers anecdotal evidence of a price of five pounds sterling; interestingly, he also observed acts of resistance of renegade Indians "hunting" the white hunters (Bridges 2007, 314–15). For the link to Menéndez, see Marchante 2014. On European encounters with the Yagán and Selk'nam, see Chapman 2010. See also Display References, "Uttermost Part of the World Exhibition: Occupation," Museo Etnográfico "Juan Bautista Ambrosetti," Buenos Aires.

21. See Moore, Pandian, and Kosek 2003.

22. Darwin 1997, 730.

23. Bridges 2007, 34.

24. Darwin 1997, 730.

25. For a thoughtful, critical account of Darwin's perception of the Yagán people, see Chapman 2010, 157–89.

26. Miller 1998.

27. Darwin 2009, 269.

28. Miller 1998, 3.

29. Ahmed 2004.

30. Ahmed 2004, 97.

31. Baretta and Markoff 1978.

32. Darwin 1997, 220.

33. Austin et al. 2013.

34. On pests, biosecurity, and race in Latin America, see Wanderer 2020.

35. Packe, "Letter to Robinson, May 23, 1867," JNCA, H/25. For a parallel with the dingo in Australia, see Trigger et al. 2008.

36. Hamley 2016.

37. See Spruce and Smith 2018, 76–77, 80.

38. Dale, "Despatches to Directors of Falkland Islands Company, 1852–1853" (Bernhardson 1989, 313–14).

39. On the relationship between sheep breeding, racial purification, and white privilege in Tierra del Fuego, see Dicenta 2022.

40. An outlier to this trend in the historical record is a landowner named John Hamilton, whose environmental management of several islands in the early twentieth century included introducing various native South American animals, including foxes, skunks, and guanacos (Andean camelids). Farmers later considered descendants of the foxes as pests. Islanders spent holidays hunting the guanacos, and their remaining population on Staats Island presents a biogeographical enigma. Franklin and Grigione 2005; Spruce and Smith 2018, 14–15; Bernhardson 2020.

41. For sea lion as native pest, see J. J. Davis, "Letter to Colonial Secretary, n.d.," JNCA, FIGA, CSO 663/19. See also Bernhardson 1989, 706–7.

42. See Penrose 1775, 37; Bernhardson 1989, 91, 94, 111; Layman and Cameron 1995; Pernety (1771) 2013.

43. See Penrose 1775; Clayton 1981; Bernhardson 1989, 107. Yagán people also collected penguin eggs at the South American Missionary Society settlement at Keppel Island and traded them between the Falklands and Patagonia (Bridges 2007, 52; Philpott 2009, 20, 62).

44. A skeptical reader may view this framing as ahistorical, but I consulted conservationists as well as farmers about the similarities between "dipping" and non-native species eradication, and they agreed that this is an appropriate historicization. See Gressier 2015 for a useful distinction between agricultural and environmental "pests."

45. "Governor D'Arcy to the Earl of Kimberley, Colonial Report 1872, Government House, Stanley, March 22, 1873," JCNA.

46. Dipping was adopted as early as 1861. Lane, "Despatch No. 122," JCNA.

47. Robert Greenshields, "Letter to Governor D'Arcy, August 9, 1875," JCNA, FIGA, H/32. See also Bernhardson 1989, 414.

48. "Important Case under the Scab Ordinance," *Falkland Islands Magazine*, January 1896, JCNA.

49. James Lewis, "Letter to F. S. Sanguinetti, May 8, 1891," JCNA, FIGA, H/46. See also Bernhardson 1989, 448.

50. Anderson, "Letter to Brooks, January 15, 1886," JCNA, FIGA, H/39. See also Bernhardson 1989, 448.

51. Bernhardson 1989, 486.

52. Franklin 2007, 46–72.

53. Haraway 2007, 16.

54. See also Robin 1997, 2007 for a comparable symbolic value of nature for nation and empire.

55. Spruce and Smith 2018, 110.

56. Melville 1997; Woods 2015.

57. Franklin 2007, 5.

58. Birdlife International considers an "important bird area" anywhere with more than 1 percent of a species' breeding population. The Falklands are home to 70 percent of the world's breeding pairs of black-browed albatross (*Thalassarche melanophrys*) and 30–40 percent of some penguin species.

59. Non-native animal eradications include foxes, cats, mice, rats, and rabbits. See Poncet et al. 2011.

60. Fogwill 2012.

61. Moore 1968, 19. On the social imaginaries of biological invasions in Patagonia, see Archibald et al. 2020.

62. Mastnak, Elyachar, and Boellstorff 2014; Davis and Todd 2017; Lennon 2017; Whyte 2017; Salazar Parreñas 2018.

63. The use of insects is also rooted in a long British imperial tradition of using winged insects for brigading and espionage. See Russell 2001; Mitchell 2002; Kosek 2010.

64. James 2016.

65. BBC 2012; Simpson 2014; *Mercopress* 2015b.

66. More explicit identification of humans with insects authorized genocide in other contexts. See Mamdani 2001; Raffles 2011b. For critical analyses on anthropomorphism, see Milton 2005; Candea 2010.

67. Leader-Williams 2009.

68. The reasoning for this was that rats and reindeer would be attracted to bait containing anticoagulant poison. The rat eradication project starting in 2000 in South Georgia was the first of its kind in the South Atlantic, but Falkland Islanders have long considered rats to be pests, as well as proxies for measuring their own civility. See *Falkland Islands Magazine*, February 1902, May 1908, JCNA. Starting in 2001, Sally Poncet, a dedicated Antipodean conservationist who has lived in the Falklands for decades, collaborating with experts from New Zealand and North America as well as Falklands Conservation, has attempted to eradicate rats from about sixty-six islands in the archipelago, ranging from very small ones of only a few hectares to First Passage Island, the biggest island ever attempted by hand-broadcast at 750 hectares. See Tabak, Poncet, Passfield, Goheen, and Martinez del Rio 2014.

69. According to a local conservationist, for similar reasons, islanders have been complacent in protecting introduced rabbits.

70. See Milton 1993; Brosius 1999; Agrawal 2005; Checker 2005; Nixon 2011; Henne 2015; Barandiarán 2018.

71. The reindeer eradication underscores how Falkland Islanders harbor resentment toward the UK for "taking away" South Georgia from the management of the FIG. Islanders also sometimes refer to how South Georgia, unlike the Falklands, does not have any "natives," questioning why the international community accepts British sovereignty over South Georgia yet still maintains doubts about the Falklands' right to self-determination.

72. Ingold 1988.

73. Blair 2019b.

74. To decrease "diminishing returns," the government did not contract the Saami for herding in later phases of the eradication. Instead, with the leader of the Saami group's consent, shooters from the Norwegian Nature Inspectorate hunted the remaining reindeer.

75. Governor D'Arcy, "Colonial Report 1871, February 27, 1872," JCNA. For discussion of the whaling stations on South Georgia, see Robertson 1954; Basberg 2004; Gordon 2004.

76. Interview with conservation officer, December 15, 2014.

77. Trouillot 1997. In spite of his pride, Bobby did regret particular moments in his whaling experience, particularly when his colleagues caught accidentally pregnant mothers carrying nearly born calves.

78. Some islanders objecting to the eradication of reindeer on South Georgia argue that they may be the only members of the species that is unaffected by radiation after

the Chernobyl nuclear accident. However, a FIG environment officer pointed out that reindeer also populate the Kerguelen Islands in the southern Indian Ocean.

79. Anthropologists have been mired in debate about whether Indigenous groups in general, and Saami people in particular, have developed harmonic relations with earth-beings, such as reindeer, that constitute either wholly different ontologies or historically produced ethnopolitical discourses (Ingold 2015; Rada 2015; Willerslev, Vitebsky, and Alekseyev 2015).

80. Groff, Hamley et al., 2020.

81. Groff, Hamley et al., 2020.

82. The controversial grazing methods of white Zimbabwean ecologist and farmer Allan Savory have been influential in the Falklands. For critical assessments of Savory's work, see Monbiot 2014; Nordborg 2016; Ketcham 2017. I thank Timothy Ivey for these suggestions.

83. Trigger and colleagues (Trigger et al. 2008; Trigger, Toussaint, and Mulcock 2010) refer to the reintroduction of native species in "postsettler" Australia as "re-naturing." These studies offer similar fascinating insights on the construction of "nativeness." However, they propose a human-centered environment, in which settler colonialism is understood as a past event rather than a structure (Wolfe 2006). Martin and Trigger (2015) have moved the research agenda on indigeneity in Australia in a more comparable direction. As I have discussed in relation to the permanent frontier, here, "colonizing" with natives may be understood not necessarily as a past event or a closed structure, but rather as an open, unstable dynamic.

84. For an anthropological approach to the political ecology of land mines, see Kim 2016.

85. Hamley 2016.

86. Buckland and Edwards 1998.

87. Hamley 2016.

88. See Groff, Williams, and Gill 2020, who note that "The term 'tussock' is used to describe the clumping growth form of P. flabellata, while the species itself is commonly known as 'tussac'" (p. 4546).

89. Hamley 2016; Hamley et al. 2021.

90. See also Gressier 2015, 46.

91. Depending on one's perspective, their settler indigeneity resembles anti-essentialist notions of "diasporic natives" in the process of "becoming Indigenous," in Clifford's (2013, 84) framing, or "perfected natives" with "degrees of imperial sovereignty," in Stoler's (1995, 2006) sense.

92. On emergent autochthony, see Geschiere and Nyamnjoh 2000; Hilgers 2011. On "green extractivism" or "green productivism" that has made the expansion of mining and other extractive industries prerequisite to sustainable development, see Mendoza 2018; Dunlap and Jakobsen 2019; Riofrancos 2019; Jerez, Garcés, and Torres 2021; Voskoboynik and Andreucci 2021; Blair and Balcázar 2022.

93. See Bruyneel 2007; Moreton-Robinson 2007; Gausset, Kenrick, and Gibb 2011. These indeterminate claims have broader implications for the challenges of analyzing ethnogenesis and indigeneity in settler colonial contexts. Despite its wide influence, Crosby's (1986) concept of "Neo-Europes" gave short shrift to Indigenous environmental knowledge and the mutually transformative relationship between ethnicity and landscape (Cronon 1983; Sillitoe 1998; Anderson 2006; Trigger 2008; Johnson and Hunn 2010; Kimmerer 2015).

94. David Trigger and colleagues refer to similar reintroduction of native species in "postsettler" Australia as "re-naturing" (Trigger et al. 2008; Trigger, Toussaint, and Mulcock 2010). However, apart from Richard Martin and David Trigger (2015), who use

multispecies ethnography to address indigeneity, these studies generally propose a human-centered environment, in which imperial formations are understood as past events, rather than durable structures and processes (Wolfe 2006; Stoler 2016). Moreover, as Tomaz Mastnak, Julia Elyachar, and Tom Boellstorff (2014) observe, ecological optimists may have oversimplified native habitat restoration as tantamount to xenophobia (Raffles 2011b).

95. Cooper and Stoler 1997, 3; Helmreich 2005; Denvir 2020.

CONCLUSION

1. Stoler 2006.

2. While islanders celebrated by flocking to the newly opened beach, some Argentine supporters of the Malvinas cause argued that it was wrong to have cleared the mines unilaterally (Fowler 2020).

3. *Penguin News* 2020a.

4. Federici 2004; Tsing 2015a; Sultana 2021.

5. Scott 2014, 70.

6. Plans for a new port to replace the interim port in Stanley Harbour have continued to unfold despite economic uncertainty (*Penguin News* 2020b).

7. Geschiere and Nyamnjoh 2000; Trigger 2008; Hilgers 2011.

8. Todd 2017b, 2020.

9. See Moreton-Robinson 2007; Gausset, Kenrick, and Gibb 2011; Povinelli 2011; Coulthard 2014; Simpson 2014. This assertion of settler indigeneity should not be confused with other forms of "race-shifting" through overt claims to Native identity (Sturm 2011), nor should it be collapsed with racialization of Indigenous peoples as "mixed race" or *mestizo* (De la Cadena 2000; Andersen 2015), but comparison with homogenizing tendencies may offer new insights.

10. Hamley et al. 2021, 6.

11. González 2022; González Calderón 2022.

12. Ogden 2021, 54.

13. Wolfe 2006.

14. Coulthard 2014; Simpson 2014.

15. Rosaldo 1989; Bissell 2005; Piot 2010; Wilder 2015; Mitchell 2021.

16. On the broader role of networked governance in sovereignty claims, see Canfield 2022.

17. Watts 2006, 2008.

18. Interestingly, both Macri and now former UK prime minister David Cameron were implicated in the offshore tax haven scandal revealed in the Panama Papers leak. See ICIJ n.d.

19. Blair 2016.

20. FIG 2020.

21. FIG 2020.

22. Bates 2020; FIG 2020.

23. *Telam* 2020.

24. *Offshore Energy Today* 2015; Krauss 2020.

25. *Mercopress* 2020; Tomic 2020.

26. Bridge and Dodge 2020.

27. Ámbito 2021.

28. Livingstone 2022.

29. Castle 2020.

30. *Mercopress* 2013a; Osborne 2020.

31. *Mercopress* 2013b.

References

Academia Nacional de la Historia. 1967. *El episodio ocurrido en Puerto de la Soledad de Malvinas el 26 de agosto de 1833. Testimonios documentales.* Buenos Aires: Author.

Acosta, Alberto. 2017. "Post-Extractivism: From Discourse to Practice—Reflections for Action." *International Development Policy | Revue internationale de politique de développement* 9 (October): 77–101. https://doi.org/10.4000/poldev.2356.

Adelman, Jeremy. 2006. *Sovereignty and Revolution in the Iberian Atlantic.* Princeton, NJ: Princeton University Press.

Agamben, Giorgio. 1998. *Homo Sacer: Sovereign Power and Bare Life.* Stanford, CA: Stanford University Press.

Agier, Michel. 2010. *Managing the Undesirables: Refugee Camps and Humanitarian Government.* Translated by David Fernbach. Cambridge, UK: Polity.

Agrawal, Arun. 2005. *Environmentality.* Durham, NC: Duke University Press.

Ahmed, Sara. 2004. *The Cultural Politics of Emotion.* New York: Routledge.

Allan, John Robert. 2010. *Sealing in the Falkland Islands.* Stanley, Falkland Islands, UK.

Allen, Theodore W. 1994. *The Invention of the White Race: Racial Oppression and Social Control.* New York: Verso.

——. 2005. "On Roediger's Wages of Whiteness." *Cultural Logic: An Electronic Journal of Marxist Theory & Practice.* http://clogic.eserver.org/4-2/allen.html.

Allinson, Jamie, China Miéville, Richard Seymour, and Rosie Warren. 2021. *The Tragedy of the Worker: Towards the Proletarocene.* London: Verso.

Allison III, James Robert. 2015. *Sovereignty for Survival: American Energy Development and Indian Self-Determination.* New Haven, CT: Yale University Press.

Alvarez León, Luis F. 2018. "A Blueprint for Market Construction? Spatial Data Infrastructure(s), Interoperability, and the EU Digital Single Market." *Geoforum* 92 (June): 45–57. https://doi.org/10.1016/j.geoforum.2018.03.013.

Ámbito. 2021, October 2. "Petrolera británica abandona la explotación de hidrocarburos en las Islas Malvinas." https://www.ambito.com/politica/malvinas/petrolera-britanica-abandona-la-explotacion-hidrocarburos-las-islas-n5290996.

Amin, Ash. 2014. "Lively Infrastructure." *Theory, Culture & Society* 31 (7–8): 137–61.

Anand, Nikhil. 2017. *Hydraulic City: Water and the Infrastructures of Citizenship in Mumbai.* Durham, NC: Duke University Press.

Anand, Nikhil, Akhil Gupta, and Hannah Appel, eds. 2018. *The Promise of Infrastructure.* Durham, NC: Duke University Press Books.

Andersen, Chris. 2015. *"Métis": Race, Recognition, and the Struggle for Indigenous Peoplehood.* Vancouver: UBC Press.

Anderson, Benedict. 1991. *Imagined Communities: Reflections on the Origin and Spread of Nationalism.* New York: Verso.

Anderson, Christian M., and Amanda Huron. 2021. "The Mixed Potential of Salvage Commoning: Crisis and Commoning Practices in Washington, DC and New York City." *Antipode.* https://doi.org/10.1111/anti.12788.

Anderson, Judith M. 2014. "A Million Little Ways: Racism and Everyday Performances of Blackness in Buenos Aires." *African and Black Diaspora: An International Journal* 7 (2): 165–76. https://doi.org/10.1080/17528631.2014.908541.

Anderson, Mark. 2019. *From Boas to Black Power: Racism, Liberalism, and American An-thropology.* Stanford, CA: Stanford University Press.

Anderson, Virginia. 2006. *Creatures of Empire.* Oxford, UK: Oxford University Press.

Andrews, George Reid. 1980. *The Afro-Argentines of Buenos Aires, 1800–1900.* Madison: Wisconsin University Press.

Anghie, Antony. 2004. *Imperialism, Sovereignty and the Making of International Law.* Cambridge: Cambridge University Press.

Anker, Peder. 2002. *Imperial Ecology: Environmental Order in the British Empire, 1895–1945.* Cambridge, MA: Harvard University Press.

Anon. 2016, August 23. "Premier Oil to Submit Proposal for Oil." *Falkland Islands News Dump.* https://www.facebook.com/note.php?note_id=1118164181571529.

Appadurai, Arjun. 1981. "The Past as a Scarce Resource." *Man* 16 (2): 201–19.

———. 2013. *The Future as Cultural Fact: Essays on the Global Condition.* London: Verso.

Appel, Hannah. 2012a. "Offshore Work: Oil, Modularity, and the How of Capitalism in Equatorial Guinea." *American Ethnologist* 39 (4): 692–709.

———. 2012b. "Walls and White Elephants: Oil Extraction, Responsibility, and Infra-structural Violence in Equatorial Guinea." *Ethnography* 13 (4): 439–65.

———. 2019. *The Licit Life of Capitalism: US Oil in Equatorial Guinea.* Durham, NC: Duke University Press Books.

Appel, Hannah, Arthur Mason, and Michael Watts, eds. 2015. *Subterranean Estates: Life Worlds of Oil and Gas.* Ithaca, NY: Cornell University Press.

Archibald, Jessica L., Christopher B. Anderson, Mara Dicenta, Catherine Roulier, Kelly Slutz, and Erik A. Nielsen. 2020. "The Relevance of Social Imaginaries to Under-stand and Manage Biological Invasions in Southern Patagonia." *Biological Invasions* 22 (11): 3307–23.

Arendt, Hannah. (1951) 1968. *Imperialism: Part Two of The Origins of Totalitarianism.* Boston: Mariner Books.

Arkhipkin, Alexander, John Barton, Stuart Wallace, and Andreas Winter. 2013. "Close Cooperation between Science, Management and Industry Benefits Sustainable Ex-ploitation of the Falkland Islands Squid Fisheries." *Journal of Fish Biology* 83 (4): 905–20.

Arkhipkin, Alexander, Paul Brickle, and Vladimir Laptikhovsky. 2010. "The Use of Is-land Water Dynamics by Spawning Red Cod, Salilota Australis (Pisces: Moridae) on the Patagonian Shelf (Southwest Atlantic)." *Fisheries Research* 105 (3): 156–62.

Armitage, David, and Michael J. Braddick. 2009. *The British Atlantic World, 1500–1800.* 2nd ed. Basingstoke, UK: Palgrave Macmillan.

Arnaquq-Baril, Alethea. 2016. *Angry Inuk.* https://www.nfb.ca/film/angry_inuk/.

Arrighi, Giovanni. 1978. *The Geometry of Imperialism.* London: NLB.

Asad, Talal. 1973. *Anthropology & the Colonial Encounter.* Ithaca, NY: Ithaca Press.

———. 2003. *Formations of the Secular: Christianity, Islam, Modernity.* Stanford, CA: Stanford University Press.

Ascui, Francisco, Marcus Haward, and Heather Lovell. 2018. "Salmon, Sensors, and Translation: The Agency of Big Data in Environmental Governance." *Environment and Planning D: Society and Space* 36 (5): 905–25. https://doi.org/10.1177/0263775818766892.

Austin, Diane E., Thomas R. McGuire, and Rylan Higgins. 2007. "Work and Change in the Gulf of Mexico Offshore Petroleum Industry." *Research in Economic Anthro-pology* 24: 89–122.

Austin, Jeremy J., Julien Soubrier, Francisco J. Prevosti, Luciano Prates, Valentina Trejo, Francisco Mena, and Alan Cooper. 2013. "The Origins of the Enigmatic Falkland Islands Wolf." *Nature Communications* 4 (March): 1552.

Auty, Richard M. 1993. *Sustaining Development in Mineral Economies: The Resource Curse Thesis*. New York: Routledge.

Auyero, Javier. 2000. *Poor People's Politics: Peronist Survival Networks and the Legacy of Evita*. Durham, NC: Duke University Press.

Babidge, Sally, Fernanda Kalazich, Manuel Prieto, and Karina Yager. 2019. "'That's the Problem with That Lake; It Changes Sides': Mapping Extraction and Ecological Exhaustion in the Atacama." *Journal of Political Ecology* 26 (1): 738–60. https://doi.org/10.2458/v26i1.23169.

Babini, José. 1954. *La evolución del pensamiento científico argentino*. Buenos Aires: La Fragua.

Bailyn, Bernard. 2005. *Atlantic History: Concepts and Contours*. Cambridge, MA: Harvard University Press.

Baker, Janelle Marie. 2016, October 23. "Harvesting Ruins: The Im/Permanence of Work Camps and Reclaiming Colonized Landscapes in the Northern Alberta Oil Sands." *Engagement* (blog). https://aesengagement.wordpress.com/2016/10/23/harvesting-ruins-the-impermanence-of-work-camps-and-reclaiming-colonized-landscapes-in-the-northern-alberta-oil-sands/.

Baker, Lee D. 1998. *From Savage to Negro: Anthropology and the Construction of Race, 1896–1954*. Berkeley: University of California Press.

——. 2021. "The Racist Anti-Racism of American Anthropology." *Transforming Anthropology* 29 (2): 127–42. https://doi.org/10.1111/traa.12222.

Ballvé, Teo. 2020. *The Frontier Effect: State Formation and Violence in Colombia*. Ithaca, NY: Cornell University Press.

Banerjee, Mukulika. 2008. "Democracy, Sacred and Everyday: An Ethnographic Case from India." In *Democracy: Anthropological Approaches*, edited by Julia Paley. Santa Fe, NM: School for Advanced Research Press.

Barandiarán, Javiera. 2018. *Science and Environment in Chile: The Politics of Expert Advice in a Neoliberal Democracy*. Cambridge, MA: The MIT Press.

——. 2019. "Lithium and Development Imaginaries in Chile, Argentina and Bolivia." *World Development* 113 (January): 381–91. https://doi.org/10.1016/j.worlddev.2018.09.019.

Baretta, Silvio R. Duncan, and John Markoff. 1978. "Civilization and Barbarism: Cattle Frontiers in Latin America." *Comparative Studies in Society and History* 20 (4): 587–620.

Barker, Joanne, ed. 2006. *Sovereignty Matters: Locations of Contestation and Possibility in Indigenous Struggles for Self-Determination*. Lincoln: University of Nebraska Press.

Barry, Andrew. 2006. "Technological Zones." *European Journal of Social Theory* 9 (2): 239–53.

——. 2013. *Material Politics: Disputes along the Pipeline*. Oxford, UK: Wiley Blackwell.

Barton, Gregory Allen. 2002. *Empire Forestry and the Origins of Environmentalism*. Cambridge: Cambridge University Press.

Basberg, Bjorn L. 2004. *The Shore Whaling Stations at South Georgia*. Oslo, Norway: Novus Press.

Basso, Keith H. 1996. *Wisdom Sits in Places: Landscape and Language among the Western Apache*. Albuquerque: University of New Mexico Press.

Bates, James. 2020, December 28. "Falklands Fishing Companies First Reaction to the UK/EU Post Brexit Agreement." *MercoPress*. https://en.mercopress.com/2020/12/28/falklands-fishing-companies-first-reaction-to-the-uk-eu-post-brexit-agreement.

Baucom, Ian. 2005. *Specters of the Atlantic: Finance Capital, Slavery, and the Philosophy of History*. Durham, NC: Duke University Press.

Bauer, Tristán. 2005. *Iluminados por el fuego*. Buenos Aires: Editorial Sudamericana.

Baylis, Alastair M. M., Megan Tierney, Rachael A. Orben, Victoria Warwick-Evans, Ewan Wakefield, W. James Grecian, Phil Trathan et al. 2019. "Important At-Sea Areas of Colonial Breeding Marine Predators on the Southern Patagonian Shelf." *Scientific Reports* 9 (1): 8517. https://doi.org/10.1038/s41598-019-44695-1.

Baylis, Alastair, Megan Tierney, Rachael Orben, Iain Staniland, and Paul Brickle. 2018. "Geographic Variation in the Foraging Behaviour of South American Fur Seals." *Marine Ecology Progress Series* 596 (May): 233–45. https://doi.org/10.3354/meps12557.

BBC. 2012. "Falklands War Cemetery Vandalised," August 1, sec. UK. http://www.bbc .co.uk/news/uk-19069597.

Bedi, Heather P. 2013. "Environmental Mis-Assessment, Development and Mining in Orissa, India." *Development and Change* 44 (1): 101–23. https://doi.org/10.1111 /dech.12000.

Belcher, Oliver, Patrick Bigger, Ben Neimark, and Cara Kennelly. 2019. "Hidden Carbon Costs of the 'Everywhere War': Logistics, Geopolitical Ecology, and the Carbon Boot-Print of the US Military." *Transactions of the Institute of British Geographers* 45 (1): 65–80. https://doi.org/10.1111/tran.12319.

Belich, James. 2009. *Replenishing the Earth: The Settler Revolution and the Rise of the Angloworld*. Oxford, UK: Oxford University Press.

Benjamin, Walter. 1968. *Illuminations*. New York: Schocken Books.

Benson, Etienne. 2010. *Wired Wilderness: Technologies of Tracking and the Making of Modern Wildlife*. Baltimore, MD: JHU Press.

Benton, Lauren. 2009. *A Search for Sovereignty: Law and Geography in European Empires, 1400–1900*. Cambridge: Cambridge University Press.

Benwell, Matthew C. 2013. "Connecting Southern Frontiers: Argentina, the South West Atlantic and 'Argentine Antarctic Territory.'" In *Polar Geopolitics? Knowledges, Resources and Legal Regimes*, edited by R. Powell and K. Dodds. Cheltenham, UK: Edward Elgar.

——. 2016a. "Encountering Geopolitical Pasts in the Present: Young People's Everyday Engagements with Memory in the Falkland Islands." *Transactions of the Institute of British Geographers* 41 (2): 121–33.

——. 2016b. "Reframing Memory in the School Classroom: Remembering the Malvinas War." *Journal of Latin American Studies* 48 (2): 273–300.

——. 2017. "Connecting Ontological (In)Securities and Generation through the Everyday and Emotional Geopolitics of Falkland Islanders." *Social & Cultural Geography* 20 (4): 485–506. https://doi.org/10.1080/14649365.2017.1290819.

——. 2020. "Going Underground: Banal Nationalism and Subterranean Elements in Argentina's Falklands/Malvinas Claim." *Geopolitics* 25 (1): 88–108. https://doi.org/10 .1080/14650045.2017.1387776.

Benwell, Matthew C., and Klaus Dodds. 2011. "Argentine Territorial Nationalism Revisited: The Malvinas/Falklands Dispute and Geographies of Everyday Nationalism." *Political Geography* 30 (8): 441–49.

Benwell, Matthew C., Alejandro F. Gasel, and Andres Núñez. 2019. "Bringing the Falklands/Malvinas Home: Young People's Everyday Engagements with Geopolitics in Domestic Space." *Bulletin of Latin American Research* 39 (4): 424–38. https:// doi.org/10.1111/blar.13018.

Benwell, Matthew C., and Alasdair Pinkerton. 2020. "Everyday Invasions: *Fuckland*, Geopolitics, and the (Re)Production of Insecurity in the Falkland Islands." *Environment and Planning C: Politics and Space* 38 (6): 998–1016. https://doi.org/10.1177 /2399654420912434.

———. 2022. "Diplomatic Memories: Remembering the Falklands/Malvinas War through the Diplomatic Practices of Argentina and the Falkland Islands." *Journal of War & Culture Studies* 15 (3): 284–308. https://doi.org/10.1080/17526272.2022.2078539.

Berkes, Fikret, and Nancy J. Turner. 2006. "Knowledge, Learning and the Evolution of Conservation Practice for Social-Ecological System Resilience." *Human Ecology* 34 (4): 479. https://doi.org/10.1007/s10745-006-9008-2.

Bernal, Federico. 2011. *Malvinas y petróleo: Una historia de piratas*. Buenos Aires: Capital Intelectual.

Bernhardson, Wayne. 1989. "Land and Life in the Falkland Islands (Islas Malvinas)." PhD diss., University of California, Berkeley.

———. 2020. "Hamilton, John (1858–1945)." Dictionary of Falklands Biography. https://falklandsbiographies.org/biographies/224.

Bessire, Lucas. 2014. *Behold the Black Caiman: A Chronicle of Ayoreo Life*. Chicago: University of Chicago Press.

Betts, Alexander. 1987. *La verdad sobre las Malvinas: Mi tierra natal (hechos reales)*. Buenos Aires: Emece Editores.

Betts, Terrence. 2012. *A Falkland Islander Till I Die*. Weymouth, UK: Dudes Publishing.

Bigger, Patrick, and Benjamin D. Neimark. 2017. "Weaponizing Nature: The Geopolitical Ecology of the US Navy's Biofuel Program." *Political Geography* 60 (September): 13–22. https://doi.org/10.1016/j.polgeo.2017.03.007.

Bingham, M. 1998. "The Distribution, Abundance and Population Trends of Gentoo, Rockhopper and King Penguins in the Falkland Islands." *Oryx* 32 (3): 223–32.

———. 2002. "The Decline of Falkland Islands Penguins in the Presence of a Commercial Fishing Industry/La disminución de los pinguinos de las Islas Falklands en la presencia de actividades de pesca comercial." *Revista chilena de historia natural* 75: 805–18.

———. 2005. *The Falklands Regime*. AuthorHouse.

Bissell, William Cunningham. 2005. "Engaging Colonial Nostalgia." *Cultural Anthropology* 20 (2): 215–48.

Bjork-James, Carwil. 2015. "Hunting Indians: Globally Circulating Ideas and Frontier Practices in the Colombian Llanos." *Comparative Studies in Society and History* 57 (1): 98–129.

Blackhawk, Ned. 2006. *Violence over the Land: Indians and Empires in the Early American West*. Cambridge, MA: Harvard University Press.

Blackhawk, Ned, and Isaiah Lorado Wilner, eds. 2018. *Indigenous Visions: Rediscovering the World of Franz Boas*. New Haven, CT: Yale University Press.

Blackwell, David. 2009, February 4. "Desire Director Quits after Insider Trading Fine." *Financial Times*. http://www.ft.com/cms/s/0/ee29ac84-f2df-11dd-abe6-0000779fd2ac .html#axzz3ya3uBR3w.

Blaikie, Piers. 1985. *The Political Economy of Soil Erosion in Developing Countries*. London: Longman.

Blaikie, Piers, and Harold Brookfield. 1987. *Land Degradation and Society*. London: Methuen.

Blair, James J. A. 2013a, March 16. "Loud and Clear." *The Economist*. http://www .economist.com/news/americas/21573581-islanders-seek-sway-world-opinion -voting-stay-british-loud-and-clear.

———. 2013b, June 21. "Referendum Rewound." *The Economist*. http://www.economist .com/blogs/americasview/2013/06/falkland-islands.

———. 2013c, August 8. "Self-Determined." *The Economist*. http://www.economist.com /blogs/americasview/2013/08/argentina-falklands-and-un.

——. 2013d, March 11. "Sending Their Message." *The Economist*. http://www.economist
   .com/blogs/americasview/2013/03/falkland-islands-referendum.
——. 2014, February 28. "Treasure Islands?" *The Economist*. http://www.economist.com
   /blogs/americasview/2014/02/oil-and-gas-falklands.
——. 2016, August 4. "Brexit's South American Ripple Effect." *NACLA Report on the
   Americas*. https://nacla.org/news/2016/08/04/brexit%E2%80%99s-south-american
   -ripple-effect.
——. 2017. "Settler Indigeneity and the Eradication of the Non-Native: Self-Determination
   and Biosecurity in the Falkland Islands (Malvinas)." *Journal of the Royal Anthropological Institute* 23 (3): 580–602. https://doi.org/10.1111/1467-9655.12653.
——. 2019a. "South Atlantic Universals: Science, Sovereignty and Self-Determination in
   the Falkland Islands (Malvinas)." *Tapuya: Latin American Science, Technology and
   Society* 2 (1): 220–36. https://doi.org/10.1080/25729861.2019.1633225.
——. 2019b. "Splintered Hinterlands: Public Anthropology, Environmental Advocacy,
   and Indigenous Sovereignty." *Journal of Ethnobiology* 39 (1): 32–49. https://doi.org
   /10.2993/0278-0771-39.1.32.
——. 2022. "Tracking Penguins, Sensing Petroleum: 'Data Gaps' and the Politics of Marine Ecology in the South Atlantic." *Environment and Planning E: Nature and
   Space* 5 (1): 60–80. https://doi.org/10.1177/2514848619882938.
Blair, James J. A., and Ramón M. Balcázar. 2022, June 23. "Plurinational Climate Action:
   Environmental Governance beyond Green Extractivism." *Cultural Anthropology—
   Hot Spots, Fieldsites*. https://culanth.org/fieldsights/plurinational-climate-action
   -environmental-governance-beyond-green-extractivism.
Blake, Sally, Jane Cameron, and Joan Spruce. 2011. *Diddle Dee to Wire Gates: A Diction-
   ary of Falklands Vocabulary*. Stanley, Falkland Islands: Jane & Alastair Cameron
   Memorial Trust.
Blanchette, Alex. 2020. *Porkopolis: American Animality, Standardized Life, and the Fac-
   tory Farm*. Durham, NC: Duke University Press.
Blaser, Mario, Harvey A. Feit, and Glenn McRae, eds. 2004. *In the Way of Development:
   Indigenous Peoples, Life Projects and Globalization*. London: Zed.
Blay, S. K. N. 1986. "Self-Determination versus Territorial Integrity in Decolonization."
   International Law and Politics 18: 441–72.
Blumenberg, Hans. 1996. *Shipwreck with Spectator*. UK ed. Cambridge, MA: The MIT
   Press.
Boas, Franz. 1938. *The Mind of Primitive Man*. New York: Macmillan. http://archive.org
   /details/jstor-533099.
——. 1989. *A Franz Boas Reader: The Shaping of American Anthropology, 1883–1911*.
   Edited by George W. Stocking Jr. Chicago: University of Chicago Press.
Bocci, Paolo. 2019. "Utopian Conservation: Scientific Humanism, Evolution, and Island
   Imaginaries on the Galápagos Islands." *Science, Technology, & Human Values*, No-
   vember. https://doi.org/10.1177/0162243919889135.
Bond, David. 2013. "Governing Disaster: The Political Life of the Environment during
   the BP Oil Spill." *Cultural Anthropology* 28 (4): 694–715.
Bonilla, Yarimar. 2013. "Ordinary Sovereignty." *Small Axe* 17 (3): 151–65.
——. 2015. *Non-Sovereign Futures: French Caribbean Politics in the Wake of Disenchant-
   ment*. Chicago: University of Chicago Press.
Bonomi, Facundo. 2014. *Las Islas del Viento*. https://www.youtube.com/watch?v=s25v2Bk
   PVaY&feature=youtube_gdata_player.
Borucki, Alex. 2011. "The Slave Trade to the Río de la Plata, 1777–1812: Trans-Imperial
   Networks and Atlantic Warfare." *Colonial Latin American Review* 20 (1): 81–107.

Bougainville, Antoine Louis de. 1772. *Voyage around the World by Lewis de Bougainville 1766–9*. Translated by John Reinhold Foster. London. http://archive.org/details/VoyageAroundTheWorldByLewisDeBougainville1766-9.

Bound, Graham. 2006. *Falkland Islanders at War*. New ed. Barnsley, UK: Pen and Sword.

———. 2012. *Fortress Falklands*. Barnsley, UK: Pen and Sword.

Bound, Mensun, ed. 1998a. *Excavating Ships of War*. Oswestry, UK: Anthony Nelson Publishers.

———. 1998b. *Lost Ships: The Discovery and Exploration of the Ocean's Sunken Treasures*. New York: Simon & Schuster.

———. 2022, March 9 *"Endurance*: 'Finest Wooden Shipwreck I've Ever Seen.'" *BBC News*, sec. Science & Environment. https://www.bbc.com/news/science-environment-60654016.

Bowden, Gary. 1985. "The Social Construction of Validity in Estimates of US Crude Oil Reserves." *Social Studies of Science* 15 (2): 207–40.

Bowen, H. V., and John G. Reid, eds. 2012. *Britain's Oceanic Empire: Atlantic and Indian Ocean Worlds c. 1550–1850*. Cambridge: Cambridge University Press.

Bowker, Geoffrey C. 1994. *Science on the Run: Information Management and Industrial Geophysics at Schlumberger, 1920–1940*. Cambridge, MA: The MIT Press.

———. 2000. "Biodiversity Datadiversity." *Social Studies of Science* 30 (5): 643–83.

Boyer, Dominic. 2011. "Energopolitics and the Anthropology of Energy." *Anthropology News* 52 (5): 5–7.

Boyson, V. F. 1924. *The Falkland Islands*. Oxford, UK: Clarendon Press.

Braslavsky, Guido. 2013, March 13. "Tras el referéndum, Cristina llamó 'consorcio de okupas' a los kelpers." *Clarín*. http://www.clarin.com/politica/referendum-Cristina-consorcio-okupas-kelpers_0_881911828.html.

Braudel, Fernand. 1982. *On History*. Chicago: University of Chicago Press.

Breglia, Lisa. 2013. *Living with Oil: Promises, Peaks, and Declines on Mexico's Gulf Coast*. Austin: University of Texas Press.

Bridge, Gavin. 2008. "Global Production Networks and the Extractive Sector: Governing Resource-Based Development." *Journal of Economic Geography* 8 (3): 389–419.

Bridge, Gavin, and Alexander Dodge. 2020, October 8. "North Sea Oil: New Owners for Twilight Years Raise Questions of National Interest." *The Conversation* (blog). http://theconversation.com/north-sea-oil-new-owners-for-twilight-years-raise-questions-of-national-interest-145658.

Bridges, E. Lucas. 2007. *The Uttermost Part of the Earth*. New York: Rookery.

Briggs, Laura. 2002. *Reproducing Empire: Race, Sex, Science, and U.S. Imperialism in Puerto Rico*. Berkeley: University of California Press.

Briones, Claudia. 2005. *Cartografías argentinas: Políticas indigenistas y formaciones provinciales de alteridad*. Buenos Aires: EA.

Britain, David, and Andrea Sudbury. 2008. "What Can the Falkland Islands Tell Us about Diphthong Shift?" Working paper. Essex Research Reports in Linguistics, University of Essex, Colchester, UK.

Brockington, Dan. 2009. *Nature Unbound: Conservation, Capitalism and the Future of Protected Areas*. London: Earthscan.

Brodkin, Karen. 1998. *How Jews Became White Folks and What That Says about Race in America*. New Brunswick, NJ: Rutgers University Press.

Brosius, J. Peter. 1999. "Analyses and Interventions: Anthropological Engagements with Environmentalism." *Current Anthropology* 40 (3): 277–310.

Brown, Jacqueline Nassy. 2005. *Dropping Anchor, Setting Sail: Geographies of Race in Black Liverpool*. Princeton, NJ: Princeton University Press.

Brown, Vincent. 2010. *The Reaper's Garden: Death and Power in the World of Atlantic Slavery*. Cambridge, MA: Harvard University Press.

Brubaker, Rogers. 2005. "The 'Diaspora' Diaspora." *Ethnic and Racial Studies* 28 (1): 1–19. https://doi.org/10.1080/0141987042000289997.

Bruyneel, Kevin. 2007. *The Third Space of Sovereignty*. Minneapolis: University of Minnesota Press.

——. 2021. *Settler Memory: The Disavowal of Indigeneity and the Politics of Race in the United States*. Chapel Hill: University of North Carolina Press.

Bryce, Benjamin. 2019. "Undesirable Britons: South Asian Migration and the Making of a White Argentina." *Hispanic American Historical Review* 99 (2): 247–73. https://doi.org/10.1215/00182168-7370225.

Buck, Pem D. 2001. *Worked to the Bone: Race, Class, Power and Privilege in Kentucky*. New York: Monthly Review Press.

Buckland, Paul C., and Kevin J. Edwards. 1998. "Palaeoecological Evidence for Possible Pre-European Settlement in the Falkland Islands." *Journal of Archaeological Science* 25 (6): 599–602.

Bukharin, Nikolai. (1929) 1973. *Imperialism and World Economy*. New York: Monthly Review Press.

Bunker, Stephen G. 1990. *Underdeveloping the Amazon: Extraction, Unequal Exchange, and the Failure of the Modern State*. Chicago: University of Chicago Press.

Burkett, J. 2013. *Constructing Post-Imperial Britain: Britishness, "Race" and the Radical Left in the 1960s*. Houndmills, UK: Palgrave Macmillan.

Büscher, Bram, Wolfram Dressler, and Robert Fletcher, eds. 2014. *Nature Inc.: Environmental Conservation in the Neoliberal Age*. Tucson: University of Arizona Press.

Bushnell, Amy Turner. 2008. "Indigenous America and the Limits of the Atlantic World, 1493–1825." In *Atlantic History: A Critical Appraisal*, edited by Jack P. Greene and Philip D. Morgan. Oxford, UK: Oxford University Press.

Bustamante, Juan. 2013. "Falklands Votes in Sovereignty Referendum Rejected by Argentina." *Reuters*, March 10. http://www.reuters.com/article/2013/03/10/us-falklands-referendum-idUSBRE9290CK20130310.

Bustos-Gallardo, Beatriz, Manuel Prieto, and Jonathan Barton. 2015. *Ecología política en Chile. naturaleza, propiedad, conocimiento y poder*. Santiago, Chile: Editorial Universitaria.

Byrd, Jodi A. 2011. *The Transit of Empire*. Minneapolis: University of Minnesota Press.

Cain, P. J., and Tony Hopkins. 2002. *British Imperialism, 1688–2000*. Harlow, UK: Longman.

Callison, Candis. 2014. *How Climate Change Comes to Matter: The Communal Life of Facts*. Durham, NC: Duke University Press.

Callon, Michel. 1986. "Some Elements of a Sociology of Translation: Domestication of the Scallops and Fishermen of St. Brieuc Bay." In *Power, Action, and Belief*, edited by John Law, 196–233. London: Routledge and Kegan Paul.

Campero, Cecilia, Nathan J. Bennett, and Nayadeth Arriagada. 2022. "Technologies of Dispossession in the Blue Economy: Socio-Environmental Impacts of Seawater Desalination in the Antofagasta Region of Chile." *The Geographical Journal*. Published online January 17, 2022. https://doi.org/10.1111/geoj.12429.

Canclini, Arnoldo. 2000. *Malvinas. Su historia en historias*. Buenos Aires: Planeta.

——. 2009. *El fueguino: La cautivante historia de Jemmy Button*. Ushuaia, Arg.: Ediciones Monte Olivia.

Candea, Matei. 2010. "'I Fell in Love with Carlos the Meerkat': Engagement and Detachment in Human–Animal Relations." *American Ethnologist* 37 (2): 241–58.

Canfield, Matthew C. 2022. *Translating Food Sovereignty: Cultivating Justice in an Age of Transnational Governance*. Stanford, CA: Stanford University Press.

Carassai, Sebastián. 2021. "'The Dagger of Dispossession Will Be Ripped Out': The Malvinas/Falkland Islands in Argentine Song (1941–82)." *Journal of Latin American Studies* 53 (4): 717–40. https://doi.org/10.1017/S002226X21000766.

Carey, Grace. 2019, April 23. "Anthropology's 'Repugnant Others.'" American Ethnological Society. http://americanethnologist.org/features/reflections/anthropologys -repugnant-others.

Carse, Ashley. 2014. *Beyond the Big Ditch: Politics, Ecology, and Infrastructure at the Panama Canal*. Cambridge, MA: The MIT Press.

Carter, Paul. 2010. *The Road to Botany Bay: An Exploration of Landscape and History*. Minneapolis: University of Minnesota Press.

Castle, Stephen. 2020, December 12. "U.K. to Halt Subsidies for Fossil Fuel Projects Abroad." *New York Times*, sec. World. https://www.nytimes.com/2020/12/11 /world/europe/UK-fossil-fuel-subsidies.html.

Castree, Noel. 2008. "Neoliberalising Nature: The Logics of Deregulation and Reregulation." *Environment and Planning A* 40 (1): 131.

Cattelino, Jessica R. 2008. *High Stakes: Florida Seminole Gaming and Sovereignty*. Durham, NC: Duke University Press.

Cawkell, M. B. R., D. H. Maling, and E. M. Cawkell. 1960. *The Falkland Islands*. London: Macmillan.

Cawkell, Mary. 1983. *The Falkland Story, 1592–1982*. Oswestry, UK: A. Nelson.

——. 2001. *History of the Falkland Islands*. Oswestry, UK: Anthony Nelson.

Cepek, Michael. 2012. "The Loss of Oil: Constituting Disaster in Amazonian Ecuador." *Journal of Latin American and Caribbean Anthropology* 17 (3): 393–412.

——. 2018. *Life in Oil: Cofán Survival in the Petroleum Fields of Amazonia*. Austin: University of Texas Press.

Chakrabarty, Dipesh. 2008. "The Climate of History: Four Theses." *Critical Inquiry* 35: 197–222.

Chamosa, Oscar. 2008. "Indigenous or Criollo: The Myth of White Argentina in Tucumán's Calchaquí Valley." *Hispanic American Historical Review* 88 (1): 71–106.

Chao, Daniel, and Rosana Guber. 2021. *¿Qué hacer con los héroes? Los veteranos de Malvinas como problema de estado*. Madrid: Sb editorial.

Chapman, Anne. 2010. *European Encounters with the Yamana People of Cape Horn, before and after Darwin*. Cambridge: Cambridge University Press.

Chari, Sharad. 2019. "Recompositions in the Subaltern Sea: Geo-graphy as Errantry." In *Subaltern Geographies: Subaltern Studies, Space and the Geographical Imagination*, edited by Tariq Jazeel and Stephen Legg. Athens: University of Georgia Press.

Chasteen, John C. 1995. *Heroes on Horseback. A Life and Times of the Last Gaucho Caudillos*. Albuquerque: University of New Mexico Press.

Chatterjee, Partha. 1993. *The Nation and Its Fragments: Colonial and Postcolonial Histories*. Princeton, NJ: Princeton University Press.

Checker, Melissa. 2005. *Polluted Promises: Environmental Racism and the Search for Justice in a Southern Town*. New York: NYU Press.

Clancy, Laura. 2021. *Running the Family Firm: How the Royal Family Manages Its Image and Our Money*. Manchester: Manchester University Press.

Clark, Jonathan L. 2015. "Uncharismatic Invasives." *Environmental Humanities* 6: 29–52.

Clarke, Cyrene M. 1854. *Glances at Life upon the Sea, or Journal of a Voyage to the Antarctic Ocean: In the Brig Parana, of Sag Harbor, L.I., in the Years '53 '54: Description of Sea-Elephant Hunting among the Icy Islands of South Shetland, Capture of Whales, Scenery in the Polar Regions, &c.* Middletown, CT: Charles H. Pelton.

Clausen, Andrea Patricia, and Klemens Pütz. 2002. "Recent Trends in Diet Composition and Productivity of Gentoo, Magellanic and Rockhopper Penguins in the Falkland Islands." *Aquatic Conservation: Marine and Freshwater Ecosystems* 12 (1): 51–61.

Clayton, William. 1981. "An Account of Falkland Islands." *Falkland Islands Journal* (Originally published in *The Philosophical Transactions of the Royal Society [1776]*): 99–108.

Clifford, James. 2013. *Returns: Becoming Indigenous in the Twenty-First Century.* Cambridge, MA: Harvard University Press.

Cohen, Robin. 2008. *Global Diasporas: An Introduction.* 2nd ed. London: Routledge.

Cohn, Bernard. 1983. "Representing Authority in Victorian India." In *The Invention of Tradition*, edited by Eric J. Hobsbawm, 165–209. Cambridge: Cambridge University Press.

Colectivo Situaciones. 2011. *19 and 20: Notes for a New Social Protagonism.* Edited by Malav Kanuga, Stevphen Shukaitis, and James Blair. Translated by Nate Holdren and Sebastián Touza. New York: Minor Compositions and Common Notions (Autonomedia).

Coles, Kimberley. 2007. *Democratic Designs: International Intervention and Electoral Practices in Postwar Bosnia-Herzegovina.* Ann Arbor: University of Michigan Press.

Collins, John. 2011. "Culture, Content, and the Enclosure of Human Being: UNESCO's 'Intangible' Heritage in the New Millennium." *Radical History Review* 109: 121–35.

Colwell, Chip. 2019. *Plundered Skulls and Stolen Spirits: Inside the Fight to Reclaim Native America's Culture.* Chicago: University of Chicago Press.

Connell, John, and Robert Aldrich. 2020. *The Ends of Empire: The Last Colonies Revisited.* Singapore: Palgrave Macmillan.

Cooper, Anderson. 2022. "Canada's Unmarked Graves: How Residential Schools Carried out 'Cultural Genocide' against Indigenous Children." *60 Minutes.* CBS News. https://www.cbsnews.com/news/canada-residential-schools-unmarked-graves -indigenous-children-60-minutes-2022-06-05/.

Cooper, Frederick. 1996. *Decolonization and African Society: The Labor Question in French and British Africa.* Cambridge: Cambridge University Press.

——. 2014. *Citizenship between Empire and Nation: Remaking France and French Africa, 1945–1960.* Princeton, NJ: Princeton University Press.

Cooper, Frederick, and Ann Laura Stoler, eds. 1997. *Tensions of Empire: Colonial Cultures in a Bourgeois World.* Berkeley: University of California Press.

Cordis, Shanya. 2019. "Settler Unfreedoms." *American Indian Culture and Research Journal* 43 (2): 9–23. https://doi.org/10.17953/aicrj.43.2.cordis.

Coronil, Fernando. 1997. *The Magical State: Nature, Money, and Modernity in Venezuela.* Chicago: University of Chicago Press.

Coulthard, Glen Sean. 2014. *Red Skin, White Masks: Rejecting the Colonial Politics of Recognition.* Minneapolis: University of Minnesota Press.

Crapanzano, Vincent. 1985. *Waiting: The Whites of South Africa.* New York: Random House.

Cronon, William. 1983. *Changes in the Land: Indians, Colonists, and the Ecology of New England*. New York: Farrar, Straus and Giroux.

———. 1987. "Revisiting the Vanishing Frontier: The Legacy of Frederick Jackson Turner." *Western Historical Quarterly* 18 (2): 157–76.

———. 1991. *Nature's Metropolis: Chicago and the Great West*. New York: W. W. Norton.

Crosby, Alfred W. 1986. *Ecological Imperialism: The Biological Expansion of Europe, 900–1900*. Cambridge: Cambridge University Press.

Croxall, John P., S. J. McInness, and P. A. Prince. 1984. "The Status and Conservation of Seabirds at the Falkland Islands." In *Status and Conservation of the World's Seabirds*, edited by J. P. Croxall, Peter G. H. Evans, and Ralph W. Schreiber. Cambridge, UK: International Council for Bird Preservation.

Dagnino, Renato, Hernán Thomas, and Amílcar Davyt. 1996. "El pensamiento en ciencia, tecnología y sociedad en Latinoamérica: Una interpretación política de su trayectoria." *Redes* 3 (7): 13–51.

Darwin, Charles. 1997. *The Voyage of the Beagle*. Ware, UK: Wordsworth Editions.

———. 2009. *The Expression of the Emotions in Man and Animals*. Edited by Francis Darwin. Cambridge: Cambridge University Press.

Darwin, Charles, David Kohn, and William Montgomery. 1985. *The Correspondence of Charles Darwin, Volume I: 1821–1836*. Edited by Frederick Burkhardt and Sydney Smith. Cambridge: Cambridge University Press.

Darwin, Charles, Robert FitzRoy, and Phillip Parker King. 2015. *Narrative of the Surveying Voyages of His Majesty's Ships Adventure and Beagle: Between the Years 1826 and 1836*. Cambridge: Cambridge University Press.

Daston, Lorraine J., and Peter Galison. 2010. *Objectivity*. New York: Zone Books.

Daversa, Fabiana. 2012. *La balsa de Malvina*. Buenos Aires: Aguilar, Altea, Taurus, Alfaguara.

Davey, Ryan, and Insa Lee Koch. 2021. "Everyday Authoritarianism: Class and Coercion on Housing Estates in Neoliberal Britain." *PoLAR: Political and Legal Anthropology Review* 44 (1): 43–59. https://doi.org/10.1111/plar.12422.

David, Andrew, Felipe Fernandez-Armesto, Carlos Novi, and Glyndwr Williams, eds. 2002. *The Malaspina Expedition 1789 to 1794: Journal of the Voyage by Alejandro Malaspina: Cadiz to Panama*. London: Hakluyt Society.

Davies, T. H. 1971. *The Sheep and Cattle Industries of the Falkland Islands: A Report*. London: Foreign and Commonwealth Office Overseas Development Administration.

Davis, Heather, and Zoe Todd. 2017. "On the Importance of a Date, or, Decolonizing the Anthropocene." *ACME: An International Journal for Critical Geographies* 16 (4): 761–80.

Davis, Janae, Alex A. Moulton, Levi Van Sant, and Brian Williams. 2019. "Anthropocene, Capitalocene, . . . Plantationocene?: A Manifesto for Ecological Justice in an Age of Global Crises." *Geography Compass* 13 (5): e12438. https://doi.org/10.1111/gec3.12438.

Davis, Mike. 2017. *Late Victorian Holocausts: El Niño Famines and the Making of the Third World*. London: Verso.

De Genova, Nicholas. 2005. *Working the Boundaries: Race, Space, and "Illegality" in Mexican Chicago*. Durham, NC: Duke University Press.

Dehnhard, Nina, Maud Poisbleau, Laurent Demongin, Katrin Ludynia, Miguel Lecoq, Juan F. Masello, and Petra Quillfeldt. 2013. "Survival of Rockhopper Penguins in Times of Global Climate Change." *Aquatic Conservation: Marine and Freshwater Ecosystems* 23 (5): 777–89.

De la Cadena, Marisol. 2000. *Indigenous Mestizos: The Politics of Race and Culture in Cuzco, Peru, 1919–1991*. Durham, NC: Duke University Press.

De la Cadena, Marisol, and Orin Starn, eds. 2007. *Indigenous Experience Today*. Oxford, UK: Bloomsbury.

De la Fuente, Ariel. 2000. *Children of Facundo: Caudillo and Gaucho Insurgency during the Argentine State-Formation Process*. Durham, NC: Duke University Press.

Delaney, Jeane. 1996. "Making Sense of Modernity: Changing Attitudes toward the Immigrant and the Gaucho in Turn-of-the-Century Argentina." *Comparative Studies in Society and History* 38 (3): 434–59.

Delbourgo, James, and Nicholas Dew, eds. 2007. *Science and Empire in the Atlantic World*. New York: Routledge.

Dempsey, Jessica. 2016. *Enterprising Nature: Economics, Markets, and Finance in Global Biodiversity Politics*. Chichester, UK: John Wiley & Sons.

Denvir, Daniel. 2020. *All-American Nativism: How the Bipartisan War on Immigrants Explains Politics as We Know It*. London: Verso.

Destefani, Laurio H. 1982. *The Malvinas, the South Georgias and South Sandwich Islands, the Conflict with Britain*. Translated by Martha Heath and Ruth James. Buenos Aires: Edipress S. A.

Diaz, Vicente M. 2006. "Creolization and Indigeneity." *American Ethnologist* 33 (4): 576–78.

Dicenta, Mara. 2022. "White Animals: Racializing Sheep and Beavers in the Argentinian Tierra del Fuego." *Latin American and Caribbean Ethnic Studies*, 1–22.

Dicenta, Mara, and Gonzalo Correa. 2021. "Worlding the End: A Story of Colonial and Scientific Anxieties over Beavers' Vitalities in the Castorcene." *Tapuya: Latin American Science, Technology and Society* 4 (1): 1973290. https://doi.org/10.1080/25729861.2021.1973290.

Docherty, Paddy. 2021. *Blood and Bronze: The British Empire and the Sack of Benin*. London: Hurst.

Dodds, Klaus. 2002. *Pink Ice: Britain and the South Atlantic Empire*. London: I. B. Tauris.

——. 2013. "White Puffs of Smoke? The Falkland Islands Referendum." *The Huffington Post*. http://www.huffingtonpost.co.uk/professor-klaus-dodds/falkland-islands-referendum_b_2908572.html.

Dodds, Klaus, and Matthew C. Benwell. 2010. "More Unfinished Business: The Falklands/Malvinas, Maritime Claims, and the Spectre of Oil in the South Atlantic." *Environment and Planning D: Society and Space* 28: 571–80.

Doel, Ronald E., Urban Wråkberg, and Suzanne Zeller. 2014. "Science, Environment, and the New Arctic." *Journal of Historical Geography* 44 (April): 2–14. https://doi.org/10.1016/j.jhg.2013.12.003.

Dominy, Michéle. 1995. "White Settler Assertions of Native Status." *American Ethnologist* 22 (2): 358–74.

——. 2000. *Calling the Station Home: Place and Identity in New Zealand's High Country*. Lanham, MD: Rowman & Littlefield.

Donaghy, Aaron. 2014. *The British Government and the Falkland Islands, 1974–79*. Houndmills, UK: Palgrave Macmillan.

Dougherty, Michael L. 2019. "Boom Times for Technocrats? How Environmental Consulting Companies Shape Mining Governance." *Extractive Industries and Society* 6 (2): 443–53. https://doi.org/10.1016/j.exis.2019.01.007.

Drayton, Richard. 2000. *Nature's Government: Science, Imperial Britain, and the "Improvement" of the World*. New Haven, CT: Yale University Press.

Du Bois, W. E. B. 1965 [1946]. *The World and Africa*. New York: International Publishers.

——. 1998. *Black Reconstruction in America 1860–1880*. New York: Free Press.

Du Plessis, Gitte, Cameron Grimm, Kyle Kajihiro, and Kenneth Gofigan Kuper. 2022. "Sustaining Empire: Conservation by Ruination at Kalama Atoll." *Environment and Planning D: Society and Space* 40 (4): 706–25. https://doi.org/10.1177/026377 58221102156.

Dunlap, Alexander, and Jostein Jakobsen. 2019. *The Violent Technologies of Extraction: Political Ecology, Critical Agrarian Studies and the Capitalist Worldeater.* Cham, Switzerland: Palgrave Pivot.

Eckert, Penelope. 1980. "Diglossia: Separate and Unequal." *Linguistics* 18: 56–64.

The Economist. 2016, July 9. "What's the Point?" http://www.economist.com/international /2016/07/07/whats-the-point.

Edelman, Marc. 1992. *The Logic of the Latifundio.* Stanford, CA: Stanford University Press.

———. 1999. *Peasants against Globalization: Rural Social Movements in Costa Rica.* Stanford, CA: Stanford University Press.

———. 2005. "Bringing the Moral Economy Back in . . . to the Study of 21st-Century Transnational Peasant Movements." *American Anthropologist* 107 (3): 331–45.

———. 2013. "Messy Hectares: Questions about the Epistemology of Land Grabbing Data." *Journal of Peasant Studies* 40 (3): 485–501.

———. 2020. "From 'Populist Moment' to Authoritarian Era: Challenges, Dangers, Possibilities." *Journal of Peasant Studies* 47 (7): 1418–44. https://doi.org/10.1080 /03066150.2020.1802250.

Eden, Sally. 2012. "Counting Fish: Performative Data, Anglers' Knowledge-Practices and Environmental Measurement." *Geoforum* 43 (5): 1014–23.

Edwards, Brent Hayes. 2003. *The Practice of Diaspora: Literature, Translation, and the Rise of Black Internationalism.* Cambridge, MA: Harvard University Press.

Edwards, Erika Denise. 2020. *Hiding in Plain Sight: Black Women, the Law, and the Making of a White Argentine Republic.* Tuscaloosa: University Alabama Press.

Edwards, Ryan C. 2021. *A Carceral Ecology: Ushuaia and the History of Landscape and Punishment in Argentina.* Oakland: University of California Press.

Eilenberg, Michael. 2014. "Frontier Constellations: Agrarian Expansion and Sovereignty on the Indonesian-Malaysian Border." *Journal of Peasant Studies* 41 (2): 157–82.

Elden, Stuart. 2006. "Contingent Sovereignty, Territorial Integrity and the Sanctity of Borders." *SAIS Review of International Affairs* 26 (1): 11–24.

Elena, Eduardo. 2011. *Dignifying Argentina: Peronism, Citizenship, and Mass Consumption.* Pittsburgh, PA: University of Pittsburgh Press.

Elkins, Caroline. 2022. *Legacy of Violence: A History of the British Empire.* New York: Knopf.

Ellerby, Clive Richard. 1990. "British Interests in the Falkland Islands: Economic Development, the Falklands Lobby and the Sovereignty Dispute 1945–1989." PhD thesis, University of Oxford.

Elliot, J. H. 2007. *Empires of the Atlantic World: Britain and Spain in America 1492–1830.* New Haven, CT: Yale University Press.

Ellner, Steve. 2007. "Toward a 'Multipolar World': Using Oil Diplomacy to Sever Venezuela's Dependence." *NACLA Report on the Americas* 40 (5): 15–22. https://doi.org /10.1080/10714839.2007.11722295.

Enns, Charis, Brock Bersaglio, and Adam Sneyd. 2019. "Fixing Extraction through Conservation: On Crises, Fixes and the Production of Shared Value and Threat." *Environment and Planning E: Nature and Space* 2 (4): 967–88. https://doi.org/10.1177 /2514848619867615.

Escobar, Arturo. 2001. "Culture Sits in Places: Reflections on Globalism and Subaltern Strategies of Localization." *Political Geography* 20: 139–74.

Fairhead, James, Melissa Leach, and Ian Scoones. 2012. "Green Grabbing: A New Appropriation of Nature?" *Journal of Peasant Studies* 39 (2): 237–61.

Falkland Islands Government. 2012. *Fisheries Department Fisheries Statistics*. Vol. 16 (2002–2011). Stanley, Falkland Islands, UK: Crown.

——. n.d. "Falkland Calamari (Loligo) Description." Accessed October 25, 2022. https://www.falklands.gov.fk/fisheries/overview/commercial-species/falkland-calamari-loligo.

Fanning, Edmund. 1924. *Voyages & Discoveries in the South Seas, 1792–1832*. Salem, MA: Marine Research Society. http://archive.org/details/voyagesroundwor00fanngoog.

Fanon, Frantz. 2007 [1963]. *The Wretched of the Earth*. New York: Grove Press.

Fassin, Didier. 2011. *Humanitarian Reason: A Moral History of the Present*. Berkeley: University of California Press.

Federici, Silvia. 2004. *Caliban and the Witch*. New York: Autonomedia.

Ferguson, Charles. 1959. "Diglossia." *Word* 15: 325–40.

——. 1985. "Diglossia." In *Language and Social Context*. New York: Penguin.

Ferguson, James. 2005. "Seeing Like an Oil Company: Space, Security, and Global Capital in Neoliberal Africa." *American Anthropologist* 107 (3): 377–82. https://doi.org/10.1525/aa.2005.107.3.377.

——. 2013. "Declarations of Dependence: Labour, Personhood, and Welfare in Southern Africa." *Journal of the Royal Anthropological Institute* 19 (2): 223–42.

Fernández, Cristina, Jorge Capitanich, Héctor Marcos Timerman, Eduardo Zuain, Daniel Filmus, and Javier Esteban Figueroa. 2014. *The International Community and the Malvinas Question*. Argentine Ministry of Foreign Affairs and Worship.

Ferns, H. S. 1960. *Britain and Argentina in the Nineteenth Century*. Oxford, UK: Clarendon Press.

Ferry, Elizabeth Emma, and Mandana E. Limbert. 2008. *Timely Assets: The Politics of Resources and Their Temporalities*. Santa Fe, NM: School for Advanced Research Press.

Fidler, Stephen. 1997, April 1. "Financial Times Survey: Falkland Islands." *Financial Times*.

FIG. 2020. "Focus on Falkland Islands Fishing Industry." BrexitFalklands. https://www.brexitfalklands.gov.fk/fishing/.

FIG Legislative Assembly. 2015. "Response to Government of Argentina's Announcement of Further Action against Hydrocarbons Exploration in the Falklands." April 17. http://www.falklands.gov.fk/response-to-government-of-argentinas-announcement-of-further-action-against-hydrocarbons-exploration-in-the-falklands/.

FIG Policy Unit. 2013. "Executive Council Publishes Hydrocarbon Development Policy Statement." July 26. http://www.falklandnews.com/public/story.cfm?get=6600&source=3.

Figueira, José Antonio da Fonseca. 1985. *David Jewett; Una biografía para la historia de las Malvinas*. Buenos Aires: Sudamericana-Planeta.

Figueroa-Vásquez, Yomaira C. 2020. *Decolonizing Diasporas: Radical Mappings of Afro-Atlantic Literature*. Evanston, IL: Northwestern University Press.

Finchelstein, Federico. 2017. *From Fascism to Populism in History*. Oakland: University of California Press.

Fishman, Joshua. 1967. "Bilingualism with and without Diglossia; Diglossia with and without Bilingualism." *Journal of Social Issues* 23 (2): 29–38.

Fitte, Ernesto J. 1966. *La agresión norteamericana a las Islas Malvinas*. Buenos Aires: Emece Editores.

FitzRoy, R. 1839. *Proceedings of the Second Expedition, 1831–36, under the Command of Captain Robert Fitz-Roy, R.N.* London: Henry Colburn.

Fogwill, Rodolfo. 2012. *Los pichiciegos*. Buenos Aires: Editorial el Ateneo.

Folch, Christine. 2013. "Surveillance and State Violence in Stroessner's Paraguay: Itaipú Hydroelectric Dam, Archive of Terror." *American Anthropologist* 115 (1): 44–57.

———. 2019. *Hydropolitics: The Itaipu Dam, Sovereignty, and the Engineering of Modern South America*. Princeton, NJ: Princeton University Press.

Forsyth, Tim. 2002. *Critical Political Ecology: Environmental Knowledge, Policy and Risk*. New York: Routledge.

Fortun, Kim. 2004. "From Bhopal to the Informating of Environmentalism: Risk Communication in Historical Perspective." *Osiris* 19: 283–96.

Foulkes, Haroldo. 1987. *Los kelpers: En las Malvinas y en la Patagonia*. Buenos Aires: Corregidor.

Fowler, John. 2012. *1982 & All That*. Stanley, Falkland Islands, UK: Amazon.

———. 2020, November 20. "A New Low for Argentine Sovereignty Claim Propagandists." *Penguin News*. https://penguin-news.com/opinion-piece/2020/a-new-low-for-argentine-sovereignty-claim-propagandists/.

Frank, Thomas. 2020. *The People, No: A Brief History of Anti-Populism*. 3rd ed. New York: Metropolitan Books.

Frankenberg, Ruth. 1993. *White Women, Race Matters: The Social Construction of Whiteness*. Minneapolis: University of Minnesota Press.

Franklin, Sarah. 2007. *Dolly Mixtures: The Remaking of Genealogy*. Durham, NC: Duke University Press.

Franklin, William L., and Melissa M. Grigione. 2005. "The Enigma of Guanacos in the Falkland Islands: The Legacy of John Hamilton." *Journal of Biogeography* 32: 661–75.

Freedman, Lawrence. 2004. *The Official History of the Falklands Campaign*. 2 vols. London: Routledge.

Freedman, Lawrence, and Virginia Gamba-Stonehouse. 1990. *Signals of War: The Falklands Conflicts of 1982*. London: Faber & Faber.

Freidberg, Susanne. 2007. "Supermarkets and Imperial Knowledge." *Cultural Geographies* 14 (3): 321–42.

Freidenberg, Judith Noemí. 2009. *The Invention of the Jewish Gaucho: Villa Clara and the Construction of Argentine Identity*. Austin: University of Texas Press.

Fujimura, Joan. 1988. "The Molecular Biological Bandwagon in Cancer Research: Where Social Worlds Meet." *Social Problems* 35 (3): 261–83.

———. 1992. "Crafting Science: Standardized Packages, Boundary Objects, and 'Translation.'" In *Science as Practice and Culture*, edited by A. Pickering, 168–214. Chicago: University of Chicago Press.

Fynn, Lino. 2014. *Falklandness*. Stanley, Falkland Islands, UK: The Printshop.

Gabrys, Jennifer. 2016. *Program Earth: Environmental Sensing Technology and the Making of a Computational Planet*. Minneapolis: University of Minnesota Press.

———. 2019. "Sensors and Sensing Practices: Reworking Experience across Entities, Environments, and Technologies." *Science, Technology, & Human Values* 44 (5): 723–36. https://doi.org/10.1177/0162243919860211.

Gal, Susan. 1987. "Codeswitching and Consciousness in the European Periphery." *American Ethnologist* 14: 637–53.

Galindo, Alberto Flores. 1993. *Buscando un Inca*. Mexico City: Consejo Nacional para la Cultura y las Artes.

Gallo, Ezequiel. 1983. *La pampa gringa*. Buenos Aires: Edhasa.

Gamerro, Carlos. 2014. *The Islands*. Translated by Ian Barnett. High Wycombe, UK: And Other Stories.

Gandy, Matthew. 2003. *Concrete and Clay: Reworking Nature in New York City*. Cambridge, MA: The MIT Press.

Garland, David. 2014. "What Is a 'History of the Present'? On Foucault's Genealogies and Their Critical Preconditions." *Punishment & Society* 16 (4): 365–84. https://doi.org/10.1177/1462474514541711.

Garnett, Emma. 2016. "Developing a Feeling for Error: Practices of Monitoring and Modelling Air Pollution Data." *Big Data & Society* 3 (2): 2053951716658061. https://doi.org/10.1177/2053951716658061.

Garrett, Paul B. 2007. "'Say It Like You See It': Radio Broadcasting and the Mass Mediation of Creole Nationhood in St. Lucia." *Identities* 14: 135–60.

Gausset, Quentin, Justin Kenrick, and Robert Gibb. 2011. "Indigeneity and Autochthony: A Couple of False Twins?" *Social Anthropology* 19 (2): 135–42.

Geertz, Clifford. 1963a. *Agricultural Involution: The Process of Ecological Change in Indonesia*. Berkeley: University of California Press.

——. 1963b. "The Integrative Revolution: Primordial Sentiments and Politics in the New States." In *Old Societies and New States: The Quest for Modernity in Asia and Africa*. New York: Free Press.

Gellner, Ernest. 1983. *Nations and Nationalism*. Ithaca, NY: Cornell University Press.

Gereffi, Gary, John Humphrey, and Timothy Sturgeon. 2005. "The Governance of Global Value Chains." *Review of International Political Economy* 12 (1): 78–104.

Gereffi, Gary, and M. Korzeniewicz. 1994. *Commodity Chains and Global Capitalism*. Westport, CT: Praeger.

Germond-Duret, Celine. 2022. "Framing the Blue Economy: Placelessness, Development and Sustainability." *Development and Change* 53 (2): 308–34. https://doi.org/10.1111/dech.12703.

Geschiere, Peter. 2009. *The Perils of Belonging: Autochthony, Citizenship, and Exclusion in Africa and Europe*. Chicago: University of Chicago Press.

Geschiere, Peter, and Francis Nyamnjoh. 2000. "Capitalism and Autochthony: The Seesaw of Mobility and Belonging." *Public Culture* 12 (2): 423–52.

Getachew, Adom. 2019. *Worldmaking after Empire: The Rise and Fall of Self-Determination*. Princeton, NJ: Princeton University Press.

Gidwani, Vinay, and Rajyashree N. Reddy. 2011. "The Afterlives of 'Waste': Notes from India for a Minor History of Capitalist Surplus." *Antipode* 43 (5): 1625–58.

Gillis, John R. 2004. *Islands of the Mind: How the Human Imagination Created the Atlantic World*. New York: Palgrave Macmillan.

Gilroy, Paul. 1987. *"There Ain't No Black in the Union Jack": The Cultural Politics of Race and Nation*. Chicago: University of Chicago Press.

——. 1993. *The Black Atlantic: Modernity and Double Consciousness*. London: Verso.

Giminiani, Piergiorgio di. 2018. *Sentient Lands: Indigeneity, Property, and Political Imagination in Neoliberal Chile*. Tucson: University of Arizona Press.

Giraldo, Omar Felipe, and Peter M. Rosset. 2018. "Agroecology as a Territory in Dispute: Between Institutionality and Social Movements." *Journal of Peasant Studies* 45 (3): 545–64. https://doi.org/10.1080/03066150.2017.1353496.

Gleyzer, Raymundo. 1966. *Nuestras Islas Malvinas*. https://www.youtube.com/watch?v=ywR9X6taVXQ&feature=youtube_gdata_player.

Glissant, Edouard. 1997. *Poetics of Relation*. Translated by Betsy Wing. Ann Arbor: University of Michigan Press.

Gluckman, Max. 1965. *The Ideas of Barotse Jurisprudence*. New Haven, CT: Yale University Press.

Godreau, Isar, and Yarimar Bonilla. 2021. "Nonsovereign Racecraft: How Colonialism, Debt, and Disaster Are Transforming Puerto Rican Racial Subjectivities." *American Anthropologist* 123 (3): 509–25. https://doi.org/10.1111/aman.13601.

Goebel, Julius. 1982. *The Struggle for the Falkland Islands*. New Haven, CT: Yale University Press.

Goldman, Mara, Paul Nadasdy, and Matt Turner, eds. 2011. *Knowing Nature: Conversations at the Intersection of Political Ecology and Science Studies*. Chicago: University of Chicago Press.

Goldman, Michael. 1998. *Privatizing Nature: Political Struggles for the Global Commons*. New Brunswick, NJ: Rutgers University Press.

Goldstein, Daniel M. 2014. "Laying the Body on the Line: Activist Anthropology and the Deportation of the Undocumented." *American Anthropologist* 116 (4): 839–42. https://doi.org/10.1111/aman.12155.

Goldstein, Jenny E. 2016. "Knowing the Subterranean: Land Grabbing, Oil Palm, and Divergent Expertise in Indonesia's Peat Soil." *Environment and Planning A* 48 (4): 754–70. https://doi.org/10.1177/0308518X15599787.

Goldstein, Jenny, and Eric Nost, eds. 2022. *The Nature of Data: Infrastructures, Environments, Politics*. Lincoln: University of Nebraska Press.

Gómez-Barris, Macarena. 2017. *The Extractive Zone: Social Ecologies and Decolonial Perspectives*. Durham, NC: Duke University Press Books.

González, Alberto. 2022, February 16. "Muere 'la abuela Cristina' Calderón, última hablante nativa Yagán." BioBioChile. https://www.biobiochile.cl/noticias/nacional/region-de-magallanes/2022/02/16/muere-la-abuela-cristina-calderon-ultima-hablante-nativa-yagan.shtml.

González, Martín Abel. 2013. *The Genesis of the Falklands (Malvinas) Conflict*. Edited by Nigel Ashton. Basingstoke, UK: Palgrave Macmillan.

González Calderón, Lidia. 2022, May 7. "La Convención, un fructífero diálogo intercultural." El Desconcierto—Prensa digital libre. https://www.eldesconcierto.cl/opinion/2022/05/07/la-convencion-un-fructifero-dialogo-intercultural.html.

Goodfellow, Maya. 2020. *Hostile Environment: How Immigrants Became Scapegoats*. London; Brooklyn, NY: Verso.

Gopal, Priyamvada. 2020. *Insurgent Empire: Anticolonial Resistance and British Dissent*. London: Verso.

Gordillo, Gastón. 2004. *Landscapes of Devils: Tensions of Place and Memory in the Argentinean Chaco*. Durham, NC: Duke University Press.

———. 2014. *Rubble: The Afterlife of Destruction*. Durham, NC: Duke University Press.

———. 2016. "The Savage outside of White Argentina." In *Rethinking Race in Modern Argentina*, edited by Eduardo Elena and Paulina Alberto. Cambridge: Cambridge University Press.

Gordillo, Gastón, and Silvia Hirsch. 2003. "Indigenous Struggles and Contested Identities in Argentina: Histories of Invisibilization and Reemergence." *Journal of Latin American Anthropology* 8 (3): 4–30.

Gordon, Tam. 2004. *Whaling Thoughts Recalled*. Glasgow: Nevisprint Limited.

Gott, Richard. 2022. *Britain's Empire: Resistance, Repression and Revolt*. London: Verso.

Gough, Barry M. 1992. *The Falkland Islands/Malvinas: The Contest for Empire in the South Atlantic*. London: Athlone Press.

Graeber, David. 2012. *Debt: The First 5,000 Years*. Brooklyn, NY: Melville House.

Gramaglia, Christelle, and François Mélard. 2019. "Looking for the Cosmopolitical Fish: Monitoring Marine Pollution with Anglers and Congers in the Gulf of Fos, Southern France." *Science, Technology, & Human Values* 44 (5): 814–42. https://doi.org/10.1177/0162243919860197.

Grandia, Liza. 2012. *Enclosed: Conservation, Cattle, and Commerce among the Q'eqchi' Maya Lowlanders*. Seattle: University of Washington Press.

Grandin, Greg. 2006. *Empire's Workshop: Latin America, the United States, and the Rise of the New Imperialism*. New York: Macmillan.

——. 2014. *The Empire of Necessity: Slavery, Freedom, and Deception in the New World*. New York: Metropolitan Books.

——. 2019. *The End of the Myth: From the Frontier to the Border Wall in the Mind of America*. New York: Metropolitan Books.

Greene, Jack P., and Philip D. Morgan, eds. 2008. *Atlantic History: A Critical Appraisal*. Oxford, UK: Oxford University Press.

Gregory, Chris. 2015. *Gifts and Commodities*. Chicago: Hau Books. http://haubooks.org/gifts-and-commodities/.

Greiff, Alexis de. 2012. *A las puertas del universo derrotado*. Bogotá: Universidad Nacional de Colombia.

Gressier, Catie. 2015. *At Home in the Okavango: White Botswana Narratives of Emplacement and Belonging*. Oxford, UK: Berghahn Books.

Groff, Dulcinea V., Kit M. Hamley, Trevor J. R. Lessard, Kayla E. Greenawalt, Moriaki Yasuhara, Paul Brickle, and Jacquelyn L. Gill. 2020. "Seabird Establishment during Regional Cooling Drove a Terrestrial Ecosystem Shift 5000 Years Ago." *Science Advances* 6 (43): eabb2788. https://doi.org/10.1126/sciadv.abb2788.

Groff, Dulcinea V., David G. Williams, and Jacquelyn L. Gill. 2020. "Modern Calibration of *Poa flabellata* (Tussac Grass) as a New Paleoclimate Proxy in the South Atlantic." *Biogeosciences* 17 (18): 4545–57. https://doi.org/10.5194/bg-17-4545-2020.

Groussac, Paul. 1934. *Las Islas Malvinas*. Buenos Aires: Comision Protectora de Bibliotecas Populares.

Guber, Rosana. 1999a. "Alfredo Lorenzo Palacios: Honor y dignidad en la nacionalización de la causa 'Malvinas.'" *Revista de ciencias sociales, Universidad Nacional de Quilmes* 10: 83–116.

——. 1999b. "'El cabecita negra' o las categorias de la investigación etnográfica en la Argentina." *Revista de investigaciones folclóricas* 14 (diciembre).

——. 2000. "Un gaucho y 18 cóndores en las Islas Malvinas: Identidad política y nación bajo el autoritarismo Argentino." *Mana* 6 (2): 97–125.

——. 2001. *¿Por qué Malvinas? De la causa nacional a la guerra absurda*. Buenos Aires: Fondo de Cultura Económica.

——. 2009. *De chicos a veteranos: Nación y memorias de la guerra de Malvinas*. Buenos Aires: Colección la Otra Ventana.

——. 2022. "Hyper-Realist Mirrors: Exequiel Martinez's Oil Paintings as Testimonies of the Air Battle over the Malvinas/Falkland Islands." *Journal of War & Culture Studies* 15 (3): 350–75. https://doi.org/10.1080/17526272.2022.2078542.

Guber, Rosana, Alejandra Barrutia, Héctor Tessey, Cecilia García Sotomayor, María Jazmín Ohanian, Laura Marina Panizo, and Hernando Flórez. 2022. *Mar de guerra: La armada de la República Argentina y sus formas de habitar el Atlántico Sur*. Madrid: Sb editorial.

Gudynas, Eduardo. 2018. "Extractivisms." In *Reframing Latin American Development*, edited by Ronaldo Munck and Raúl Delgado Wise, 61–76. New York: Routledge Critical Development Studies.

Guillebaud, C. W. 1967. "Report on an Economic Survey of the Falkland Islands [with Corrections]." Stanley: Falkland Islands Government.

Gupta, Akhil. 2015, September 24. "Suspension." *Cultural Anthropology: Theorizing the Contemporary* (blog). http://www.culanth.org/fieldsights/722-suspension.

Gurr, Andrew. 1996. "The Future of Agriculture in the Falkland Islands: Recommendations regarding Agricultural Policy."

Gusinde, Martin. 1961. *The Yámana: The Life and Thought of the Water Nomads of Cape Horn*. Austin: University of Texas.

Gustafson, Bret. 2020. *Bolivia in the Age of Gas*. Durham, NC: Duke University Press Books.

Gustafson, Lowell S. 1984. "The Principle of Self-Determination and the Dispute about Sovereignty over the Falkland Islands." *Inter-American Economic Affairs* 37 (4): 81–99.

———. 1988. *The Sovereignty Dispute over the Falkland (Malvinas) Islands*. New York: Oxford University Press.

———. 2015. "Colonial versus National Self-Determination in the Falkland (Malvinas) Islands." In *Nationalism and Intra-State Conflicts in the Postcolonial World*, edited by Fonkem Achankeng, 453–74. Lanham, MD: Lexington Books.

Guy, Donna J., and Thomas E. Sheridan. 1998. *Contested Ground: Comparative Frontiers on the Northern and Southern Edges of the Spanish Empire*. Tucson: University of Arizona Press.

Guyer, Jane. 2007. "Prophecy and the Near Future." *American Ethnologist* 34 (3): 409–21.

———. 2015. "Oil Assemblages and the Production of Confusion: Price Fluctuations in Two West African Oil-Producing Economies." In *Subterranean Estates: Life Worlds of Oil and Gas*, edited by Hannah Appel, Arthur Mason, and Michael Watts, 237–52. Ithaca, NY: Cornell University Press.

Hall, Stuart. 1980. "Popular-Democratic vs Authoritarian Populism: Two Ways of 'Taking Democracy Seriously.'" In *Marxism and Democracy*, edited by Alan Hunt. London: Lawrence and Wishart.

———. 1985. "Authoritarian Populism: A Reply." *New Left Review*, no. 151.

———. 1988. *The Hard Road to Renewal: Thatcherism and the Crisis of the Left*. London: Verso.

———. 2002. "Race, Articulation, and Societies Structured in Dominance." In *Race Critical Theories: Text and Context*, edited by Philomena Essed and David Theo Goldberg, 38–68. Malden, MA: Blackwell Publishers.

———. 2016. "The Empire Strikes Back 1982." In *Selected Political Writings: The Great Moving Right Show and Other Essays*. Durham, NC: Duke University Press.

———. 2021. "Cultural Identity and Diaspora [1990]." In *Selected Writings on Race and Difference*, edited by Paul Gilroy and Ruth Wilson Gilmore, 257–71. Durham, NC: Duke University Press Books.

Halperín Donghi, Tulio. 2012. *La larga agonia de la Argentina peronista*. Buenos Aires: Ariel.

Hamley, Catherine. 2016. "Humans and the Falkland Islands Warrah: Investigating the Origins of an Extinct Endemic Canid." PhD diss., The University of Maine. https://digitalcommons.library.umaine.edu/etd/2538.

Hamley, Kit M., Jacquelyn L. Gill, Kathryn E. Krasinski, Dulcinea V. Groff, Brenda L. Hall, Daniel H. Sandweiss, John R. Southon, Paul Brickle, and Thomas V. Lowell. 2021. "Evidence of Prehistoric Human Activity in the Falkland Islands." *Science Advances* 7 (44): eabh3803. https://doi.org/10.1126/sciadv.abh3803.

Hancock, David. 1997. *Citizens of the World: London Merchants and the Integration of the British Atlantic Community, 1735–1785*. Cambridge: Cambridge University Press.

Hansen, Thomas Blom. 2021. "Sovereignty in a Minor Key." *Public Culture* 33 (1): 41–61. https://doi.org/10.1215/08992363-8742160.

Hansen, Thomas Blom, and Finn Stepputat. 2005. *Sovereign Bodies: Citizens, Migrants, and States in the Postcolonial World*. Princeton, NJ: Princeton University Press.

Haraway, Donna. 1988. "Situated Knowledges: The Science Question in Feminism and the Privilege of Partial Perspective." *Feminist Studies* 14 (3): 575–99.

——. 2007. *When Species Meet*. Minneapolis: University of Minnesota Press.

Harding, Luke. 2014, October 27. "British Drilling in Falklands Risks Eco-Disaster, Says Argentina." *The Guardian*, sec. UK News. http://www.theguardian.com/uk-news/2014/oct/27/british-drilling-falklands-eco-disaster-argentina-malvinas-daniel-filmus-oil-gas-sea-bed.

Harding, Sandra. 2008. *Sciences from Below: Feminisms, Postcolonialities, and Modernities*. Durham, NC: Duke University Press.

——. 2015. *Objectivity and Diversity: Another Logic of Scientific Research*. Chicago: University of Chicago Press.

Harding, Susan. 1991. "Representing Fundamentalism: The Problem of the Repugnant Cultural Other." *Social Research* 58 (2): 373–93.

Hardt, Michael, and Antonio Negri. 2001. *Empire*. Cambridge, MA: Harvard University Press.

Harris, Cheryl I. 1993. "Whiteness as Property." *Harvard Law Review* 106: 1707–91.

Hart, Gillian. 2019. "From Authoritarian to Left Populism?: Reframing Debates." *South Atlantic Quarterly* 118 (2): 307–23. https://doi.org/10.1215/00382876-7381158.

Hartigan, John. 2017. *Care of the Species: Races of Corn and the Science of Plant Biodiversity*. Minneapolis: University of Minnesota Press.

Hartman, Saidiya. 2008. *Lose Your Mother: A Journey along the Atlantic Slave Route*. New York: Farrar, Straus and Giroux.

Harvey, David. 1995. "Militant Particularism and Global Ambition: The Conceptual Politics of Place, Space, and Environment in the Work of Raymond Williams." *Social Text* 42: 69–98.

——. 2005. *The New Imperialism*. Oxford, UK: Oxford University Press.

——. 2009. "Places, Regions, Territories." In *Cosmopolitanism and the Geographies of Freedom*. New York: Columbia University Press.

——. 2011. *The Enigma of Capital: And the Crises of Capitalism*. Oxford, UK: Oxford University Press.

Harvey, Penelope. 2014. "Infrastructures of the Frontier in Latin America." *Journal of Latin American and Caribbean Anthropology* 19 (2): 280–83.

Harvey, Penny, and Hannah Knox. 2015. *Roads: An Anthropology of Infrastructure and Expertise*. Ithaca, NY: Cornell University Press.

Hastings, Max, and Simon Jenkins. 2012. *The Battle for the Falklands*. London: Pan Macmillan.

Hawkes, Steve. 2012. "US Firm's £1bn for Falklands Oil." *The Sun*. http://www.thesun.co.uk/sol/homepage/news/money/4077559/US-firms-1bn-for-Falklands-oil.html.

Hazlewood, Nick. 2001. *Savage: The Life and Times of Jemmy Button*. New York: Thomas Dunne Books.

Healey, Mark A. 2011. *The Ruins of the New Argentina: Peronism and the Remaking of San Juan after the 1944 Earthquake*. Durham, NC: Duke University Press.

Heatherington, Tracey. 2010. *Wild Sardinia: Indigeneity and the Global Dreamtimes of Environmentalism*. Seattle: University of Washington Press.

Hecht, Gabrielle. 2012. *Being Nuclear: Africans and the Global Uranium Trade*. Cambridge, MA: MIT Press.

Helmreich, Stefan. 2005. "How Scientists Think; About 'Natives,' for Example. A Problem of Taxonomy among Biologists of Alien Species in Hawaii." *Journal of the Royal Anthropological Institute* 11 (1): 107–28.

——. 2009. *Alien Ocean: Anthropological Voyages in Microbial Seas.* Berkeley: University of California Press.

——. 2015. *Sounding the Limits of Life: Essays in the Anthropology of Biology and Beyond.* Princeton, NJ: Princeton University Press.

Henne, Adam. 2015. *Environmentalism, Ethical Trade, and Commodification: Technologies of Value and the Forest Stewardship Council in Chile.* New York: Routledge.

Herrscher, Roberto. 2010. *The Voyages of the Penelope.* Translated by John Fowler. Ushuaia, Arg.: Editorial Sudpol.

Herzfeld, Michael. 1985. *The Poetics of Manhood: Contest and Identity in a Cretan Mountain Village.* Princeton, NJ: Princeton University Press.

——. 1987. *Anthropology through the Looking-Glass: Critical Ethnography in the Margins of Europe.* Cambridge: Cambridge University Press.

Hicks, Dan. 2021. *The Brutish Museums: The Benin Bronzes, Colonial Violence and Cultural Restitution.* London: Pluto Press.

Hilgers, Mathieu. 2011. "Autochthony as Capital in a Global Age." *Theory, Culture & Society* 28 (1): 34–54.

Hinchliffe, Steve. 2008. "Reconstituting Nature Conservation: Towards a Careful Political Ecology." *Geoforum* 39 (1): 88–97.

Hirsch, Silvia María. 2003. "Bilingualism, Pan-Indianism and Politics in Northern Argentina: The Guaraní's Struggle for Identity and Recognition." *Journal of Latin American Anthropology* 8: 84–103.

Ho, Engseng. 2004. "Empire through Diasporic Eyes: A View from the Other Boat." *Society for Comparative Study of Society and History* 46 (2): 210–46.

Hobson, John Atkinson. 1902. *Imperialism: A Study.* London: James Nisbet.

Hobsbawm, Eric. 1983a, January. "Falklands Fallout." *Marxism Today*, 13–19.

——. 1983b. *The Invention of Tradition.* Cambridge: Cambridge University Press.

——. 1990. *Industry and Empire: From 1750 to the Present Day.* London: Penguin Books.

Hodgson, Dorothy L. 2009. "Becoming Indigenous in Africa." *African Studies Review* 52 (3): 1–32.

Holmes, George, and Connor J. Cavanagh. 2016. "A Review of the Social Impacts of Neoliberal Conservation: Formations, Inequalities, Contestations." *Geoforum* 75 (October): 199–209. https://doi.org/10.1016/j.geoforum.2016.07.014.

Holt, Thomas C. 1991. *The Problem of Freedom: Race, Labor, and Politics in Jamaica and Britain, 1832–1938.* Baltimore, MD: Johns Hopkins University Press.

Hoogeveen, Dawn. 2016. "Fish-Hood: Environmental Assessment, Critical Indigenous Studies, and Posthumanism at Fish Lake (Teztan Biny), Tsilhqot'in Territory." *Environment and Planning D: Society and Space* 34 (2): 355–70. https://doi.org/10.1177/0263775815615123.

Horne, Gerald. 2018. *The Apocalypse of Settler Colonialism: The Roots of Slavery, White Supremacy, and Capitalism in 17th Century North America and the Caribbean.* New York: Monthly Review Press.

Horowitz, Leah S. 2015. "Culturally Articulated Neoliberalisation: Corporate Social Responsibility and the Capture of Indigenous Legitimacy in New Caledonia." *Transactions of the Institute of British Geographers* 40 (1): 88–101. https://doi.org/10.1111/tran.12057.

Howe, Cymene. 2019. "Sensing Asymmetries in Other-than-Human Forms." *Science, Technology, & Human Values*, 44 (5): 900–910. https://doi.org/10.1177/0162243919852675.

Howe, Cymene, Jessica Lockrem, Hannah Appel, Edward Hackett, Dominic Boyer, Randal Hall, Matthew Schneider-Mayerson et al. 2016. "Paradoxical Infrastructures: Ruins, Retrofit, and Risk." *Science, Technology, & Human Values* 41 (3): 547–65.

Huber, Matthew T. 2013. *Lifeblood: Oil, Freedom, and the Forces of Capital.* Minneapolis: University of Minnesota Press.

Hughes, David McDermott. 2006. "Hydrology of Hope: Farm Dams, Conservation, and Whiteness in Zimbabwe." *American Ethnologist* 33 (2): 269–87.

——. 2010. *Whiteness in Zimbabwe: Race, Landscape, and the Problem of Belonging.* New York: Palgrave Macmillan.

——. 2013. "Climate Change and the Victim Slot: From Oil to Innocence." *American Anthropologist* 115 (4): 570–81.

——. 2017. *Energy without Conscience: Oil, Climate Change, and Complicity.* Durham, NC: Duke University Press.

Humphreys, Macartan, Jeffrey D. Sachs, and Joseph Eugene Stiglitz, eds. 2007. *Escaping the Resource Curse.* New York: Columbia University Press.

Hunt, Sarah. 2014. "Ontologies of Indigeneity: The Politics of Embodying a Concept." *Cultural Geographies* 21 (1): 27–32. https://doi.org/10.1177/1474474013500226.

Hyades, Paul, and Joseph Deniker. 1891. *Mission scientifique du Cap Horn, 1882–1883. Tome VII. Anthropologie, ethnographie.* Paris: Gauthier Villars.

Hymes, Dell, ed. 1972. *Reinventing Anthropology.* Ann Arbor: University of Michigan Press.

ICIJ. n.d. "The Panama Papers: The Power Players." Accessed April 10, 2016. https://panamapapers.icij.org/the_power_players/.

Ignatiev, Noel. 1995. *How the Irish Became White.* London: Routledge.

Ingold, Tim. 1988. *Hunters, Pastoralists and Ranchers: Reindeer Economies and Their Transformations.* Cambridge: Cambridge University Press.

——. 2013. "Anthropology beyond Humanity." *Suomen antropologi: Journal of the Finnish Anthropological Society* 38 (3): 5–23.

——. 2015. "From the Master's Point of View: Hunting Is Sacrifice." *Journal of the Royal Anthropological Institute* 21 (1): 24–27. https://doi.org/10.1111/1467-9655.12145.

Isla, Alejandro. 1998. "Terror, Memory and Responsibility in Argentina." *Critique of Anthropology* 18 (2): 134–56.

Ivanov, Lyubomir. 2003. *The Future of the Falkland Islands and Its People.* Sofia: Double T Publishers.

Jackson, Ben, and Robert Saunders, eds. 2012. *Making Thatcher's Britain.* Cambridge: Cambridge University Press.

Jackson, Robert H. 1993. *Quasi-States: Sovereignty, International Relations and the Third World.* Cambridge: Cambridge University Press.

Jackson, Shona N. 2012. *Creole Indigeneity: Between Myth and Nation in the Caribbean.* Minneapolis: University of Minnesota Press.

Jacobson, Matthew Frye. 1999. *Whiteness of a Different Color: European Immigrants and the Alchemy of Race.* Cambridge, MA: Harvard University Press.

Jaffe, Alexandra. 1999. *Ideologies in Action: Language Politics on Corsica.* Berlin: Mouton de Gruyter.

James, Daniel. 1988. *Resistance and Integration, Peronism and the Argentine Working Class 1946–1976.* Cambridge: Cambridge University Press.

James, Ross. 2016, May 16. "Falkland Islands' Biosecurity Officers Thwart Argentine Ant Army Invasion." *MercoPress* and *Penguin News.* http://en.mercopress.com/2016/05/16/falkland-islands-biosecurity-officers-thwart-argentine-ant-army-invasion.

Jasanoff, Sheila. 2006. *States of Knowledge: The Co-Production of Science and Social Order.* London: Routledge.

Jasanoff, Sheila, and Sang-Hyun Kim. 2009. "Containing the Atom: Sociotechnical Imaginaries and Nuclear Power in the United States and South Korea." *Minerva* 47: 119–46.

——, eds. 2015. *Dreamscapes of Modernity: Sociotechnical Imaginaries and the Fabrication of Power*. Chicago: University of Chicago.

Jeffery, Laura. 2011. *Chagos Islanders in Mauritius and the UK: Forced Displacement and Onward Migration*. Manchester, UK: Manchester University Press.

Jegathesan, Mythri. 2021. "Black Feminist Plots before the Plantationocene and Anthropology's 'Regional Closets.'" *Feminist Anthropology* 2 (1): 78–93. https://doi.org /10.1002/fea2.12037.

Jenkings, K. Neil, and John Beales. 2022. "Pilgrimage Respecified: Falklands War Veterans' Accounts of Their Returns to the Falkland Islands." *Journal of War & Culture Studies* 15 (3): 328–49. https://doi.org/10.1080/17526272.2022.2078540.

Jerez, Bárbara, Ingrid Garcés, and Robinson Torres. 2021. "Lithium Extractivism and Water Injustices in the Salar de Atacama, Chile: The Colonial Shadow of Green Electromobility." *Political Geography* 87 (May). https://doi.org/10.1016/j.polgeo.2021.102382.

Jessop, Bob. 2019. "Authoritarian Neoliberalism: Periodization and Critique." *South Atlantic Quarterly* 118 (2): 343–61. https://doi.org/10.1215/00382876-7381182.

Jobson, Ryan Cecil. 2018. "Road Work: Highways and Hegemony in Trinidad and Tobago." *Journal of Latin American and Caribbean Anthropology* 23 (3): 457.

Johnson, Adrienne, Anna Zalik, Sharlene Mollett, Farhana Sultana, Elizabeth Havice, Tracey Osborne, Gabriela Valdivia, Flora Lu, and Emily Billo. 2021. "Extraction, Entanglements, and (Im)Materialities: Reflections on the Methods and Methodologies of Natural Resource Industries Fieldwork." *Environment and Planning E: Nature and Space* 4 (2): 383–428. https://doi.org/10.1177/2514848620907470.

Johnson, Leslie Main, and Eugene S. Hunn, eds. 2010. *Landscape Ethnoecology: Concepts of Biotic and Physical Space*. Oxford, UK: Berghahn Books.

Johnson, Lyman. 2011. *Workshop of a Revolution: Plebeian Buenos Aires and the Atlantic World, 1776–1810*. Durham, NC: Duke University Press.

Johnson, Samuel. 1771. *Thoughts on the Late Transactions Respecting Falkland's Islands*. London: T. Cadell.

Joseph, Galen. 2000. "Taking Race Seriously: Whiteness in Argentina's National and Transnational Imaginary." *Identities* 7: 333–71.

Jusionyte, Ieva. 2015. *Savage Frontier: Making News and Security on the Argentine Border*. Berkeley: University of California Press.

Kamin, Bebe. 1984. *Los chicos de la guerra*. Buenos Aires: K Films.

Karl, Terry Lynn. 1997. *The Paradox of Plenty*. Berkeley: University of California Press.

Karush, Matthew B., and Oscar Chamosa, eds. 2010. *The New Cultural History of Peronism: Power and Identity in Mid-Twentieth-Century Argentina*. Durham, NC: Duke University Press.

Kauanui, J. Kēhaulani. 2016. "'A Structure, Not an Event': Settler Colonialism and Enduring Indigeneity." *Lateral* 5 (1).

Kauanui, J. Kēhaulani, and Patrick Wolfe. 2012. "Settler Colonialism Then and Now: A Conversation between J. Kēhaulani Kauanui and Patrick Wolfe." *Politica & società* 2: 235–58.

Kaur Hundle, Anneeth. 2022. "On Blue Economies: Afro-Asianism, Imperial Entanglements, Geopolitics." *American Anthropologist* website. August 4. https://www .americananthropologist.org/geopolitical-lives/hundle.

Kelley, Robin D. G. 2017. "The Rest of Us: Rethinking Settler and Native." *American Quarterly* 69 (2): 267–76. https://doi.org/10.1353/aq.2017.0020.

Kemp, Alex. 2011. *The Official History of North Sea Oil and Gas*. 2 vols. New York: Routledge.

Ketcham, Christopher. 2017, February 14. "Allan Savory's Holistic Management Theory Falls Short on Science." *Sierra Magazine*. https://www.sierraclub.org/sierra/2017-2

-march-april/feature/allan-savory-says-more-cows-land-will-reverse-climate
-change.

Khalili, Laleh. 2021. *Sinews of War and Trade: Shipping and Capitalism in the Arabian Peninsula*. London; New York: Verso.

Kim, Eleana J. 2016. "Toward an Anthropology of Landmines: Rogue Infrastructure and Military Waste in the Korean DMZ." *Cultural Anthropology* 31 (2): 162–87.

Kimmerer, Robin Wall. 2015. *Braiding Sweetgrass: Indigenous Wisdom, Scientific Knowledge and the Teachings of Plants*. Minneapolis, MN: Milkweed Editions.

King, Charles. 2019. *Gods of the Upper Air: How a Circle of Renegade Anthropologists Reinvented Race, Sex, and Gender in the Twentieth Century*. 3rd ed. New York: Doubleday.

King, Tiffany Lethabo. 2019. *The Black Shoals: Offshore Formations of Black and Native Studies*. Durham, NC: Duke University Press Books.

Kingsbury, Benedict. 2000. "Self-Determination: A Relational Approach." In *Operationalizing the Rights of Indigenous Peoples to Self-Determination*, edited by Pekka Aikio and Martin Scheinin. Turku/Abo, Finland: Institute for Human Rights, Abo Akademi University.

Kirksey, Eben. 2015. "Species: A Praxiographic Study." *Journal of the Royal Anthropological Institute* 21 (4): 758–780.

Kirksey, S. Eben, and Stefan Helmreich. 2010. "The Emergence of Multispecies Ethnography." *Cultural Anthropology* 25 (4): 545–76.

Kirsch, Stuart. 2018. *Engaged Anthropology: Politics beyond the Text*. Oakland: University of California Press.

Knorr-Cetina, Karin. 1999. *Epistemic Cultures: How the Sciences Make Knowledge*. Cambridge, MA: Harvard University Press.

Koch, Natalie, and Tom Perreault. 2019. "Resource Nationalism." *Progress in Human Geography* 43 (4): 611–31. https://doi.org/10.1177/0309132518781497.

Kohen, Marcelo G., and Facundo D. Rodríguez. 2013. *Las Malvinas entre el derecho y la historia: refutación del folleto británico "Más allá de la historia oficial. La verdadera historia de las Falklands/Malvinas."* Buenos Aires: EUDEBA.

Kohen, Marcelo. 2014, October 22. "'Las Islas Malvinas y la libre determinación de los pueblos.'" *MercoPress*. http://es.mercopress.com/2014/10/22/las-islas-malvinas-y-la-libre-determinacion-de-los-pueblos.

Kon, Daniel. 1982. *Los chicos de la guerra: Hablan los soldados que estuvieron en Malvinas*. Buenos Aires: Editorial Galerna.

Koram, Kojo. 2022. *Uncommon Wealth: Britain and the Aftermath of Empire*. London: John Murray.

Kosek, Jake. 2010. "Ecologies of Empire: On the New Uses of the Honeybee." *Cultural Anthropology* 25 (4): 650–78.

Koselleck, Reinhart. 2004. *Futures Past: On the Semantics of Historical Time*. New York: Columbia University Press.

Koutonin, Mawuna Remarque. 2015, March 13. "Why Are White People Expats When the Rest of Us Are Immigrants?" *The Guardian*. http://www.theguardian.com/global-development-professionals-network/2015/mar/13/white-people-expats-immigrants-migration.

Krauss, Clifford. 2020, July 20. "Chevron Deal for Oil and Gas Fields May Set Off New Wave of Mergers." *New York Times*, sec. Business. https://www.nytimes.com/2020/07/20/business/energy-environment/chevron-noble-oil-mergers.html.

Kreimer, Pablo. 2007. "Social Studies of Science and Technology in Latin America: A Field in Process of Consolidation." *Science, Technology and Society* 11 (1): 1–9.

Kröger, Markus, and Anja Nygren. 2020. "Shifting Frontier Dynamics in Latin America." *Journal of Agrarian Change* 20 (3): 364–86. https://doi.org/10.1111/joac.12354.

Kunz, Sarah. 2019. "A Business Empire and Its Migrants: Royal Dutch Shell and the Management of Racial Capitalism." *Transactions of the Institute of British Geographers* 45 (2): 377–91. https://doi.org/10.1111/tran.12366.

Kuper, Adam. 2003. "The Return of the Native." *Current Anthropology* 44 (3): 389–402.

La Nación. 2012, February 23. "Una visión alternativa sobre la causa de Malvinas." *La nación.* http://www.lanacion.com.ar/1450787-una-vision-alternativa-sobre-la-causa -de-malvinas.

Laclau, Ernesto. 1977. *Politics and Ideology in Marxist Theory: Capitalism, Fascism, Populism.* London: New Left Books.

Larkin, Brian. 2008. *Signal and Noise: Media, Infrastructure, and Urban Culture in Nigeria.* Durham, NC: Duke University Press.

———. 2013. "The Politics and Poetics of Infrastructure." *Annual Review of Anthropology* 42 (1): 327–43.

Larson, Edward J. 2012. *An Empire of Ice: Scott, Shackleton, and the Heroic Age of Antarctic Science.* New Haven, CT: Yale University Press.

Latour, Bruno. 1987. *Science in Action: How to Follow Scientists and Engineers through Society.* Cambridge, MA: Harvard University Press.

———. 1999. "Circulating Reference: Sampling Soil in the Amazon Forest." In *Pandora's Hope: Essays on the Reality of Science Studies.* Cambridge, MA: Harvard University Press.

———. 2004. *Politics of Nature: How to Bring the Sciences into Democracy.* Translated by Catherine Porter. Cambridge, MA: Harvard University Press.

———. 2007. *Reassembling the Social: An Introduction to Actor-Network-Theory.* Oxford, UK: Oxford University Press.

Latour, Bruno, and Steve Woolgar. 1986. *Laboratory Life: The Construction of Scientific Facts.* Princeton, NJ: Princeton University Press.

Lattimore, Owen. 1940. *Inner Asian Frontiers of China.* New York: American Geographical Society.

Law, John, and Annemarie Mol, eds. 2002. *Complexities: Social Studies of Knowledge Practices.* Durham, NC: Duke University Press.

Layman, Kit, and Jane Cameron. 1995. *The Falklands and The Dwarf 1881–1882.* Milton Keynes, UK: Back to Front.

Lazar, Sian. 2017. *The Social Life of Politics: Ethics, Kinship, and Union Activism in Argentina.* Stanford, CA: Stanford University Press.

Le Guin, Ursula K. 1972. *The Word for World Is Forest.* London: Granada.

Leader-Williams, N. 2009. *Reindeer on South Georgia: The Ecology of an Introduced Population.* Cambridge: Cambridge University Press.

Leivestad, Hege Høyer, and Johanna Markkula. 2021. "Inside Container Economies." *Focaal* 2021 (89): 1–11. https://doi.org/10.3167/fcl.2021.890101.

Lenin, V. I. (1917) 1939. *Imperialism: The Highest Stage of Capitalism.* New York: International Publishers.

Lennon, Myles. 2017. "Decolonizing Energy: Black Lives Matter and Technoscientific Expertise amid Solar Transitions." *Energy Research & Social Science* 30 (August): 18–27. https://doi.org/10.1016/j.erss.2017.06.002.

Leonard, Lori. 2016. *Life in the Time of Oil: A Pipeline and Poverty in Chad.* Bloomington: Indiana University Press.

Leonardo, Micaela di. 1994. "White Ethnicities, Identity Politics, and Baby Bear's Chair." *Social Text* 41: 165–91.

Levick, R. B. M. 1982. "Chief Events and Episodes in the History of the Falkland Islands and the Gibraltar Parallel." Margin Print. JCNA.

Levinson, Marc. 2008. *The Box: How the Shipping Container Made the World Smaller and the World Economy Bigger*. Princeton, NJ: Princeton University Press.

Levitsky, Steven. 2003. "From Labor Politics to Machine Politics: The Transformation of Party-Union Linkages in Argentine Peronism, 1983–1999." *Latin American Research Review* 38 (3): 3–36.

Li, Fabiana. 2015. *Unearthing Conflict: Corporate Mining, Activism, and Expertise in Peru*. Durham, NC: Duke University Press.

Li, Tania Murray. 1998. "Working Separately but Eating Together: Personhood, Property, and Power in Conjugal Relations." *American Ethnologist* 25 (4): 675–94.

——. 2007. *The Will to Improve: Governmentality, Development, and the Practice of Politics*. Durham, NC: Duke University Press.

Liboiron, Max. 2021. *Pollution Is Colonialism*. Durham, NC: Duke University Press Books.

Liboiron, Max, Manuel Tironi, and Nerea Calvillo. 2018. "Toxic Politics: Acting in a Permanently Polluted World." *Social Studies of Science* 48 (3): 331–49. https://doi.org /10.1177/0306312718783087.

Limbert, Mandana. 2010. *In the Time of Oil: Piety, Memory, and Social Life in an Omani Town*. Stanford, CA: Stanford University Press.

Linebaugh, Peter, and Marcus Rediker. 2000. *The Many-Headed Hydra: Sailors, Slaves, Commoners, and the Hidden History of the Revolutionary Atlantic*. Boston: Beacon Press.

Lins Ribeiro, Gustavo. 1994. *Transnational Capitalism and Hydropolitics in Argentina: The Yacyreta High Dam*. Gainesville: University of Florida Press.

Lipsitz, George. 2006. *The Possessive Investment in Whiteness: How White People Profit from Identity Politics*. Philadelphia: Temple University Press.

Livingstone, Grace. 2018. *Britain and the Dictatorships of Argentina and Chile, 1973–82: Foreign Policy, Corporations and Social Movements*. Cham, Switzerland: Palgrave MacMillan.

——. 2022. "Oil and the Falklands/Malvinas: Oil Companies, Governments and Islanders." *The Round Table* 111 (1): 91–103. https://doi.org/10.1080/00358533.2022 .2037235.

Locke, John. 1964. "On Property." In *Two Treatises of Government*, edited by Peter Laslett. Cambridge: Cambridge University Press.

Lorenz, Federico. 2006. *Las guerras por Malvinas*. Buenos Aires: Edhasa.

——. 2009. *Malvinas: Una guerra argentina*. Buenos Aires: Sudamericana.

——. 2012. *Montoneros o la Ballena Blanca*. Buenos Aires: Tusquets Editores.

——. 2014. *Todo lo que necesitás saber sobre Malvinas*. Buenos Aires: Paidós.

Lorenz, Federico, and Julio Vezub. 2022. *Malvinas. Historia, conflictos, perspectivas*. Buenos Aires: Sb editorial.

Lowe, Celia. 2006. *Wild Profusion: Biodiversity Conservation in an Indonesian Archipelago*. Princeton, NJ: Princeton University Press.

——. 2010. "Viral Clouds: Becoming H5N1 in Indonesia." *Cultural Anthropology* 25 (4): 625–49.

Löwy, Michael. 2016. *Fire Alarm: Reading Walter Benjamin's "On the Concept of History."* Translated by Chris Turner. London: Verso.

Lutz, Catherine. 2008. "Empire Is in the Details." *American Ethnologist* 33 (4): 593–611. https://doi.org/10.1525/ae.2006.33.4.593.

Luxemburg, Rosa. (1913) 2015. *The Accumulation of Capital*. London: Routledge.

MacFarquhar, Larissa. 2020, July 6. "How Prosperity Transformed the Falklands." *The New Yorker*. https://www.newyorker.com/magazine/2020/07/06/how-prosperity-transformed-the-falklands.

MacKenzie, John M. 1997. *The Empire of Nature: Hunting, Conservation and British Imperialism*. Manchester, UK: Manchester University Press.

Macpherson, C. B. 1962. *The Political Theory of Possessive Individualism: Hobbes to Locke*. Oxford, UK: Oxford University Press.

Madeley, John. 1982. "Diego Garcia: A Contrast to the Falklands." Report No. 54. Minority Rights Group.

Mainwaring, Michael. 1983. *From the Falklands to Patagonia*. London: Allison & Busby.

Makihara, Miki. 2004. "Linguistic Syncretism and Language Ideologies: Transforming Sociolinguistic Hierarchy on Rapa Nui (Easter Island)." *American Anthropologist* 106: 529–40.

Malinowski, Bronislaw. 1922. *Argonauts of the Western Pacific: An Account of Native Enterprise and Adventure in the Archipelagoes of Melanesian New Guinea*. London: Routledge.

———. 1935. *Coral Gardens and Their Magic*. London: Allen & Unwin.

Malm, Andreas. 2016. *Fossil Capital: The Rise of Steam Power and the Roots of Global Warming*. 3rd ed. London: Verso.

Mamdani, Mahmood. 1996. *Citizen and Subject: Contemporary Africa and the Legacy of Late Colonialism*. Princeton, NJ: Princeton University Press.

———. 2001. *When Victims Become Killers: Colonialism, Nativism and the Genocide in Rwanda*. Princeton, NJ: Princeton University Press.

———. 2020. *Neither Settler nor Native: The Making and Unmaking of Permanent Minorities*. Cambridge, MA: Harvard University Press.

Manela, Erez. 2009. *The Wilsonian Moment: Self-Determination and the International Origins of Anticolonial Nationalism*. Oxford, UK: Oxford University Press.

Mansfield, Becky. 2003. "From Catfish to Organic Fish: Making Distinctions about Nature as Cultural Economic Practice." *Geoforum* 34 (3): 329–42.

———. 2004. "Neoliberalism in the Oceans: 'Rationalization,' Property Rights, and the Commons Question." *Geoforum* 35 (3): 313–26.

Marchante, José Luis Alonso. 2014. *Menéndez, Rey de la Patagonia*. Santiago, Chile: Catalonia.

Markkula, Johanna Sofia Kristina. 2022. "The Ship." *History and Anthropology* 33 (2): 188–95. https://doi.org/10.1080/02757206.2022.2066097.

Marks, Richard Lee. 1991. *Three Men of the Beagle*. New York: Avon Books.

Marsh, J. W., and W. H. Stirling. 1867. *The Story of Commander Allen Gardiner, R.N., with Sketches of Missionary Work in South America*. London: James Nisbet.

Martin, Richard J., and David Trigger. 2015. "Negotiating Belonging: Plants, People, and Indigeneity in Northern Australia." *Journal of the Royal Anthropological Institute* 21 (2): 276–95. https://doi.org/10.1111/1467-9655.12206.

Martinez, Diego. 2015. "Piratas con licencia de Malvinas." *Página/12*. http://www.pagina12.com.ar/diario/elpais/1-273944-2015-06-01.html.

Martinic, Mateo. 2002. "Brief History of the Selk'nam." In *12 Perspectives. Essays on the Selknam, Yahgan and Kawesqar*, edited by Peter Mason and Carolina Odone, 231–59. Santiago, Chile: Taller Experimental Cuerpos Pintados.

Martínez-Reyes, José E. 2016. *Moral Ecology of a Forest: The Nature Industry and Maya Post-Conservation*. Tucson: University of Arizona Press.

Marx, Karl. 1976. "Part Eight: So-Called 'Primitive Accumulation.'" In *Capital*. Vol. 1. London: Penguin Books.

——. 1993. *Grundrisse: Foundations of the Critique of Political Economy (Rough Draft)*. Translated by Martin Nicolaus. New York: Penguin Books.

Mason, Arthur. 2007. "The Rise of Consultant Forecasting in Liberalized Gas Markets." *Public Culture* 19 (2): 367–79.

——. 2016, July 29. "Arctic Abstractive Industry." Hot Spots, Cultural Anthropology. https://culanth.org/fieldsights/945-arctic-abstractive-industry.

Mason, Rowena. 2010, December 7. "Desire Petroleum's Falklands Oil 'Find' Is Actually Water." *The Telegraph*. http://www.telegraph.co.uk/finance/newsbysector/energy/oilandgas/8185019/Desire-Petroleums-Falklands-oil-find-is-actually-water.html.

Massey, Doreen. 1993. "Power-Geometry and a Progressive Sense of Place." In *Mapping the Futures: Local Cultures, Global Change*. London: Routledge.

Mastnak, Tomaz, Julia Elyachar, and Tom Boellstorff. 2014. "Botanical Decolonization: Rethinking Native Plants." *Environment and Planning D: Society and Space* 32 (2): 363–80. https://doi.org/10.1068/d13006p.

Mattei, Ugo, and Laura Nader. 2008. *Plunder: When the Rule of Law Is Illegal*. Malden, MA: John Wiley & Sons.

Maurer, Bill. 2000. *Recharting the Caribbean: Land, Law, and Citizenship in the British Virgin Islands*. Ann Arbor: University of Michigan Press.

Mauss, Marcel. 1950. *The Gift: The Form and Reason for Exchange in Archaic Societies*. New York: W. W. Norton.

McAfee, Kathleen. 1999. "Selling Nature to Save It? Biodiversity and Green Developmentalism." *Environment and Planning D: Society and Space* 17 (2): 133–54. https://doi.org/10.1068/d170133.

McCarthy, James. 2019. "Authoritarianism, Populism, and the Environment: Comparative Experiences, Insights, and Perspectives." *Annals of the American Association of Geographers* 109 (2): 301–13. https://doi.org/10.1080/24694452.2018.1554393.

McCay, Bonnie J., and James M. Acheson. 1987. *The Question of the Commons: The Culture and Ecology of Communal Resources*. Tucson: University of Arizona.

McClintock, Anne. 1995. *Imperial Leather: Race, Gender, and Sexuality in the Colonial Contest*. London: Routledge.

McCook, Stuart George. 2002. *States of Nature: Science, Agriculture, and Environment in the Spanish Caribbean, 1760–1940*. Austin: University of Texas Press.

McGranahan, Carole. 2016. "Theorizing Refusal: An Introduction." *Cultural Anthropology* 31 (3): 319–25. https://doi.org/10.14506/ca31.3.01.

McGranahan, Carole, and John F. Collins, eds. 2018. *Ethnographies of U.S. Empire*. Durham, NC: Duke University Press Books.

McIntosh, Janet. 2016. *Unsettled: Denial and Belonging among White Kenyans*. Oakland: University of California Press.

McKee, Alexander. 1985. *Tarquin's Ship: The Etruscan Wreck in Campese Bay*. London: Souvenir Press.

McKittrick, Katherine, and Clyde Woods, eds. 2007. *Black Geographies and the Politics of Place*. Toronto: South End Press.

Mehta, Uday. 1999. *Liberalism and Empire: A Study in Nineteenth Century British Legal Thought*. Chicago: University of Chicago Press.

Melchionne, Thomas Louis. 1985. "Alcohol Beverage Use in the Falkland Islands Camp: An Ethnography of a Cultural Domain." New Brunswick, NJ: Rutgers. http://www.worldcat.org/title/alcohol-beverage-use-in-the-falkland-islands-camp-an-ethnography-of-a-cultural-domain/oclc/22286526.

Melville, Elinor. 1997. *A Plague of Sheep: Environmental Consequences of the Conquest of Mexico*. Cambridge: Cambridge University Press.

Mendoza, Marcos. 2018. *The Patagonian Sublime: The Green Economy and Post-Neoliberal Politics*. New Brunswick, NJ: Rutgers University Press.

Mendoza, Marcos, Maron Greenleaf, and Eric H. Thomas. 2021. "Green Distributive Politics: Legitimizing Green Capitalism and Environmental Protection in Latin America." *Geoforum* 126 (November): 1–12. https://doi.org/10.1016/j.geoforum.2021.07.012.

Mentz, Steve. 2015. *Shipwreck Modernity: Ecologies of Globalization, 1550–1719*. Minneapolis: University of Minnesota Press.

Mercau, Ezequiel. 2019. *The Falklands War: An Imperial History*. Cambridge: Cambridge University Press.

MercoPress. 2013a, May 20. "Falkland Islands: Argentina Blasts UK's 'Double Standard' on Self Determination Right." *MercoPress*. https://en.mercopress.com/2013/05/20/falkland-islands-argentina-blasts-uk-s-double-standard-on-self-determination-right.

———. 2013b, June 20. "Falklands' People Happy Exercising Self-Determination and with 'Current Relationship with UK.'" *MercoPress*. https://en.mercopress.com/2013/06/20/falklands-people-happy-exercising-self-determination-and-with-current-relationship-with-uk.

———. 2014a, December 1. "A 'Malvinas "Invasion" with Women,' Mujica Proposes to the Argentines." *MercoPress*. http://en.mercopress.com/2014/12/01/a-malvinas-invasion-with-women-mujica-proposes-to-the-argentines.

———. 2014b, April 2. "'We Planted between 15.000 and 20.000 Antipersonnel and Antitank Mines' in Malvinas." *MercoPress*. http://en.mercopress.com/2014/04/02/we-planted-between-15.000-and-20.000-antipersonnel-and-antitank-mines-in-malvinas.

———. 2014c, September 4. "Malvinas: Argentina Recalls Fifty Years since Falklands Claim Statement before C24." *MercoPress*. http://en.mercopress.com/2014/09/04/malvinas-argentina-recalls-fifty-years-since-falklands-claim-statement-before-c24.

———. 2015a, October 16. "Argentine Yacht Salvaged in the Falklands, Effectively Had the Purpose of Showing the Flag." *MercoPress*. http://en.mercopress.com/2015/10/16/argentine-yacht-salvaged-in-the-falklands-effectively-had-the-purpose-of-showing-the-flag.

———. 2015b, April 22. "Falklands' Argentine Cemetery Remains' Identification Turns into Controversy." *MercoPress*. http://en.mercopress.com/2015/04/22/falklands-argentine-cemetery-remains-identification-turns-into-controversy.

———. 2015c, October 19. "Falklands' Salvage of la Sanmartiniana, Triggered Minor Air Incident, Reports Argentine Media." *MercoPress*. http://en.mercopress.com/2015/10/19/falklands-salvage-of-la-sanmartiniana-triggered-minor-air-incident-reports-argentine-media.

———. 2020, January 8. "Falklands' Sea Lion Project Reaches Farm-in Deal with Tel Aviv Listed Navitas." *MercoPress*, https://en.mercopress.com/2020/01/08/falklands-sea-lion-project-reaches-farm-in-deal-with-tel-aviv-listed-navitas.

Milanesio, Natalia. 2013. *Workers Go Shopping in Argentina: The Rise of Popular Consumer Culture*. Albuquerque: University of New Mexico Press.

Miller, David. 1995. *Wreck of the Isabella*. Annapolis, MD: Naval Institute Press.

Miller, Sydney. 1988. *A Life of Our Choice*. Stanley, Falkland Islands, UK: Private.

Miller, William Ian. 1998. *The Anatomy of Disgust*. Cambridge, MA: Harvard University Press.

Milne, Seumas. 2014. *The Enemy Within*. Updated ed. London: Verso.

Milton, Kay. 1993. *Environmentalism: The View from Anthropology*. London: Routledge.

———. 2005. "Anthropomorphism or Egomorphism? The Perception of Non-Human Persons by Human Ones." In *Animals in Person: Cultural Perspectives on Human-Animal Intimacy*, edited by John Knight, 255–71. Oxford, UK: Berg.

Milun, Kathryn. 2011. *The Political Uncommons: The Cross-Cultural Logic of the Global Commons*. Farnham, UK: Ashgate Publishing.

Mitchell, Peter. 2021. *Imperial Nostalgia: How the British Conquered Themselves*. Manchester: Manchester University Press.

Mitchell, Timothy. 2002. *Rule of Experts: Egypt, Techno-Politics, Modernity*. Berkeley: University of California Press.

——. 2011. *Carbon Democracy: Political Power in the Age of Oil*. New York: Verso.

Mitman, Gregg. 1996. "When Nature Is the Zoo: Vision and Power in the Art and Science of Natural History." *Osiris* 11: 117–43.

Mohanram, Radhika. 2007. *Imperial White: Race, Diaspora, and the British Empire*. Minneapolis: University of Minnesota Press.

Monbiot, George. 2014, August 4. "Eat More Meat and Save the World: The Latest Implausible Farming Miracle." *The Guardian*, sec. Environment. http://www.theguardian.com/environment/georgemonbiot/2014/aug/04/eat-more-meat-and-save-the-world-the-latest-implausible-farming-miracle.

Moore, Charles. 2015. *Margaret Thatcher: The Authorized Biography: Volume I: From Grantham to the Falklands*. 3rd ed. New York: Vintage.

Moore, D. M. 1968. "The Vascular Flora of the Falkland Islands." *British Antarctic Survey Scientific Reports* 60: 19.

Moore, Donald S., Anand Pandian, and Jake Kosek, eds. 2003. *Race, Nature, and the Politics of Difference*. Durham, NC: Duke University Press.

Moreton-Robinson, Aileen, ed. 2007. *Sovereign Subjects: Indigenous Sovereignty Matters*. Crows Nest, Australia: Allen & Unwin.

——. 2015. *The White Possessive: Property, Power, and Indigenous Sovereignty*. Minneapolis: University of Minnesota Press.

Moro, Rubén O. 2005. *La trampa de Malvinas*. Buenos Aires: Edivern.

Mosko, Mark. 2010. "Partible Penitents: Dividual Personhood and Christian Practice in Melanesia and the West." *Journal of the Royal Anthropological Institute* 16 (2): 215–40.

Moyo, Sam, and Walter Chambati, eds. 2013. *Land and Agrarian Reform in Zimbabwe: Beyond White-Settler Capitalism*. Dakar, Senegal: CODESRIA & AIAS.

Muehlebach, Andrea. 2003. "What Self in Self-Determination? Notes from the Frontiers of Transnational Indigenous Activism." *Identities: Global Studies in Culture and Power* 10: 241–68.

Muir, Sarah. 2021. *Routine Crisis: An Ethnography of Disillusion*. Chicago: University of Chicago Press.

Mukerji, Chandra. 2010. "The Territorial State as a Figured World of Power: Strategics, Logistics, and Impersonal Rule." *Sociological Theory* 28 (4): 402–24.

Mullings, Leith. 2005. "Interrogating Racism: Towards an Anti-Racist Anthropology." *Annual Review of Anthropology* 43: 667–93.

Muñoz Azpiri, José Luis, ed. 1966. *Historia completa de las Malvinas*. 3 vols. Buenos Aires: Ediciones Oriente.

Munro, Hugh. 1924. *Report of an Investigation into the Conditions and Practice of Sheep Farming in the Falkland Islands*. London: Waterlow and Sons.

Munro, Richard. 1998. *Place Names of the Falkland Islands*. London: Bluntisham Books.

Muzio, Tim di. 2015. *Carbon Capitalism: Energy, Social Reproduction and World Order*. London: Rowman & Littlefield Publishers.

Nader, Laura. 1972. *Up the Anthropologist: Perspectives Gained from Studying Up*. ERIC Clearinghouse. https://eric.ed.gov/?id=ED065375.

——. 1981. "Barriers to Thinking New about Energy." *Physics Today* 34 (3): 99–102.

——, ed. 2010. *The Energy Reader*. Malden, MA: Wiley-Blackwell.

Nairn, Tom. 2015. *The Break-Up of Britain: Crisis and Neo-Nationalism*. Melbourne, Australia: Common Ground Publishing.

Natale, Eleonora. 2022. "From Campo de Mayo to Malvinas, and Back: The Falklands/Malvinas War from the Perspective of Argentine Veterans Accused of Crimes against Humanity." *Journal of War & Culture Studies* 15 (3): 309–27. https://doi.org/10.1080/17526272.2022.2078541.

Neale, Timothy. 2017. *Wild Articulations: Environmentalism and Indigeneity in Northern Australia*. Honolulu: University of Hawaii Press.

Newsinger, John. 2013. *The Blood Never Dried*. 2nd UK ed. London: Bookmarks.

Nichols, Robert. 2019. *Theft Is Property!: Dispossession and Critical Theory*. Durham, NC: Duke University Press Books.

Niebieskikwiat, Natasha. 2012. *Lagrimas de hielo*. Buenos Aires: Kapelusz.

——. 2014. *Kelpers. Ni ingleses ni argentinos: Cómo es la nación que crece frente a nuestras costas*. Buenos Aires: Sudamericana.

——. 2018, May 14. "Cierran un acuerdo antártico con Gran Bretaña y avanzan con la pesca." *Clarín*. https://www.clarin.com/politica/cierran-acuerdo-antartico-gran -bretana-avanzan-pesca_0_S10kFKwRG.html.

Nielsen, Paul Maersk. 2003. "English in Argentina." *World Englishes* 22: 199–209.

Nieto-Olarte, Mauricio. 2006. *Remedios para el Imperio. Historia natural y la apropiación del nuevo mundo*. Bogotá: Universidad de los Andes.

Niezen, Ron. 2003. *The Origins of Indigenism: Human Rights and the Politics of Identity*. Berkeley: University of California Press.

Nixon, Rob. 2011. *Slow Violence and the Environmentalism of the Poor*. Cambridge, MA: Harvard University Press.

Nkrumah, Kwame. (1965) 1987. *Neo-Colonialism: The Last Stage of Imperialism*. London: Panaf Books.

Nordborg, Maria. 2016. "Holistic Management—A Critical Review of Allan Savory's Grazing Method." Uppsala, Sweden: SLU/EPOK—Centre for Organic Food & Farming & Chalmers.

Nost, Eric. 2015. "Performing Nature's Value: Software and the Making of Oregon's Ecosystem Service Markets." *Environment and Planning A* 47 (12): 2573–90.

Nuttall, Mark. 2010. *Pipeline Dreams: People, Environment, and the Arctic Energy Frontier*. Copenhagen: International Work Group for Indigenous Affairs.

Nyerges, A. Endre. 1992. "The Ecology of Wealth-in-People: Agriculture, Settlement, and Society on the Perpetual Frontier." *American Anthropologist* 94 (4): 860–81.

NZASA, Hal Levine, Stephen Levine, David H. Turner, Anthony P. Cohen, Richard Handler, and John Sharp. 1990. "Cultural Politics in New Zealand." *Anthropology Today* 6 (3): 3–9.

O'Brien, Mary H. 1993. "Being a Scientist Means Taking Sides." *BioScience* 43 (10): 706–8. https://doi.org/10.2307/1312342.

O'Faircheallaigh, Ciaran. 2017. "Shaping Projects, Shaping Impacts: Community-Controlled Impact Assessments and Negotiated Agreements." *Third World Quarterly* 38 (5): 1181–97. https://doi.org/10.1080/01436597.2017.1279539.

O'Reilly, Jessica. 2017. *The Technocratic Antarctic: An Ethnography of Scientific Expertise and Environmental Governance*. Ithaca, NY: Cornell University Press.

Offshore Energy Today. 2015, October 29. "Noble Energy Abandons Falkland Islands Offshore Well." https://www.offshoreenergytoday.com/noble-energy-abandons -falkland-island-offshore-well/.

Ogden, Laura A. 2021. *Loss and Wonder at the World's End*. Durham, NC: Duke University Press Books.

Oka, Cynthia Dewi. 2017. *Salvage: Poems*. Evanston, IL: TriQuarterly.

Olusoga, David. 2023. *Black and British: A Forgotten History*. London: Picador.

Ong, Aihwa. 2000. "Graduated Sovereignty in South-East Asia." *Theory, Culture & Society* 17 (4): 55–75.

Onley, James. 2004. "Britain's Native Agents in Arabia and Persia in the Nineteenth Century." *Comparative Studies of South Asia, Africa and the Middle East* 24 (1): 129–37.

Orta-Martinez, Martí, and Matt Finer. 2010. "Oil Frontiers and Indigenous Resistance in the Peruvian Amazon." *Ecological Economics* 70: 207–18.

Osborne, Simon. 2020, October 29. "Falklands Islands Row Erupts as Britain Confronts Argentina over Fresh Sovereignty Claims." *Express*, sec. UK. https://www.express .co.uk/news/uk/1353860/falklands-islands-latest-uk-argentina-foreign-secretary -dominic-raab-argentine-ambassador.

Ostrom, Elinor. 1990. *Governing the Commons: The Evolution of Institutions for Collective Action*. Cambridge: Cambridge University Press.

Padmore, George. (1936) 1969. *How Britain Rules Africa*. New York: Negro Universities Press.

Palermo, Vicente. 2014. *Sal en las heridas: Las Malvinas en la cultura argentina contemporánea*. Buenos Aires: Sudamericana.

Paley, Julia. 2001. *Marketing Democracy: Power and Social Movements in Post-Dictatorial Chile*. Berkeley: University of California Press.

Panizo, Laura Marina. 2015. "Los héroes santos: Muerte y sacralización en el caso de los caídos en la guerra de Malvinas." *Páginas* 13: 11–32.

Pascoe, Graham. 2014. *The Battle of the Falklands, 1914: A Falklands Perspective*. Self-published.

Pascoe, Graham, and Peter Pepper. 2012. "False Falklands History at the United Nations: How Argentina Misled the UN in 1964—And Still Does." http://www.falklands history.org/.

Pastorino, Ana. 2013. *Malvinas: El derecho de libre determinación de los pueblos y la población de las islas*. Buenos Aires: Eudeba.

Patterson, Tiffany Ruby, and Robin D. G. Kelley. 2000. "Unfinished Migrations: Reflections on the African Diaspora and the Making of the Modern World." *African Studies Review* 43 (1): 11–45. https://doi.org/10.2307/524719.

Peck, James. 2013. *Malvinas: Una guerra privada*. Buenos Aires: Emecé.

——. 2015. *Behind the Door Lurks a Monster: Punk, Politics and the President; A Life Growing Up in the Falkland Islands*. Self-published, CreateSpace Independent Publishing Platform.

Peel, Michael. 2010. *A Swamp Full of Dollars: Pipelines and Paramilitaries at Nigeria's Oil Frontier*. Chicago: Chicago Review Press.

Peet, Richard, and Michael Watts, eds. 1996. *Liberation Ecologies*. London: Routledge.

Penguin News. 2020a, December 17. "Government to Legislate on Discrimination in Falklands." https://penguin-news.com/headlines/community/2020/government-to-legis late-on-discrimination-in-falklands/.

——. 2020b, July 15. "New Falklands Port Operational by Early 2024." https://penguin -news.com/headlines/oil/2020/new-falklands-port-operational-by-early-2024/.

Penguin News via *MercoPress*. 2020, August 8. "Falklands Targets a 3% Yearly Population Growth and Doubles Permanent Residence Permits." *MercoPress*. https://en .mercopress.com/2020/08/08/falklands-targets-a-3-yearly-population-growth -and-doubles-permanent-residence-permits.

Penrose, Bernard. 1775. *An Account of the Last Expedition to Port Egmont, in Falkland's Islands, in the Year 1772*. London: J. Johnson.

Perelman, Michael. 2000. *The Invention of Capitalism: Classical Political Economy and the Secret History of Primitive Accumulation.* Durham, NC: Duke University Press.

Pernety, Antoine-Joseph. (1771) 2013. *The History of a Voyage to the Malouine (or Falkland) Islands.* Cambridge: Cambridge University Press.

Phillips, Barnaby. 2022. *Loot: Britain and the Benin Bronzes.* London: Oneworld Publications.

Philpott, Robert A. 2009. *Keppel: The South American Missionary Society Settlement 1855–1911. An Archaeological and Historical Survey.* Stanley, Falkland Islands, UK: Falkland Islands Museum and National Trust and National Museums Liverpool.

Phipps, Colin. 1977. *What Future for the Falklands?* London: Fabian Society.

Pineda, Baron, ed. 2006. *Shipwrecked Identities: Navigating Race on Nicaragua's Mosquito Coast.* New Brunswick, NJ: Rutgers University Press.

Piot, Charles. 2010. *Nostalgia for the Future: West Africa after the Cold War.* Chicago: University of Chicago Press.

Pocock, J. G. A. 1985. *In Virtue, Commerce, and History: Essays on Political Thought and History, Chiefly in the Eighteenth Century.* New York: Cambridge University Press.

Política Argentina. 2016, April 2. "Filmus: 'Echaron a todos los que trabajaban en la secretaría de Malvinas.'" http://www.politicargentina.com/notas/201604/12891-filmus-echaron-a-todos-los-que-trabajaban-en-la-secretaria-de-malvinas8207.html.

Pompert Robertson, Sophie. 2014. "Sub-Division in the Falkland Islands—Case Study: Port Stephens." *Falkland Islands Journal*, 203–10.

Poncet, S., L. Poncet, D. Poncet, D. Christie, C. Dockrill, and D. Brown. 2011. "Introduced Mammal Eradications in the Falkland Islands and South Georgia." In *Island Invasives: Eradication and Management: Proceedings of the International Conference on Island Invasives*, edited by C. R. Veitch, M. N. Clout, and D. R. Towns, 332–36. Gland, Switzerland: IUCN.

Pondal, Martiniano Leguizamón. 1956. *Toponimia criolla en las Malvinas.* Buenos Aires: Editorial Raigal.

Pope, Frank. 2007. *Dragon Sea: A True Tale of Treasure, Archeology, and Greed off the Coast of Vietnam.* Orlando, FL: Harcourt.

Povinelli, Elizabeth A. 2002. *The Cunning of Recognition: Indigenous Alterities and the Making of Australian Multiculturalism.* Durham, NC: Duke University Press.

——. 2011. "The Governance of the Prior." *Interventions* 13 (1): 13–30.

Premier Oil Exploration & Production Limited, Falkland Islands Business Unit. 2014. "2015 Falkland Islands Exploration Campaign Environmental Impact Statement." Document No: FK-BU-PMO-EV-REP-0003.

——. 2018. "Sea Lion Development - Phase 1 Environmental Impact Statement." Document No: FK-SL-PMO-EV-REP-0008 REV. B03.

Pritchard, Helen V. 2013. "Thinking with the Animal-Hacker: Articulation in Ecologies of Earth Observation." APRJA. http://www.aprja.net/thinking-with-the-animal-hackerarticulation-in-ecologies-of-earth-observation/.

Pütz, Klemens, Andrea P. Clausen, Nic Huin, and John P. Croxall. 2003. "Re-Evaluation of Historical Rockhopper Penguin Population Data in the Falkland Islands." *Waterbirds* 26 (2): 169–75.

Rada, Ángel Díaz de. 2015. "Discursive Elaborations of 'Saami' Ethnos: A Multi-Source Model of Ethnic and Ethnopolitical Structuration." *Anthropological Theory* 15 (4): 472–96. https://doi.org/10.1177/1463499615609067.

Radcliffe-Brown, Alfred Reginald. 1922. *The Andaman Islanders: A Study in Social Anthropology.* Princeton, NJ: The University Press.

Raffles, Hugh. 2011a. *Insectopedia.* New York: Random House.

——. 2011b, April 3. "Mother Nature's Melting Pot." *New York Times*. http://www
    .nytimes.com/2011/04/03/opinion/03Raffles.html.

——. 2020. *The Book of Unconformities: Speculations on Lost Time*. 3rd ed. New York:
    Pantheon.

Rana, Junaid. 2011. *Terrifying Muslims: Race and Labor in the South Asian Diaspora*. Dur-
    ham, NC: Duke University Press Books.

Rappaport, Roy A. 1968. *Pigs for the Ancestors*. New Haven, CT: Yale University Press.

Rasmussen, Mattias Borg. 2021. "Institutionalizing Precarity: Settler Identities, National
    Parks and the Containment of Political Spaces in Patagonia." *Geoforum* 119 (Feb-
    ruary): 289–97. https://doi.org/10.1016/j.geoforum.2019.06.005.

Ratier, Hugo E. 1971. *El cabecita negra*. Buenos Aires: Centro Editor de America
    Latina.

Rausch, Jane M. 1993. *The Llanos Frontier in Colombian History, 1830–1930*. Albuquer-
    que: University of New Mexico Press.

Ray, Sarah Jaquette. 2014. "Rub Trees, Crittercams, and GIS: The Wired Wilderness of
    Leanne Allison and Jeremy Mendes' Bear 71." *Green Letters* 18 (3): 236–53. https://
    doi.org/10.1080/14688417.2014.964282.

Rediker, Marcus. 1987. *Between the Devil and the Deep Blue Sea: Merchant Seamen, Pi-
    rates and the Anglo-American Maritime World, 1700–1750*. Cambridge: Cambridge
    University Press.

——. 2007. *The Slave Ship: A Human History*. New York: Penguin Books.

——. 2010. "The Poetics of History from Below." *American Historical Association Per-
    spectives*. https://www.historians.org/publications-and-directories/perspectives
    -on-history/september-2010/the-poetics-of-history-from-below.

Reed, Kristin. 2009. *Crude Existence: Environment and the Politics of Oil in Northern An-
    gola*. Berkeley: University of California Press.

Regeneris Consulting Ltd. 2013. "Socio-Economic Study of Oil and Gas Development
    in the Falklands." Stanley: Falkland Islands Government.

Renfrew, Daniel. 2018. *Life without Lead: Contamination, Crisis, and Hope in Uruguay*.
    Oakland: University of California Press.

Richards, Philip C. 1995. *Oil and the Falkland Islands: An Introduction to the October 1995
    Offshore Licensing Round*. Keyworth, UK: British Geological Survey for Falkland
    Islands Government.

Rifkin, Mark. 2014. "The Frontier as (Movable) Space of Exception." *Settler Colonial Stud-
    ies* 4 (2): 176–80.

Rindstedt, Camilla, and Karin Aronsson. 2002. "Growing up Monolingual in a Bilin-
    gual Community: The Quichua Revitalization Paradox." *Language in Society* 31:
    721–42.

Riofrancos, Thea. 2017. "*Extractivismo* Unearthed: A Genealogy of a Radical Discourse."
    Cultural Studies 31 (2–3): 277–306. https://doi.org/10.1080/09502386.2017.1303429.

——. 2019, December 7. "What Green Costs." *Logic Magazine*. https://logicmag.io
    /nature/what-green-costs/.

——. 2020. *Resource Radicals: From Petro-Nationalism to Post-Extractivism in Ecuador*.
    Durham, NC: Duke University Press Books.

Robbins, Paul. 2003. "Beyond Ground Truth: GIS and the Environmental Knowledge of
    Herders, Professional Foresters, and Other Traditional Communities." *Human
    Ecology* 31 (2): 233–53. https://doi.org/10.1023/A:1023932829887.

Robertson, R. B. 1954. *Of Whales and Men*. New York: Alfred A. Knopf.

Robin, Libby. 1997. "Ecology: A Science of Empire?" In *Ecology and Empire: Environ-
    mental History of Settler Societies*, edited by Tom Griffiths and Libby Robin, 63–
    75. Seattle: University of Washington Press.

——. 2007. *How a Continent Created a Nation*. Kensington, Australia: UNSW Press.

Rocheleau, Dianne E. 2015. "Networked, Rooted and Territorial: Green Grabbing and Resistance in Chiapas." *The Journal of Peasant Studies* 42 (3–4): 695–723. https://doi.org/10.1080/03066150.2014.993622.

Rock, David, ed. 1975. *Argentina in the Twentieth Century*. Pittsburgh, PA: University of Pittsburgh Press.

Rodney, Walter. (1972) 1981. *How Europe Underdeveloped Africa*. Washington, DC: Howard University Press.

Rodríguez, Yliana. 2022. "Spanish Place Names of the Falkland Islands: A Novel Classification System." *Names* 70 (March): 1–8. https://doi.org/10.5195/names.2022.2376.

Rodríguez, Yliana, Adolfo Elizaincín, and Paz González. 2022. "The Spanish Component of Falkland Islands English: a Micro-Corpus Approach to the Study of Loanwords." *English World-Wide* (published online in November).

Rodríguez, Yliana, Paz González, and Adolfo Elizaincín. (forthcoming). "Los préstamos lingüísticos como registro de la historia: Indigenismos en el inglés de las Islas Malvinas/Falkland."

Rodriguez-Medina, Leandro. 2013. *Centers and Peripheries in Knowledge Production*. New York: Routledge.

Roediger, David R. 2007. *The Wages of Whiteness: Race and the Making of the American Working Class*. London: Verso.

Rogers, Douglas. 2012. "The Materiality of the Corporation: Oil, Gas, and Corporate Social Technologies in the Remaking of a Russian Region." *American Ethnologist* 39: 284–96.

——. 2014. "Petrobarter: Oil, Inequality, and the Political Imagination in and after the Cold War." *Current Anthropology* 55 (2): 131–53.

——. 2015a. "Oil and Anthropology." *Annual Review of Anthropology* 44 (1): 365–80.

——. 2015b. *The Depths of Russia: Oil, Power, and Culture after Socialism*. Ithaca, NY: Cornell University Press.

Rolston, Jessica Smith. 2013. "Specters of Syndromes and the Everyday Lives of Wyoming Energy Workers." In *Cultures of Energy: Power, Practices, Technologies*, edited by Sarah Strauss, Stephanie Rupp, and Thomas Love. Walnut Creek, CA: Left Coast Press.

Rosaldo, Renato. 1989. "Imperialist Nostalgia." *Representations* 26: 107–22.

Rosenberg, Andrew S. 2022. *Undesirable Immigrants: Why Racism Persists in International Migration*. Princeton, NJ: Princeton University Press.

Ross, Michael Lewin. 2012. *The Oil Curse: How Petroleum Wealth Shapes the Development of Nations*. Princeton, NJ: Princeton University Press.

Rouse, Roger. 1992. "Making Sense of Settlement: Class Transformation, Cultural Struggle, and Transnationalism among Mexican Migrants in the United States." *Annals of the New York Academy of Sciences* 645 (1): 25–52.

Rozitchner, León. 2005. *Malvinas: De la guerra sucia a la guerra limpia*. Buenos Aires: Losada.

Ruiz, Rafico. 2021. *Slow Disturbance: Infrastructural Mediation on the Settler Colonial Resource Frontier*. Durham, NC: Duke University Press.

Russell, Edmund. 2001. *War and Nature: Fighting Humans and Insects with Chemicals from World War I to Silent Spring*. Cambridge: Cambridge University Press.

Rutledge, Ian. 1977. "The Integration of the Highland Peasantry into the Sugar Cane Economy of Northern Argentina, 1930–43." In *Land and Labour in Latin America*, edited by Kenneth Duncan and Ian Rutledge. Cambridge: Cambridge University Press.

Ruzza, Alice. 2011. "The Falkland Islands and the UK v. Argentina Oil Dispute: Which Legal Regime?" *Goettingen Journal of International Law* 3 (1): 71–99.

Sabato, Hilda. 1987. "La cuestión agraria pampeana: Un debate inconcluso." *Desarrollo económico* 27 (106): 291–301.

Sábato, Jorge, ed. 1975. *El pensamiento latinoamericano en la problemática ciencia-tecnología-desarrollo-dependencia*. Buenos Aires: Paidós.

Saez de Vernet, Maria. 1829. *Cronista de nuestra soberania en Malvinas*. Buenos Aires: Ediciones Puerto Luis.

Salazar Parreñas, Juno. 2018. *Decolonizing Extinction: The Work of Care in Orangutan Rehabilitation*. Durham, NC: Duke University Press.

Salvatore, Ricardo D. 2008. "The Unsettling Location of a Settler Nation: Argentina, from Settler Economy to Failed Developing Nation." *South Atlantic Quarterly* 107 (4): 755–89.

Sands, Philippe. 2022. *The Last Colony: A Tale of Exile, Justice and Britain's Colonial Legacy*. London: Orion.

Sarlo, Beatriz. 2013a. "En Gilbert House, el chalecito de gobierno." 15 2013. http://www.lanacion.com.ar/1563297-en-gilbert-house-el-chalecito-de-gobierno.

——. 2013b. "La experiencia inconmensurable." 10 2013. http://www.lanacion.com.ar/1561779-la-experiencia-inconmensurable.

——. 2013c. "Let's go pasear." 16 2013. http://www.lanacion.com.ar/1563749-lets-go-pasear.

——. 2013d. "Manda la identidad, pero también la necesidad." 11 2013. http://www.lanacion.com.ar/1562012-tapa-manda-la-identidad-pero-tambien-la-necesidad.

——. 2013e. "Un gesto con más de un mensaje." 12 2013. http://www.lanacion.com.ar/1562320-un-gesto-con-mas-de-un-mensaje.

Sarmiento, Domingo Faustino. 2010. *Facundo*. Buenos Aires: Losada.

Sathnam, Sanghera. 2021. *Empireland: How Imperialism Has Shaped Modern Britain*. London: Viking.

Sawyer, Suzana. 2004. *Crude Chronicles: Indigenous Politics, Multinational Oil, and Neoliberalism in Ecuador*. Durham, NC: Duke University Press.

——. 2022. *The Small Matter of Suing Chevron*. Durham, NC: Duke University Press Books.

Sawyer, Suzana, and Edmund Terence Gomez, eds. 2012. *The Politics of Resource Extraction: Indigenous Peoples, Multinational Corporations and the State*. Basingstoke, UK: Palgrave Macmillan.

Saxton, Alexander. 1990. *The Rise and Fall of the White Republic: Class Politics and Mass Culture in Nineteenth-Century America*. New York: Verso.

Scanlan, Padraic X. 2022. *Slave Empire: How Slavery Built Modern Britain*. London: Robinson.

Scheper-Hughes, Nancy. 1995. "The Primacy of the Ethical: Propositions for a Militant Anthropology." *Current Anthropology* 36 (3): 409–40. https://doi.org/10.1086/204378.

Schumpeter, Joseph. 1951. "The Sociology of Imperialisms." In *Imperialism, Social Classes: Two Essays*, 1–98. New York: New American Library.

Scoones, Ian, ed. 2010. *Zimbabwe's Land Reform: Myths and Realities*. Martlesham, UK: James Currey.

Scott, David. 2014. *Omens of Adversity: Tragedy, Time, Memory, Justice*. Durham, NC: Duke University Press.

Scott, James. 1977. *The Moral Economy of the Peasant: Rebellion and Subsistence in Southeast Asia*. New Haven, CT: Yale University Press.

Scott, Julius S. 2020. *The Common Wind: Afro-American Currents in the Age of the Haitian Revolution*. London: Verso.

Segato, Rita Laura. 2007. *La nación y sus otros*. Buenos Aires: Prometeo Libros Editorial.

Semán, Ernesto. 2017. *Ambassadors of the Working Class: Argentina's International Labor Activists & Cold War Democracy in the Americas*. Durham, NC: Duke University Press.

Serje, Margarita. 2011. *El revés de la nación: Territorios salvajes, fronteras y tierras de nadie*. Bogotá: Ediciones Uniandes.

Sewell, William H. 2005. *Logics of History: Social Theory and Social Transformation*. Chicago: University of Chicago Press.

Shackleton KG PC OBE, The Rt. Hon. Lord. 1976a. *Economic Survey of the Falkland Islands: Resources and Development Potential*. Vol. 1. London: Crown.

——. 1976b. *Economic Survey of the Falkland Islands: Strategy, Recommendations & Implementation*. Vol. 2. London: Crown.

——. 1982. *Falkland Islands Economic Study*. Vol. 3. Reports—Fiscal—Economics. London: Her Majesty's Stationery Office.

Sharp, Lesley A. 2002. "Bodies, Boundaries, and Territorial Disputes: Investigating the Murky Realm of Scientific Authority." *Medical Anthropology* 21 (3–4): 369–79. https://doi.org/10.1080/01459740214075.

Sharpe, Christina. 2016. *In the Wake: On Blackness and Being*. Durham, NC: Duke University Press Books.

Shever, Elana. 2012. *Resources for Reform: Oil and Neoliberalism in Argentina*. Stanford, CA: Stanford University Press.

Shickell, Mark. 2009. *Don't Call Them Bennys*. Peterborough, UK: Fastprint Gold.

Shukin, Nicole. 2009. *Animal Capital: Rendering Life in Biopolitical Times*. Minneapolis: University of Minnesota Press.

Shumway, Nicolas. 1993. *The Invention of Argentina*. Berkeley: University of California Press.

Silenzi de Stagni, Adolfo. 1983. *Las Malvinas y el petróleo*. Buenos Aires: Teoría SRL.

Sillitoe, Paul. 1998. "The Development of Indigenous Knowledge: A New Applied Anthropology." *Current Anthropology* 39 (2): 223–52.

Silverwood, James, and Craig Berry. 2022. "The Distinctiveness of State Capitalism in Britain: Market-Making, Industrial Policy and Economic Space." *Environment and Planning A: Economy and Space*, May. https://doi.org/10.1177/0308518X221102960.

Simpson, Audra. 2014. *Mohawk Interruptus: Political Life across the Borders of Settler States*. Durham, NC: Duke University Press.

——. 2016. "Consent's Revenge." *Cultural Anthropology* 31 (3): 326–33. https://doi.org/10.14506/ca31.3.02.

——. 2017. "The Ruse of Consent and the Anatomy of 'Refusal': Cases from Indigenous North America and Australia." *Postcolonial Studies* 20 (1): 18–33.

——. 2018. "Why White People Love Franz Boas; Or, the Grammar of Indigenous Dispossession." In *Indigenous Visions: Rediscovering the World of Franz Boas*, edited by Ned Blackhawk and Isaiah Lorado Wilner. New Haven, CT: Yale University Press.

——. 2020. "Empire of Feeling." *General Anthropology* 27 (1): 1–8. https://doi.org/10.1111/gena.12063.

Simpson, Jack. 2014, November 24. "Archaeologists to Exhume the Bodies of over 100 Unidentified Argentine Soldiers on Falkland Islands." *The Independent*. http://www.independent.co.uk/news/world/americas/archaeologists-to-exhume-the-bodies-of-over-100-unidentified-argentine-soldiers-on-falkland-islands-9880080.html.

Sitrin, Marina A. 2012. *Everyday Revolutions: Horizontalism and Autonomy in Argentina*. London: Zed Books.

Slatta, Richard W. 1992. *Gauchos & the Vanishing Frontier*. Lincoln: University of Nebraska Press.

Sluyter, Andrew. 2012. *Black Ranching Frontiers: African Cattle Herders of the Atlantic, 1500–1900*. New Haven, CT: Yale University Press.

Smith, John. 1973. *Condemned at Stanley: Notes and Sketches on the Hulks and Wrecks at Port Stanley*. New York: National Maritime Historical Society.

——. 1984. *74 Days: An Islander's Diary of the Falklands Occupation*. London: Century Publishing House.

——. 2013. *An Historical Scrapbook of Stanley*. Stanley, Falkland Islands, UK: Falkland Islands Company.

Smith, Neil. 1984. *Uneven Development: Nature, Capital, and the Production of Space*. Athens: University of Georgia Press.

——. 1992. "Contours of a Spatialized Politics: Homeless Vehicles and the Production of Geographical Scale." *Social Text* 33: 54–81.

——. 2004. *American Empire: Roosevelt's Geographer and the Prelude to Globalization*. Berkeley: University of California Press.

——. 2006. "Nature as Accumulation Strategy." *Socialist Register* 43: 16–36.

Smith, Sherry, and Brian Frehner. 2010. *Indians & Energy: Exploitation and Opportunity in the American Southwest*. Santa Fe, NM: School for Advanced Research Press.

Snow, W. P. 1857. *A Two Years' Cruise off Tierra del Fuego, the Falkland Islands, Patagonia and in the River Plate: A Narrative of Life in the Southern Seas*. 2 vols. London: Longman.

Solari Yrigoyen, Hipólito. 1998. *Malvinas: Lo que no cuentan los ingleses*. Buenos Aires: El Ateneo.

Spice, Anne. 2018. "Fighting Invasive Infrastructures: Indigenous Relations against Pipelines." *Environment and Society* 9 (1): 40–56. https://doi.org/10.3167/ares.2018.090104.

Spiegel, Samuel J. 2017. "EIAs, Power and Political Ecology: Situating Resource Struggles and the Techno-Politics of Small-Scale Mining." *Geoforum* 87 (December): 95–107. https://doi.org/10.1016/j.geoforum.2017.10.010.

Spruce, Joan, and Natalie Smith. 2018. *Falkland Rural Heritage: Sites, Structures and Snippets of Historical Interest*. Fox Bay, Falkland Islands, UK: Falklands Publications.

Star, Susan Leigh. 1999. "The Ethnography of Infrastructure." *American Behavioral Scientist* 43 (3): 377–91.

——. 2010. "This Is Not a Boundary Object: Reflections on the Origin of a Concept." *Science, Technology, & Human Values* 35 (5): 601–17. https://doi.org/10.1177/0162243910377624.

Star, Susan Leigh, and James R. Griesemer. 1989. "Institutional Ecology, 'Translation,' and Boundary Objects: Amateurs and Professionals in Berkeley's Museum of Vertebrate Zoology, 1907–39." *Social Studies of Science* 19: 387–420.

Steedman, Carolyn. 1986. *Landscape for a Good Woman: A Story of Two Lives*. New Brunswick, NJ: Virago.

Stepan, Nancy. 1991. *The Hour of Eugenics: Race, Gender, and Nation in Latin America*. Ithaca, NY: Cornell University Press.

Stepputat, Finn. 2015. "Formations of Sovereignty at the Frontier of the Modern State." *Conflict and Society* 1 (1): 129–43.

Steward, Julian. 1955. *Theory of Culture Change: The Methodology of Multilinear Evolution*. Chicago: University of Illinois Press.

Stokland, Hakon B. 2015. "Field Studies in Absentia: Counting and Monitoring from a Distance as Technologies of Government in Norwegian Wolf Management (1960s–2010s)." *Journal of the History of Biology* 48: 1–36.

Stoler, Ann Laura. 1995. *Race and the Education of Desire: Foucault's History of Sexuality and the Colonial Order of Things*. Durham, NC: Duke University Press.

——. 2006. "On Degrees of Imperial Sovereignty." *Public Culture* 18 (1): 125–46.

——, ed. 2013. *Imperial Debris: On Ruins and Ruination*. Durham, NC: Duke University Press.

——. 2016. *Duress: Imperial Durabilities in Our Times*. Durham, NC: Duke University Press.

Stoler, Ann Laura, Carole McGranahan, and Peter C. Perdue, eds. 2007. *Imperial Formations*. Santa Fe, NM: School for Advanced Research Press.

StoryFutures Creative Cluster. 2020. *Your Scans. Your Stories. Your Heritage: Enabling Young People to Tell Their Own Heritage Stories*. https://www.youtube.com/watch?v=3cIDgJwmh00&feature=youtu.be.

Strange, Ian J. 1972. *The Falkland Islands*. Newton Abbot, UK: David & Charles.

Stratford, Elaine, Godfrey Baldacchino, Elizabeth McMahon, Carol Farbotko, and Andrew Harwood. 2011. "Envisioning the Archipelago." *Island Studies Journal* 6 (2): 113–30.

Strathern, Marilyn. 1984. "Subject or Object? Women and the Circulation of Valuables in Highlands New Guinea." In *Women and Property, Women as Property*, edited by Renee Hirschon, 158–75. London: Croom Helm.

——. 1988. *The Gender of the Gift*. Berkeley: University of California Press.

Strauss, Sarah, Stephanie Rupp, and Thomas Love, eds. 2013. *Cultures of Energy: Power, Practices, Technologies*. Walnut Creek, CA: Left Coast Press.

Sturm, Circe. 2011. *Becoming Indian: The Struggle over Cherokee Identity in the Twenty-First Century*. Santa Fe, NM: School for Advanced Research Press.

Subramaniam, Banu. 2001. "The Aliens Have Landed! Reflections on the Rhetoric of Biological Invasions." *Meridians* 2 (1): 26–40.

Sudbury, Andrea. 2001. "Falkland Islands English; A Southern Hemisphere Variety?" *English World-Wide* 22: 55–80.

——. 2004. "English on the Falklands." In *Legacies of Colonial English*. Cambridge: Cambridge University Press.

Sultana, Farhana. 2021. "Political Ecology 1: From Margins to Center." *Progress in Human Geography* 45 (1): 156–65. https://doi.org/10.1177/0309132520936751.

Sundberg, Juanita. 2003. "Conservation and Democratization: Constituting Citizenship in the Maya Biosphere Reserve, Guatemala." *Political Geography* 22 (7): 715–40. https://doi.org/10.1016/S0962-6298(03)00076-3.

Svampa, Maristella. 2013. "Consenso de los commodities y lenguajes de valorización en América Latina." *Nueva sociedad* 244 (April).

Swanson, Heather Anne, Marianne Elisabeth Lien, and Gro B. Ween, eds. 2018. *Domestication Gone Wild: Politics and Practices of Multispecies Relations*. Durham, NC: Duke University Press Books.

Swint, Brian. 2013, October 2. "Falklands Tax Dispute Clouding Oil Dream for Investors." *BusinessWeek*. https://www.bloomberg.com/news/articles/2013-10-01/falklands-tax-dispute-clouding-oil-dream-for-investors-energy.

Swyngedouw, Erik. 1997. "Neither Global nor Local: 'Glocalization' and the Politics of Scale." In *Spaces of Globalization: Reasserting the Power of the Local*, edited By Kevin R. Cox. New York: The Guilford Press.

——. 2004. *Social Power and the Urbanization of Water: Flows of Power*. Oxford, UK: Oxford University Press.

——. 2015. *Liquid Power: Contested Hydro-Modernities in Twentieth-Century Spain*. Cambridge, MA: MIT Press.

——. 2019. "The Perverse Lure of Autocratic Postdemocracy." *South Atlantic Quarterly* 118 (2): 267–86. https://doi.org/10.1215/00382876-7381134.

Szasz, Paul C. 1999. "The Irresistible Force of Self-Determination Meets the Impregnable Fortress of Territorial Integrity: A Cautionary Fairy Tale about Clashes in Kosovo and Elsewhere." *Georgia Journal of International and Comparative Law* 28 (1).

Tabak, Michael A., Sally Poncet, Ken Passfield, Jacob R. Goheen, and Carlos Martinez del Rio. 2014. "Rat Eradication and the Resistance and Resilience of Passerine Bird Assemblages in the Falkland Islands." *Journal of Animal Ecology* (October).

Tabak, Michael A., Sally Poncet, Ken Passfield, and Carlos Martinez del Rio. 2014. "Invasive Species and Land Bird Diversity on Remote South Atlantic Islands." *Biological Invasions* 16 (2): 341–52.

Taber, Peter. 2016. "Taxonomic Government: Ecuador's National Herbarium and the Institution of Biodiversity, 1986–1996." *Science & Technology Studies* (September). https://sciencetechnologystudies.journal.fi/article/view/59197.

TallBear, Kim. 2013. *Native American DNA: Tribal Belonging and the False Promise of Genetic Science*. Minneapolis: University of Minnesota Press.

Tamarkin, Noah. 2020. *Genetic Afterlives: Black Jewish Indigeneity in South Africa*. Durham, NC: Duke University Press.

Tatham, David, ed. 2008. *The Dictionary of Falklands Biography: Including South Georgia*. Ledbury, UK: David Tatham.

Taussig, Michael. 1993. *Mimesis and Alterity: A Particular History of the Senses*. New York: Routledge.

——. 2004. *My Cocaine Museum*. Chicago: University of Chicago Press.

Taylor, Christopher. 2018. *Empire of Neglect: The West Indies in the Wake of British Liberalism*. Durham, NC: Duke University Press Books.

Taylor, Jean Gelman. 1983. *The Social World of Batavia: Europeans and Eurasians in Colonial Indonesia*. Madison: University of Wisconsin Press.

Taylor, Margaret Stewart. 1971. *Focus on the Falkland Islands*. Plymouth, UK: Robert Hale Ltd.

Telam. 2020, December 28. "Filmus destacó la decisión de la UE de excluir a Malvinas de su acuerdo posbrexit con reino unido." *Telam*. https://www.telam.com.ar/notas /202012/539791-filmus-destaco-la-decision-de-la-ue-de-excluir-a-malvinas-de -su-acuerdo-posbrexit-con-reino-unido.html.

The Telegraph. 2009, February 5. "Colin Phipps Obituary." http://www.telegraph.co.uk /news/obituaries/4528103/Colin-Phipps.html.

Tella, Guido di. 1982. "The Economics of the Frontier." In *Economics in the Long View: Essays in Honor of W. W. Rostow*, edited by Charles P. Kindleberger and Guido di Tella, 1: 210–27. New York: New York University.

Telles, Edward, and René Flores. 2013. "Not Just Color: Whiteness, Nation, and Status in Latin America." *Hispanic American Historical Review* 93 (3): 411–49.

Tesler, Mario. 1971. *El Gaucho Antonio Rivero: La mentira en la historiografía académica*. Buenos Aires: Peña Lillo.

——. 2013. *Agresión militar de los EE.UU. a las Islas Malvinas y el Gaucho Antonio Rivero*. Buenos Aires: Editorial Dunken.

Thatcher, Jim, David O'Sullivan, and Dillon Mahmoudi. 2016. "Data Colonialism through Accumulation by Dispossession: New Metaphors for Daily Data." *Environment and Planning D: Society and Space* 34 (6): 990–1006. https://doi.org/10.1177 /0263775816633195.

Thatcher, Margaret. 1978. "TV Interview for Granada World in Action ('Rather Swamped')." Thatcher Archive. http://www.margaretthatcher.org/document/103485.

——. 2015. "Memoir of the Falklands War." Margaret Thatcher Foundation. http://www .margaretthatcher.org/archive/1982retpap2.asp.

Thomas, Deborah A. 2009. "The Violence of Diaspora: Governmentality, Class Cultures, and Circulations." *Radical History Review* 2009 (103): 83–104. https://doi.org/10.1215/01636545-2008-032.

——. 2016. "Time and the Otherwise: Plantations, Garrisons and Being Human in the Caribbean." *Anthropological Theory* 16 (2–3): 177–200.

Thompson, E. P. 1971. "The Moral Economy of the English Crowd in the Eighteenth Century." *Past & Present*, no. 50: 76–136.

——. 1975. *Whigs and Hunters*. New York: Pantheon Books.

——. 1982. *Zero Option*. London: Merlin Press.

——. 1991. "The Moral Economy Reviewed." In *Customs in Common*, 259–351. New York: New Press.

Tilzey, Mark. 2019. "Authoritarian Populism and Neo-Extractivism in Bolivia and Ecuador: The Unresolved Agrarian Question and the Prospects for Food Sovereignty as Counter-Hegemony." *Journal of Peasant Studies* 46 (3): 626–52. https://doi.org/10.1080/03066150.2019.1584191.

Tinker Salas, Miguel. 2009. *The Enduring Legacy: Oil, Culture, and Society in Venezuela*. Durham, NC: Duke University Press.

Tinsley, Omise'eke Natasha. 2008. "BLACK ATLANTIC, QUEER ATLANTIC: Queer Imaginings of the Middle Passage." *GLQ: A Journal of Lesbian and Gay Studies* 14 (2–3): 191–215. https://doi.org/10.1215/10642684-2007-030.

Tironi, Manuel, and Javiera Barandiarán. 2014. "Neoliberalism as Political Technology: Expertise, Energy, and Democracy in Chile." In *Beyond Imported Magic: Essays on Science, Technology, and Society in Latin America*, edited by Eden Medina, Ivan da Costa Marques, and Christina Holmes, 305–30. London: MIT Press.

Todd, Zoe. 2016. "An Indigenous Feminist's Take on the Ontological Turn: 'Ontology' Is Just Another Word for Colonialism." *Journal of Historical Sociology* 29 (1): 4–22. https://doi.org/10.1111/johs.12124.

——. 2017a, April 11. "Commentary: The Environmental Anthropology of Settler Colonialism, Part I." *Engagement* (blog). https://aesengagement.wordpress.com/2017/04/11/commentary-the-environmental-anthropology-of-settler-colonialism-part-i/.

——. 2017b. "Fish, Kin and Hope: Tending to Water Violations in Amiskwaciwâskahikan and Treaty Six Territory." *Afterall: A Journal of Art, Context and Enquiry* 43 (March): 102–7. https://doi.org/10.1086/692559.

——. 2022. "Fossil Fuels and Fossil Kin: An Environmental Kin Study of Weaponised Fossil Kin and Alberta's So-Called 'Energy Resources Heritage.'" *Antipode* (published online November 8, 2022). https://doi.org/10.1111/anti.12897.

Tomic, Bartolomej. 2020, May 13. "Premier Oil Shelves $1.8B Falkland Islands Offshore Project." *Offshore Engineer Magazine*. https://www.oedigital.com/news/478448-premier-oil-shelves-1-8b-falkland-islands-offshore-project.

Tomich, Dale W., ed. 2020. *Atlantic Transformations: Empire, Politics, and Slavery during the Nineteenth Century*. Albany, NY: SUNY Press.

Torre, Juan Carlos. 1990. *La vieja guardia sindical y Perón: Sobre los orígenes del peronismo*. Buenos Aires: Editorial Sudamericana-Instituto Torcuato di Tella.

Trigger, David S. 2008. "Indigeneity, Ferality, and What 'Belongs' in the Australian Bush: Aboriginal Responses to 'Introduced' Animals and Plants in a Settler-Descendant Society." *Journal of the Royal Anthropological Institute* 14 (3): 628–46.

Trigger, David, Jane Mulcock, Andrea Gaynor, and Yann Toussaint. 2008. "Ecological Restoration, Cultural Preferences and the Negotiation of 'Nativeness' in Australia." *Geoforum* 39 (3): 1273–83.

Trigger, David S., Yann Toussaint, and Jane Mulcock. 2010. "Ecological Restoration in Australia: Environmental Discourses, Landscape Ideals, and the Significance of Human Agency." *Society & Natural Resources* 23 (11): 1060–74.

Trouillot, Michel-Rolph. 1991. "Anthropology and the Savage Slot." In *Recapturing Anthropology*, edited by Richard G. Fox. Santa Fe, NM: School of American Research Press.

———. 1997. *Silencing the Past: Power and the Production of History*. Boston: Beacon Press.

———. 2002. "North Atlantic Universals: Analytical Fictions, 1492–1945." *The South Atlantic Quarterly* 101 (4): 839–58.

———. 2003. *Global Transformations*. New York: Palgrave Macmillan.

Trudgill, Peter. 1986. *Dialects in Contact*. New York: Basil Blackwell.

Tsing, Anna Lowenhaupt. 2005. *Friction: An Ethnography of Global Connection*. Princeton, NJ: Princeton University Press.

———. 2015a, March 30. "Salvage Accumulation, or the Structural Effects of Capitalist Generativity." *Cultural Anthropology*. http://culanth.org/fieldsights/656-salvage-accumulation-or-the-structural-effects-of-capitalist-generativity.

———. 2015b. *The Mushroom at the End of the World: On the Possibility of Life in Capitalist Ruins*. Princeton, NJ: Princeton University Press.

Tsing, Anna L., Jennifer Deger, Alder Keleman Saxena, and Feifei Zhou. 2020. *Feral Atlas: The More-Than-Human Anthropocene*. Stanford, CA: Stanford University Press. https://doi.org/10.21627/2020fa.

Tully, James. 1993. *An Approach to Political Philosophy: Locke in Contexts*. New York: Cambridge University Press.

———. 2000. "The Struggle of Indigenous Peoples for and of Freedom." In *Political Theory and the Rights of Indigenous Peoples*, edited by Duncan Ivison, Paul Patton, and Will Sanders. Cambridge: Cambridge University Press.

Turner, Frederick Jackson. 1920. "The Significance of the Frontier in American History." In *The Frontier in American History*, 1–38. New York: Henry Holt.

UN Division of Ocean Affairs and Law of the Sea. 2016, March 28. "Commission on Limits of Continental Shelf Concludes Fortieth Session." United Nations, New York. http://www.un.org/press/en/2016/sea2030.doc.htm.

United Nations. 2022, May 10. "Non-Self-Governing Territories | The United Nations and Decolonization." United Nations, New York. https://www.un.org/dppa/decolonization/en/nsgt.

Urbina, Ian. 2015, July 27. "'Sea Slaves': The Human Misery That Feeds Pets and Livestock." *New York Times*. http://www.nytimes.com/2015/07/27/world/outlaw-ocean-thailand-fishing-sea-slaves-pets.html.

Ureta, Sebastian. 2020. "Ruination Science: Producing Knowledge from a Toxic World." *Science, Technology, & Human Values* 46 (1): 29–52. https://doi.org/10.1177/0162243919900957.

Van Dooren, Thom. 2014. *Flight Ways: Life and Loss at the Edge of Extinction*. New York: Columbia University Press.

Van Wyhe, John, ed. 2002. *The Complete Work of Charles Darwin Online*. http://darwin-online.org.uk.

Varvavsky, Oscar. 1969. *Ciencia, política y cientificismo*. Buenos Aires: Centro Editor de América Latina.

Vázquez, Juan José Ponce. 2021. *Islanders and Empire*. Cambridge: Cambridge University Press.

Verbitsky, Horacio. 2002. *Malvinas: La ultima batalla de la tercera guerra mundial*. Buenos Aires: Sudamericana.

Verdery, Katherine. 2003. *The Vanishing Hectare: Property and Value in Postsocialist Transylvania*. Ithaca, NY: Cornell University Press.

Vergès, Françoise. 1999. *Monsters and Revolutionaries: Colonial Family Romance and Métissage*. Durham, NC: Duke University Press.

Vessuri, Hebe. 1983. "Consideraciones acerca del estudio social de la ciencia." In *La ciencia periférica*, edited by Elena Díaz, Yolanda Texera, and Hebe Vessuri, 9–35. Caracas: Monte Ávila Editores.

Viatori, Maximilian. 2019. "Uncertain Risks: Salmon Science, Harm, and Ignorance in Canada." *American Anthropologist* 121 (2): 325–37. https://doi.org/10.1111/aman.13241.

Vindal Ødegaard, Cecilie, and Juan Javier Rivera Andía, eds. 2018. *Indigenous Life Projects and Extractivism: Ethnographies from South America*. Cham, Switzerland: Palgrave MacMillan.

Vine, David. 2011. *Island of Shame: The Secret History of the U.S. Military Base on Diego Garcia*. Princeton, NJ: Princeton University Press.

"A Visit to the Falkland Islands." 1832. *United Service Journal and Naval and Military Magazine*, part 3.

Vitalis, Robert. 2009. *America's Kingdom: Mythmaking on the Saudi Oil Frontier*. New updated ed. New York: Verso.

Voskoboynik, Daniel Macmillen, and Diego Andreucci. 2021. "Greening Extractivism: Environmental Discourses and Resource Governance in the 'Lithium Triangle.'" *Environment and Planning E: Nature and Space* 5 (2): 787–809. https://doi.org/10.1177/25148486211006345.

Wagner, Jay Paul, Murray Jones, and Susan Dowse. 2014, March 17. "Managing Socio-Economic Risk in Frontier Areas: A Case Study from the Falkland Islands." Presented at the Society of Petroleum Engineers International Conference on Health, Safety, and Environment, Long Beach, CA.

Wallerstein, Immanuel. 1987. "The Construction of Peoplehood: Racism, Nationalism, Ethnicity." *Sociological Forum* 2 (2): 373–88.

Wanderer, Emily. 2020. *The Life of a Pest: An Ethnography of Biological Invasion in Mexico*. Oakland: University of California Press.

Wannop, A. R. 1961. *Report on Visits to Falkland Islands Sheep Stations*. Edinburgh.

Waterhouse, G. 1838. *Canis antarcticus*. In *The Zoology of the Voyage of HMS Beagle*, edited by C. Darwin. London: Smith Elder and Co.

Waterton, Claire. 2003. "Performing the Classification of Nature." *The Sociological Review* 51 (October): 111–29. https://doi.org/10.1111/j.1467-954X.2004.00454.x.

Watson, Lisa. 2012. *Waking up to War*. Stanley, Falkland Islands, UK: Amazon.

——. 2014, May 2. "£20 Million from Fishing for Falklands Government but Crewmen Welfare Still a Concern." *Penguin News*. http://pnews.falklands.info/index.php/news/fishing/item/699-p20-million-from-fishing-for-falklands-government-but-crewmen-welfare-still-a-concern.

Watts, Michael. 1983. *Silent Violence: Food, Famine, & Peasantry in Northern Nigeria*. Berkeley: University of California Press.

——, ed. 1987. *State, Oil, and Agriculture in Nigeria*. Berkeley: Institute of International Studies, University of California Press.

——. 1992. "The Shock of Modernity: Petroleum, Protest, and Fast Capitalism in an Industrializing Society." In *Reworking Modernity: Capitalism and Symbolic Discontent*, 21–63. New Brunswick, NJ: Rutgers University Press.

——. 2004. "Resource Curse? Governmentality, Oil and Power in the Niger Delta, Nigeria." *Geopolitics* 9 (1): 50–80.

——. 2006. "Empire of Oil: Capitalist Dispossession and the Scramble for Africa." *Monthly Review* 58 (4): 1–17.

——. 2008. "Blood Oil: An Anatomy of a Petro-Insurgency in the Niger Delta." *Focaal—European Journal of Anthropology* 52: 18–38.

——. 2012. "A Tale of Two Gulfs: Life, Death, and Dispossession along Two Oil Frontiers." *American Quarterly* 64 (3): 437–67.

——. 2014. "Oil Frontiers: The Niger Delta and the Gulf of Mexico." In *Oil Culture*, edited by Ross Barrett and Daniel Worden, 189–210. Minneapolis: University of Minnesota Press.

——. 2015. "Securing Oil: Frontiers, Risks, and Spaces of Accumulated Insecurity." In *Subterranean Estates: Life Worlds of Oil and Gas*, edited by Hannah Appel, Arthur Mason, and Michael Watts, 211–36. Ithaca, NY: Cornell University Press.

Weiner, Annette. 1985. "Inalienable Wealth." *American Ethnologist* 12 (2): 210–27.

Wells, John. 1982. *Accents of English*. Vols. 1–3. Cambridge: Cambridge University Press.

West, Paige. 2006. *Conservation Is Our Government Now: The Politics of Ecology in Papua New Guinea*. Durham, NC: Duke University Press.

Westman, Clinton N. 2013. "Social Impact Assessment and the Anthropology of the Future in Canada's Tar Sands." *Human Organization* 72 (2): 111–20.

Weszkalnys, Gisa. 2008. "Hope & Oil: Expectations in São Tomé e Príncipe." *Review of African Political Economy* 35 (117): 473–82.

——. 2011. "Cursed Resources, or Articulations of Economic Theory in the Gulf of Guinea." *Economy and Society* 40 (3): 345–72.

——. 2015. "Geology, Potentiality, Speculation: On the Indeterminacy of First Oil." *Cultural Anthropology* 30 (4): 611–39.

——. 2016. "A Doubtful Hope: Resource Affect in a Future Oil Economy." *Journal of the Royal Anthropological Institute* 22 (S1): 127–46. https://doi.org/10.1111/1467-9655.12397.

Weszkalnys, Gisa, and Tanya Richardson, eds. 2014. "Resource Materialities: New Anthropological Perspectives on Natural Resource Environments." *Anthropological Quarterly* 87 (1): 5–30.

Whitney, Kristoffer. 2014. "Domesticating Nature?: Surveillance and Conservation of Migratory Shorebirds in the 'Atlantic Flyway.'" *Studies in History and Philosophy of Science Part C: Studies in History and Philosophy of Biological and Biomedical Sciences* 45 (March): 78–87. https://doi.org/10.1016/j.shpsc.2013.10.008.

Whyte, Kyle Powys. 2013. "On the Role of Traditional Ecological Knowledge as a Collaborative Concept: A Philosophical Study." *Ecological Processes* 2 (7).

——. 2017. "Indigenous Climate Change Studies: Indigenizing Futures, Decolonizing the Anthropocene." *English Language Notes* 55 (1): 153–62.

Wigglesworth, Angela. 1992. *Falkland People*. London: Peter Owen.

Wilder, Gary. 2015. *Freedom Time: Negritude, Decolonization, and the Future of the World*. Durham, NC: Duke University Press.

Willerslev, Rane, Piers Vitebsky, and Anatoly Alekseyev. 2015. "Sacrifice as the Ideal Hunt: A Cosmological Explanation for the Origin of Reindeer Domestication." *Journal of the Royal Anthropological Institute* 21 (1): 1–23. https://doi.org/10.1111/1467-9655.12142.

Williams, Eric. 1994. *Capitalism and Slavery*. Chapel Hill: UNC Press Books.

Wimmer, Andreas, and Nina Glick Schiller. 2002. "Methodological Nationalism and Beyond: Nation–State Building, Migration and the Social Sciences." *Global Networks* 2 (4): 301–34. https://doi.org/10.1111/1471-0374.00043.

Wolf, Eric. 1972. "Ownership and Political Ecology." *Anthropological Quarterly* 45 (3): 201–5.

——. 1982. *Europe and the People Without History*. Berkeley: University of California Press.

Wolfe, Patrick. 2006. "Settler Colonialism and the Elimination of the Native." *Journal of Genocide Research* 8: 387–409.

Wood, Ellen Meisken. 2002. *The Origin of Capitalism: A Longer View*. New York: Verso.

Woodman, Paul. 2006. *The Toponymy of the Falkland Islands as Recorded on Maps and in Gazetteers*. UK Permanent Committee on Geographical Names. http://eastsea1994.org/data/bbsData/14911231941.pdf

Woods, Rebecca J. H. 2015. "From Colonial Animal to Imperial Edible: Building an Empire of Sheep in New Zealand, ca. 1880–1900." *Comparative Studies of South Asia, Africa and the Middle East* 35 (1): 117–36.

Woodward, Guy H., and Grace Steele Woodward. 1973. *The Secret of Sherwood Forest: Oil Production in England During World War II*. Norman: University of Oklahoma Press.

Woodward, Rachel, Matthew C. Benwell, K. Neil Jenkings, Eleonora Natale, and Helen Parr. 2022. "Reflections on Conflict and Culture on the 40th Anniversary of the Falklands/Malvinas War." *Journal of War & Culture Studies* 15 (3): 261–65. https://doi.org/10.1080/17526272.2022.2078544.

Woolard, Kathryn A., and Bambi B. Schieffelin. 1994. "Language Ideology." *Annual Review of Anthropology* 23 (1): 55–82.

Wright, Michael. 2006. *The Company: The Story of the Falkland Islands Company*. Dorchester, UK: Nisbet Media.

Wright, Winthrop R. 1975. *British-Owned Railways in Argentina: Their Effect on the Growth of Economic Nationalism, 1854–1948*. Austin: University of Texas Press.

Wylie, Sara Ann. 2018. *Fractivism: Corporate Bodies and Chemical Bonds*. Durham, NC: Duke University Press.

Yarrow, Thomas. 2017. "Remains of the Future: Rethinking the Space and Time of Rumination through the Volta Resettlement Project, Ghana." *Cultural Anthropology* 32 (4): 566–91. https://doi.org/10.14506/ca32.4.06.

Yon, Daniel A. 2007. "Race-Making/Race-Mixing: St. Helena and the South Atlantic World." *Social Dynamics* 33 (2): 144–63.

Zalik, Anna. 2009. "Zones of Exclusion: Offshore Extraction, the Contestation of Space and Physical Displacement in the Nigerian Delta and the Mexican Gulf." *Antipode* 41 (3): 557–82. https://doi.org/10.1111/j.1467-8330.2009.00687.x.

Zenker, Olaf. 2011. "Autochthony, Ethnicity, Indigeneity and Nationalism: Time-Honouring and State-Oriented Modes of Rooting Individual-Territory-Group Triads in a Globalizing World." *Critique of Anthropology* 31 (1): 63–81.

Zoomers, Annelies, Alex Gekker, and Mirko Tobias Schäfer. 2016. "Between Two Hypes: Will 'Big Data' Help Unravel Blind Spots in Understanding the 'Global Land Rush?'" *Geoforum* 69: 147–59.

Index

Note: Figures are indicated by an italicized *f* following the page numbers.